AF389499

Marine Skeletal Biopolymers and Proteins, and Their Biomedical Application

Marine Skeletal Biopolymers and Proteins, and Their Biomedical Application

Editor

Azizur Rahman

MDPI • Basel • Beijing • Wuhan • Barcelona • Belgrade • Manchester • Tokyo • Cluj • Tianjin

Editor
Azizur Rahman
Centre for Climate Change
Research
University of Toronto
(ONRamp@UTE) and
A.R. Environmental Solutions
University of Toronto
Mississauga (ICUBE)
Ontario
Canada

Editorial Office
MDPI
St. Alban-Anlage 66
4052 Basel, Switzerland

This is a reprint of articles from the Special Issue published online in the open access journal *Marine Drugs* (ISSN 1660-3397) (available at: www.mdpi.com/journal/marinedrugs/special_issues/ Skeletal_Proteins_Biomedical_Applications).

For citation purposes, cite each article independently as indicated on the article page online and as indicated below:

LastName, A.A.; LastName, B.B.; LastName, C.C. Article Title. *Journal Name* **Year**, *Volume Number*, Page Range.

ISBN 978-3-0365-2134-3 (Hbk)
ISBN 978-3-0365-2133-6 (PDF)

Contents

About the Editor

Azizur Rahman

Azizur Rahman, Ph.D., is currently serving as a Research Head/Director of the Center for Climate Change Research (CCCR), Toronto, Canada. He received his M.S. and Ph.D. degrees from the University of Ryukyus, Japan, and conducted his postdoctoral studies at the Department of Chemistry, University of Ryukyus (funded by JSPS). He also served as Humboldt Fellow at the University of Munich, Germany for two years. He joined the University of Toronto in 2013. He pursues a broad range of research interests that include marine biology, drug discovery, climate change, biomineralization, and protein biochemistry. His interdisciplinary research has been recognized with over 20 major national and international awards, including several best paper awards and a gold medal. He is a recipient of the Presidential Honorary Award, the highest honor given by the University of Ryukyus in Japan for outstanding novel research findings. Dr. Rahman also received the Top Peer-Reviewers Award (2017-2019).

Preface to "Marine Skeletal Biopolymers and Proteins, and Their Biomedical Application"

Marine skeletal proteins and biopolymers have a great potential application in the biomedical field. The skeletal biopolymers include chitin and chitosan, collagen, cellulose, and various polysaccharides. The marine skeletal proteins include calcium-binding proteins, marine enzymes, and various candidate proteins for drug discovery from the calcifying marine organisms. Due to their broad spectrum of biological functions in biopolymer and protein-based drugs and bioactivities, such as anticancer, antimicrobial, bone tissue regeneration, antioxidant, and anti-aging activities, bioactive skeletal proteins and biopolymer have recently gained great interest in the pharmaceutical, nutraceutical, and cosmeceutical industries. Marine skeletal proteins are also a very rich source of amino acids, which are essential for building good health.

Researchers around the world have found that marine resources are the most convenient and safest sources of biopolymers, proteins, and peptides. The advantages of this source are the huge availability and abundance in the shallow, mid-level, and deep-sea waters. This source includes marine invertebrates and related calcifiers, for example, soft and hard corals, mollusks/bivalves, sponges, sea urchins, coralline red algae, and other calcifying marine organisms. Marine proteins have developed significantly over several stages in both fundamental research and industrial fields. In this book, some recent innovations of these proteins have been incorporated that could be potentially applied in scientific and industrial research. This book covers recent trends in all aspects of basic and applied scientific research on marine skeletal proteins and biopolymers and their derivatives.

Azizur Rahman
Editor

Editorial

Marine Skeletal Biopolymers and Proteins and Their Biomedical Application

M. Azizur Rahman [1,2]

1   Centre for Climate Change Research, Toronto, ON M4P 1J4, Canada; aziz@climatechangeresearch.ca or mazizur.rahman@utoronto.ca

2   A.R. Environmental Solutions, ICUBE-University of Toronto Mississauga, Toronto, ON L5L 1C6, Canada

Citation: Rahman, M.A. Marine Skeletal Biopolymers and Proteins and Their Biomedical Application. *Mar. Drugs* **2021**, *19*, 389. https://doi.org/10.3390/md19070389

Received: 28 June 2021
Accepted: 5 July 2021
Published: 12 July 2021

Skeletal biopolymers and proteins in marine organisms are present as complex mixtures and have great potential applications in the biomedical field. The organic matrices of marine calcifiers are the main source of skeletal proteins [1–6] and they have very interesting structural formation. Marine skeletal proteins are also a very rich source of amino acids, which are essential for building good health. Similarly, biopolymers from marine resources have a variety of structural characteristics that make them useful for different biomedical applications. However, due to their broad array of biological functions in biopolymer- and protein-based drugs, such as anticancer, antimicrobial, bone tissue regeneration, antioxidant, and anti-aging activities, bioactive skeletal proteins and biopolymer have recently attracted a great amount of interest in the pharmaceutical, nutraceutical, and cosmeceutical industries.

The advantages of marine sources are their huge availability and abundance in the shallow, mid-level, and deep-sea waters. These sources include marine invertebrates and related calcifiers, for example, soft and hard corals, mollusks/bivalves, sponges, sea urchins, coralline red algae, and other calcifying marine organisms [7–14].

This Special Issue of Marine Drugs on marine skeletal biopolymers and proteins contains 13 high-quality original articles on different interesting topics related to biomedical and other applications. In the following sections, a short overview of the research findings contributed by the authors is provided, which could help readers to find their relevant articles.

Zheng et al. investigated the antioxidant activity of SNNH-1 in vitro and their findings showed that SNNH-1 can be used as a marine antioxidant and provide a basis for its application in the food and pharmaceutical fields. Five different proteases were used to hydrolyze the swim bladders of *Nibea japonica*, and the hydrolysate treated with neutrase (collagen peptide named SNNHs) showed the highest scavenge 2, 2-diphenyl-1-picrylhydrazyl (DPPH) radical scavenging activity. These results imply that collagen peptides from *Nibea japonica* can significantly reduce the oxidative stress caused by H_2O_2, providing a basis for the application of collagen peptides in the food industry, pharmaceuticals, and cosmetics [15].

Benayahu et al. analyzed the biocompatibility of a marine collagen-based scaffold applied both in vitro and in vivo in a rat animal model [16]. The experiment demonstrated the healing of a rotator cuff tear, the most common musculoskeletal injury occurring in the shoulder. The designed biomaterial will allow the future development of bio composite-based products with optimal mechanical properties that will fully integrate with the natural tissue, contributing to its healing processes. This study also demonstrated that the 3D structure facilitates cell migration and new blood vessel formation needed for tissue repair.

The application of sponges for water purification and collagen production was demonstrated by Gökalp et al. [17]. The main goal of this study was to investigate the effect of depth on the filtration capacity (measured as in situ bacterial clearance rates), metabolism (respiration rate as oxygen consumption), morphology (density and size of oscula), growth, and collagen/biomass production of *Chondrosia reniformis*. This study represents an important step forward, both in understanding the morphological plasticity and performance of sponges and in the application of *C. reniformis* for combined bioremediation and collagen production.

Machalowski et al. investigated chitin-based 3D scaffolds which were isolated from the cultivated under the farming *Aplysina aerophoba* marine demosponge. This biomaterial was modified by silver nanoparticle deposition using chemical reduction of silver nitrate and the antibacterial action was investigated. The results were used for the first time as a basic construct for the fabrication of an antibacterial water filter. This group of marine sponges represents a unique, renewable source of specialized chitin due to their ability to grow under marine farming conditions and which could have a high industrial potential [18].

Hermann Ehrlich and his group, Nowacki et al. [8], reported the isolation technique of chitin from the skeleton of black coral *Cirrhipathes* sp. (Antipatharia, Antipathidae) for the first time. In this study, the authors report the stepwise isolation and identification of chitin from this species and this novel method allows the isolation of α-chitin in the form of a microporous membrane-like material. Additionally, the extracted chitinous scaffold, with a well-preserved, unique pore distribution, was extracted in an amazingly short time (12 h).

Zhang et al. explored the structure of pepsin-solubilized collagen (PSC) from the skin of *Lophius litulon*. The authors used a variety of techniques such as sodium dodecylsulphate polyacrylamide gel electrophoresis (SDS-PAGE), Fourier transform infrared spectroscopy (FTIR), and scanning electron microscopy (SEM). The protein analysis results by SDS-PAGE revealed that PSC from *Lophius litulon* skin was collagen type I and had collagen-specific α1, α2, β, and γ chains. The overall results suggest that collagen from the skin of *Lophius litulon* has potential applications in wound healing along with physicochemical and antioxidant properties due to its good biocompatibility [19].

In the article by Gaspar-Pintiliescu [20], gelatin from the soft tissue of *Rapana venosa* was extracted and characterized using acidic and enzymatic methods. The main purpose of this study was to use the results in the pharmaceutical and cosmetic fields. Its physicochemical and ultrastructural properties were analyzed and compared to those of commercial pig skin gelatin. In addition, gelatins were tested on human keratinocyte cells for their cytocompatibility, cell adhesion capacity, and irritant potential.

Kovalchuk et al. published an interesting work entitled "Naturally Drug-Loaded Chitin: Isolation and Applications". In this study, the authors report the demosponge *Ianthella flabelliformis* (Linnaeus, 1759) for concurrent extraction of both naturally occurring ("ready-to-use") chitin scaffolds, and biologically active bromotyrosines, which are recognized as a potential antibiotic, antitumor, and marine antifouling substances. The results demonstrated that sponge-derived chitin scaffolds, impregnated with decamethoxine, effectively inhibit the growth of human pathogen *Staphylococcus aureus* in an agar diffusion assay [21].

Chen et al. reported the presence of type 1 collagen in red stingray skin. In this study, the authors extracted acid-soluble collagen (ASC) and pepsin-soluble collagen (PSC) from the skin of red stingray *Dasyatis akajei*, and subsequently, its physicochemical and functional properties were investigated. The authors suggested that the PSC from red stingray skin could be useful (instead of terrestrial animal collagen) in drugs, foods, cosmetics, and biological functional materials, and as scaffolds for bone regeneration [22].

The article by Jin et al. reports on the kinetic resolution of racemic styrene oxide (SO) and benzyl glycidyl ether (BGE) obtained by the variant of epoxide hydrolase from *Agromyces mediolanus* (vEH-Am). This study describes the theoretical foundation for the application of vEH-Am in the preparation of enantiopure SO and BGE [23].

The article by Chen et al. evaluated the in vitro anti-proliferative mechanism between Nereis Active Protease (NAP) and human lung cancer H1299 cells. Their experimental results indicate that colony formation and migration of cells were significantly lowered following NAP treatment. Flow cytometry results suggested that NAP-induced growth inhibition of H1299 cells is linked to apoptosis, and that NAP can arrest the cells at the G0/G1 phase [24].

Pan et al. report four antioxidant peptides (RSHP-A, RSHP-B, RSHP-C, and RSHP-D) from protein hydrolysate of red stingray (*Dasyatis akajei*) cartilages. In the work, water-soluble proteins of red stingray (*Dasyatis akajei*) cartilages were extracted by guanidine hydrochloride and hydrolyzed using trypsin [25].

The article by Lin et al. reports the extraction of collagen from bigeye tuna (*Thunnus obesus*) skins by salting-out (PSC-SO) and isoelectric precipitation (PSC-IP) methods. Their results suggest that PSC-IP could be used to rapidly extract collagen from marine by-products instead of traditional salting-out methods. Moreover, collagen from bigeye tuna skin may have strong potential for cosmetic and biomedical applications [26].

As a guest editor, I appreciate the endeavors provided by all of the authors who contributed their excellent results to this Special Issue. I would also like to thank all of the reviewers who carefully evaluated the submitted manuscripts and the editorial board of Marine Drugs for their support and kind help.

Funding: This research received no external funding.

Conflicts of Interest: The author declares no conflict of interest.

References

1. Green, D.W.; Padula, M.P.; Santos, J.; Chou, J.; Milthorpe, B.; Ben-Nissan, B. A therapeutic potential for marine skeletal proteins in bone regeneration. *Mar. Drugs* **2013**, *11*, 1203–1220. [CrossRef] [PubMed]
2. Rahman, M.A. Collagen of Extracellular Matrix from Marine Invertebrates and Its Medical Applications. *Mar. Drugs* **2019**, *17*, 118. [CrossRef]
3. Rahman, M.A.; Fujimura, H.; Shinjo, R.; Oomori, T. Extracellular matrix protein in calcified endoskeleton: A potential additive for crystal growth and design. *J. Cryst. Growth* **2011**, *324*, 177–183. [CrossRef]
4. Rahman, M.A.; Isa, Y.; Uehara, T. Proteins of calcified endoskeleton: II partial amino acid sequences of endoskeletal proteins and the characterization of proteinaceous organic matrix of spicules from the alcyonarian, synularia polydactyla. *Proteomics* **2005**, *5*, 885–893. [CrossRef]
5. Drake, J.L.; Mass, T.; Haramaty, L.; Zelzion, E.; Bhattacharya, D.; Falkowski, P.G. Proteomic analysis of skeletal organic matrix from the stony coral stylophora pistillata. *Proc. Natl. Acad. Sci. USA* **2013**, *110*, 3788–3793. [CrossRef]
6. Rahman, M.A.; Karl, K.; Nonaka, M.; Fujimura, H.; Shinjo, R.; Oomori, T.; Worheide, G. Characterization of the proteinaceous skeletal organic matrix from the precious coral *Corallium konojoi*. *Proteomics* **2014**, *14*, 2600–2606. [CrossRef]
7. Cooper, E.L.; Hirabayashi, K.; Strychar, K.B.; Sammarco, P.W. Corals and their potential applications to integrative medicine. *Evid. Based Complement. Altern. Med.* **2014**, *2014*, 184959. [CrossRef]
8. Nowacki, K.; Stępniak, I.; Langer, E.; Tsurkan, M.; Wysokowski, M.; Petrenko, I.; Khrunyk, Y.; Fursov, A.; Bo, M.; Bavestrello, G.; et al. Electrochemical Approach for Isolation of Chitin from the Skeleton of the Black Coral *Cirrhipathes* sp. (Antipatharia). *Mar. Drugs* **2020**, *18*, 297. [CrossRef]
9. Rahman, M.A.; Halfar, J. First evidence of chitin in calcified coralline algae: New insights into the calcification process of clathromorphum compactum. *Sci. Rep.* **2014**, *4*, 6162. [CrossRef]
10. Mann, K.; Poustka, A.J.; Mann, M. In-depth, high-accuracy proteomics of sea urchin tooth organic matrix. *Proteome Sci.* **2008**, *6*, 33. [CrossRef]
11. Laurienzo, P. Marine polysaccharides in pharmaceutical applications: An overview. *Mar. Drugs* **2010**, *8*, 2435–2465. [CrossRef]
12. Benayahu, D.; Sharabi, M.; Pomeraniec, L.; Awad, L.; Haj-Ali, R.; Benayahu, Y. Unique Collagen Fibers for Biomedical Applications. *Mar. Drugs* **2018**, *16*, 102. [CrossRef]
13. Rahman, M.A. An Overview of the Medical Applications of Marine Skeletal Matrix Proteins. *Mar. Drugs* **2016**, *14*, 167. [CrossRef] [PubMed]
14. Latire, T.; Legendre, F.; Bigot, N.; Carduner, L.; Kellouche, S.; Bouyoucef, M.; Carreiras, F.; Marin, F.; Lebel, J.-M.; Galéra, P.; et al. Shell Extracts from the Marine Bivalve *Pecten maximus* Regulate the Synthesis of Extracellular Matrix in Primary Cultured Human Skin Fibroblasts. *PLoS ONE* **2014**, *9*, e99931. [CrossRef]
15. Zheng, J.; Tian, X.; Xu, B.; Yuan, F.; Gong, J.; Yang, Z. Collagen Peptides from Swim Bladders of Giant Croaker (*Nibea japonica*) and Their Protective Effects against H_2O_2-Induced Oxidative Damage toward Human Umbilical Vein Endothelial Cells. *Mar. Drugs* **2020**, *18*, 430. [CrossRef]
16. Benayahu, D.; Pomeraniec, L.; Shemesh, S.; Heller, S.; Rosenthal, Y.; Rath-Wolfson, L.; Benayahu, Y. Biocompatibility of a Marine Collagen-Based Scaffold In Vitro and In Vivo. *Mar. Drugs* **2020**, *18*, 420. [CrossRef]

17. Gökalp, M.; Kooistra, T.; Rocha, M.S.; Silva, T.H.; Osinga, R.; Murk, A.J.; Wijgerde, T. The Effect of Depth on the Morphology, Bacterial Clearance, and Respiration of the Mediterranean Sponge *Chondrosia reniformis* (Nardo, 1847). *Mar. Drugs* **2020**, *18*, 358. [CrossRef]
18. Machałowski, T.; Czajka, M.; Petrenko, I.; Meissner, H.; Schimpf, C.; Rafaja, D.; Ziętek, J.; Dzięgiel, B.; Adaszek, Ł.; Voronkina, A.; et al. Functionalization of 3D Chitinous Skeletal Scaffolds of Sponge Origin Using Silver Nanoparticles and Their Antibacterial Properties. *Mar. Drugs* **2020**, *18*, 304. [CrossRef] [PubMed]
19. Zhang, W.; Zheng, J.; Tian, X.; Tang, Y.; Ding, G.; Yang, Z.; Jin, H. Pepsin-Soluble Collagen from the Skin of *Lophius litulo*: A Preliminary Study Evaluating Physicochemical, Antioxidant, and Wound Healing Properties. *Mar. Drugs* **2019**, *17*, 708. [CrossRef]
20. Gaspar-Pintiliescu, A.; Stefan, L.M.; Anton, E.D.; Berger, D.; Matei, C.; Negreanu-Pirjol, T.; Moldovan, L. Physicochemical and Biological Properties of Gelatin Extracted from Marine Snail *Rapana venosa*. *Mar. Drugs* **2019**, *17*, 589. [CrossRef] [PubMed]
21. Kovalchuk, V.; Voronkina, A.; Binnewerg, B.; Schubert, M.; Muzychka, L.; Wysokowski, M.; Tsurkan, M.V.; Bechmann, N.; Petrenko, I.; Fursov, A.; et al. Naturally Drug-Loaded Chitin: Isolation and Applications. *Mar. Drugs* **2019**, *17*, 574. [CrossRef] [PubMed]
22. Chen, J.; Li, J.; Li, Z.; Yi, R.; Shi, S.; Wu, K.; Li, Y.; Wu, S. Physicochemical and Functional Properties of Type I Collagens in Red Stingray (*Dasyatis akajei*) Skin. *Mar. Drugs* **2019**, *17*, 558. [CrossRef] [PubMed]
23. Jin, H.; Li, Y.; Zhang, Q.; Lin, S.; Yang, Z.; Ding, G. Enantioselective Hydrolysis of Styrene Oxide and Benzyl Glycidyl Ether by a Variant of Epoxide Hydrolase from *Agromyces mediolanus*. *Mar. Drugs* **2019**, *17*, 367. [CrossRef] [PubMed]
24. Chen, Y.; Tang, Y.; Tang, Y.; Yang, Z.; Ding, G. Serine Protease from *Nereis virens* Inhibits H1299 Lung Cancer Cell Proliferation via the PI3K/AKT/mTOR Pathway. *Mar. Drugs* **2019**, *17*, 366. [CrossRef]
25. Pan, X.-Y.; Wang, Y.-M.; Li, L.; Chi, C.-F.; Wang, B. Four Antioxidant Peptides from Protein Hydrolysate of Red Stingray (*Dasyatis akajei*) Cartilages: Isolation, Identification, and In Vitro Activity Evaluation. *Mar. Drugs* **2019**, *17*, 263. [CrossRef]
26. Lin, X.; Chen, Y.; Jin, H.; Zhao, Q.; Liu, C.; Li, R.; Yu, F.; Chen, Y.; Huang, F.; Yang, Z.; et al. Collagen Extracted from Bigeye Tuna (*Thunnus obesus*) Skin by Isoelectric Precipitation: Physicochemical Properties, Proliferation, and Migration Activities. *Mar. Drugs* **2019**, *17*, 261. [CrossRef]

Article

Four Antioxidant Peptides from Protein Hydrolysate of Red Stingray (*Dasyatis akajei*) Cartilages: Isolation, Identification, and In Vitro Activity Evaluation

Xiao-Yang Pan [1], Yu-Mei Wang [2], Li Li [1], Chang-Feng Chi [1,*]  and Bin Wang [2,*]

[1] National and Provincial Joint Laboratory of Exploration and Utilization of Marine Aquatic Genetic Resources, National Engineering Research Center of Marine Facilities Aquaculture, School of Marine Science and Technology, Zhejiang Ocean University, Zhoushan 316022, China; 18368091610@163.com (X.-Y.P.); wenwenlili@163.com (L.L.)

[2] Zhejiang Provincial Engineering Technology Research Center of Marine Biomedical Products, School of Food and Pharmacy, Zhejiang Ocean University, Zhoushan 316022, China; wangym731@126.com

* Correspondence: chichangfeng@hotmail.com (C.-F.C.); wangbin4159@hotmail.com (B.W.); Tel./Fax: +86-580-255-4818 (C.-F.C.); +86-580-255-4781 (B.W.)

Received: 4 April 2019; Accepted: 30 April 2019; Published: 3 May 2019

Abstract: In the work, water-soluble proteins of red stingray (*Dasyatis akajei*) cartilages were extracted by guanidine hydrochloride and hydrolyzed using trypsin. Subsequently, four antioxidant peptides (RSHP-A, RSHP-B, RSHP-C, and RSHP-D) were isolated from the water-soluble protein hydrolysate while using ultrafiltration and chromatographic techniques, and the amino acid sequences of RSHP-A, RSHP-B, RSHP-C, and RSHP-D were identified as Val-Pro-Arg (VPR), Ile-Glu-Pro-His (IEPH), Leu-Glu-Glu–Glu-Glu (LEEEE), and Ile-Glu-Glu-Glu-Gln (IEEEQ), with molecular weights of 370.46 Da, 494.55 Da, 647.64 Da, and 646.66 Da, respectively. VPR, IEPH, LEEEE, and IEEEQ exhibited good scavenging activities on the DPPH radical (EC_{50} values of 4.61, 1.90, 3.69, and 4.01 mg/mL, respectively), hydroxyl radical (EC_{50} values of 0.77, 0.46, 0.70, and 1.30 mg/mL, respectively), superoxide anion radical (EC_{50} values of 0.08, 0.17, 0.15, and 0.16 mg/mL, respectively), and ABTS cation radical (EC_{50} values of 0.15, 0.11, 0.19, and 0.18 mg/mL, respectively). Among the four isolated antioxidant peptides, IEPH showed the strongest reducing power and lipid peroxidation inhibition activity, but LEEEE showed the highest Fe^{2+}-chelating ability. The present results suggested that VPR, IEPH, LEEEE, and IEEEQ might have the possibility of being an antioxidant additive that is used in functional food and pharmaceuticals.

Keywords: red stingray (*Dasyatis akajei*); cartilage; peptide; antioxidant activity

---

1. Introduction

Superfluous reactive oxygen species (ROS) are produced under oxidative stress conditions and they can destroy all types of macromolecules, which further lead to many health disorders, such as cancer, diabetes mellitus, and inflammatory diseases [1–3]. In addition, oxidative decomposition of unsaturated lipids causes food rancidity and shortens the shelf-life of food products [4–7]. Antioxidant peptides (APs) from food resources usually contain 2–20 amino acid residues and they can donate hydrogen, trap lipid-derived free radicals, and/or activate antioxidant enzymes in cells [5–10]. Therefore, APs can convert ROS, including superoxide anion radicals ($O_2^-\bullet$) and hydroxyl radicals (HO•) into stable molecular structure to inhibit the propagation of peroxidizing chain reaction [5–10]. FWKVV and FMPLH were isolated from the protein hydrolysate of miiuy croaker (*Miichthys miiuy*) muscle and exhibited strong reducing power, lipid peroxidation inhibition ability, and radical scavenging activity [9]. The AP fractions of CPF1 (molecular weight (MW) < 1 kDa) and CPF2 (1 < MW <

3 kDa) from corn gluten exhibited cytoprotective effect and intracellular ROS scavenging activity in H_2O_2-damaged HepG2 cells because they could positively affect the activities of the intracellular antioxidant enzymes [11]. These studies suggested that APs could effectively enhance human health and/or protect food quality by reducing the oxidative stress [12]. In addition, APs can serve as ingredients of functional foods due to their nutritional and biological properties [1,13].

Collagens and peptides from porcine and cow cartilages can provide an adequate nutrient for repairing cartilage, maintain the overall health of subchondral bone and articular cartilage, and improve the joint flexibility and mobility in Osteoarthritis (OA) patients [14,15]. However, these ingredients that were used in food and medical products have caused anxiety among some consumers due to frequent outbreaks of infectious disease [15]. Therefore, cartilages from Chondrichthyes, such as sharks, skates, and rays, are thought to be one of the alternatives of porcine and cow cartilages, because they contain analogous bioactive substances, such as active proteins (angiogenesis inhibitors), proteoglycan, collagen, bioactive peptides, and elastin fibers [16–23]. Furthermore, peptides from protein and collagen hydrolysates of fish cartilages showed a variety of biological activities [15,18,19]. The polypeptide (PG155) with molecular weight (MW) of 15.500 kDa from the cartilages of *Prionace glauca* can significantly reduce the growth of subintestinal vessels (SIVs) in the zebrafish embryos model. PG155 also can inhibit vascular endothelial growth factor (VEGF) induced the migration and tubulogenesis of HUVECs [17]. FIMGPY, IVAGPQ, and IVAGPQ from *Raja porosa* cartilages exhibited high lipid peroxidation inhibiting power and radical scavenging activities [20,21]. In addition, FIMGPY could induce the apoptosis of HeLa cells by upregulating the Bax/Bcl-2 ratio and activating caspase-3 [21]. GAERP, GEREANVM, and AEVG from protein hydrolysate of *Sphyrna lewini* cartilages exhibited good radical scavenging activities, similar inhibiting ability on lipid peroxidation with butylated hydroxytoluene (BHT), and protection on H_2O_2-damaged HepG2 cells by decreasing the content of malonaldehyde (MDA) and increasing the levels of antioxidant enzymes [22]. The cartilage collagens and peptides showing significant pharmacological effects on repairing cartilage and maintaining the overall health of subchondral bones and articular cartilage collagens is more noteworthy [15,19,23].

Red stingray (*Dasyatis akajei*), which mainly resides in the northwestern Pacific Ocean, is a common species of stingray in the Dasyatidae family. In previous studies, some active substances, including proteins (angiogenesis inhibitors), polysaccharides, collagens, and collagen hydrolysates, were prepared from its cartilages [16,24,25]. Luo et al. reported that an angiogenesis inhibitor (DCAI-1) with a MW of 62 kDa from red stingray cartilages could inhibit the formation of the blood vessels of the chick embryo chorioallantoic membrane in a dose-dependent manner [24,26]. The mucopolysaccharides from red stingray cartilages could significantly reduce the values of total cholesterol, triglyceride, and low-density lipoprotein of blood serum, but increase the value of high-density lipoprotein cholesterol in hyperlipidaemia rabbits [27]. Chi et al. isolated and characterized the acid-soluble collagen (ASC) from red stingray cartilages, and the results indicated that the ASC was comprised of type I and II collagens. In addition, the properties of cartilage collagen hydrolysates from *D. akjei*, *S. lewini*, and *R. porosa* indicated that the antioxidant activities of collagen hydrolysates were negatively correlated with the logarithm of their average MWs [25]. However, there is no report on the APs from red stingray cartilages. Therefore, in the present study, peptides from water-soluble protein hydrolysates of red stingray cartilages were purified and characterized, and their *in vitro* antioxidant activities were then evaluated.

2. Results and Discussion

2.1. Isolation of APs from Water-Soluble Protein Hydrolysate (RSH) of Red Stingray Cartilages

2.1.1. Fractionation of RSH by Ultrafiltration

Bioactive peptides are inactive in the sequence of their parent proteins and they can be released by enzymatic hydrolysis either during gastrointestinal digestion in the body or during food processing [2].

Bioactive peptides may act as regulatory compounds with diverse biological activities once they are liberated from protein sources by proteolysis, such as antioxidant, antihypertensive, and enhancing immunity activities [1]. Subsequently, water-soluble proteins of red stingray cartilages were hydrolyzed using trypsin in the experiment.

Three fractions RSH-I (MW < 3 kDa), RSH-II (3 < MW < 10 kDa), and RSH-III (MW > 10 kDa) were prepared from protein hydrolysate (RSH) of red stingray cartilages using 3 and 10 kDa ultrafiltration membranes, and their radical scavenging activities are shown in Figure 1. The 2,2-diphenyl-1-picrylhydrazyl radical (DPPH•) and HO• scavenging rates of RSH-I were 40.12 ± 1.57% and 51.43 ± 1.88% at the concentration of 10 mg protein/mL, which were significantly ($p < 0.05$) higher than those of RSH, RSH-II, and RSH-III at the same concentration. Short chain peptides easily access free radicals and act as electron donors to convert them into the stable state and inhibit the chain reactions [25,28]. The results were in agreement with previous reports that fractions from protein hydrolysates with low molecular size had high antioxidant activities, such as fractions from protein hydrolysates of miiuy croaker muscle [9], cod muscle [29], skate cartilage [21], bluefin leatherjacket head [30], and blue mussel [2,31]. Therefore, RSH-I was chosen for the preparation of APs with high activity.

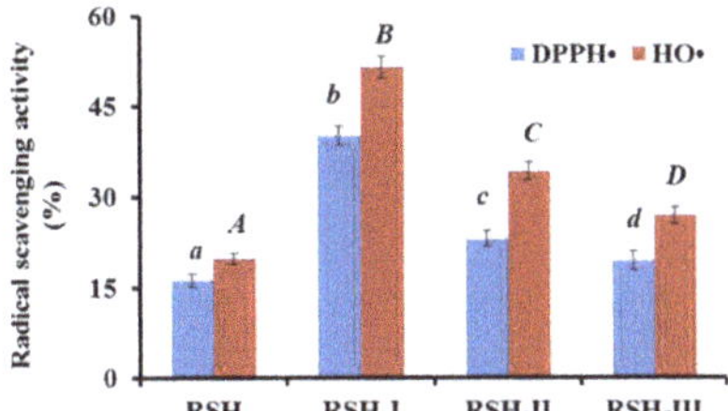

Figure 1. 2,2-diphenyl-1-picrylhydrazyl radical (DPPH•) and hydroxyl radical (HO•) scavenging activity of protein hydrolysate (RSH) and its three fractions by ultrafiltration at the concentration of 10 mg protein/mL. All data were presented as mean ± standard deviation (SD, $n = 3$). a–d or A–D, Values with same letters indicated no significant difference of different samples on DPPH• or HO• scavenging activity ($p > 0.05$).

2.1.2. Anion-Exchange Chromatography of RSH-I

Figure 2A showed that four fractions (RSH-I-1 to RSH-I-4) were separated from RSH-I using a DEAE-52 cellulose column. Among them, RSH-I-1 was eluted using deionized water (DW); RSH-I-2 and RSH-I-3 were eluted using 0.1 M NaCl; and, RSH-I-4 was eluted using 0.5 M NaCl. Figure 2B indicated that the DPPH• and HO• scavenging activities of RSH-I-4 were 70.2 ± 1.8% and 82.4 ± 3.20% at the concentration of 10 mg protein/mL, which were significantly ($p < 0.05$) higher than those of RSH-I, RSH-I-1, RSH-I-2, and RSH-I-3 at the same concentration. Acidic and hydrophobic amino acid residues in peptides can be easily adsorbed to the anion exchange resins, and the interaction strength is closely related to the number and location of the charges on the molecule structure [25,30]. The results indicated that RSH-I-4 might contain some APs, with acidic and hydrophobic amino acid residues.

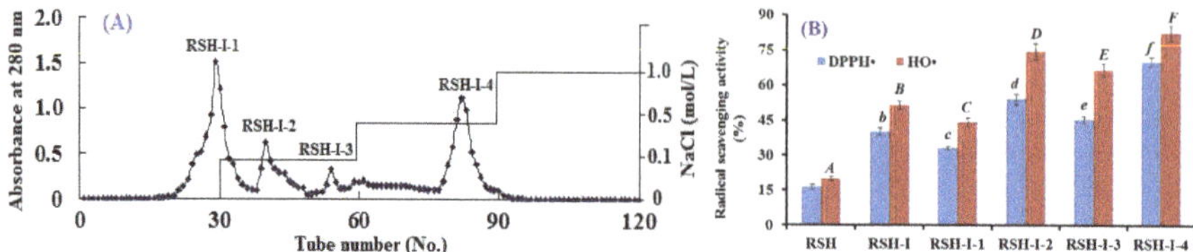

Figure 2. Elution profile of RSH-I through DEAE-52 cellulose anion-exchange chromatography (**A**) and DPPH• and HO• scavenging activities of RSH-I and its fractions at the concentration of 10 mg protein/mL (**B**). All data were presented as mean ± SD (n = 3). a–f or A–F, Values with same letters indicated no significant difference of different samples on DPPH• and HO• scavenging activity ($p > 0.05$).

2.1.3. Gel filtration Chromatography of RSH-I-4

As shown in Figure 3A, RSH-I-4 was subsequently separated into two subfractions (Frac.1 and Frac.2) by a Sephadex G-15 column. The radicals scavenging rates of Frac.1 on DPPH• and OH• were 81.1 ± 1.98% and 89.5 ± 2.37% at the concentration of 10 mg protein/mL, which were significantly ($p < 0.05$) higher than those of RSH-I-4 and Frac.2 (Figure 3B). Smaller peptide fractions generally exhibited a stronger antioxidant activity than their larger counterparts [32,33]. However, Frac.1 with bigger MW showed a higher activity than Frac.2 did, which indicated that some other factors, such as amino acid residue composition and spatial conformation, also played important roles in the bioactivity of APs.

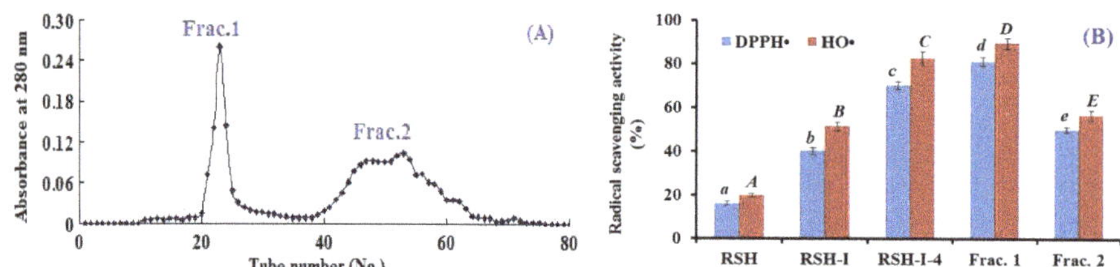

Figure 3. Elution profile of RSH-I-4 using Sephadex G-15 column (**A**) and DPPH• and OH• scavenging activities of subfractions (Frac.1 and Frac.2) from RSH-I-4 at the concentration of 10 mg protein/mL (**B**). All data were presented as mean ± SD (n = 3). a–e or A–E, Values with same letters indicated no significant difference of different samples on DPPH• and HO• scavenging activity ($p > 0.05$).

2.1.4. Isolation of APs from Frac.1 by RP-HPLC

Finally, Frac.1 with high DPPH• and HO• scavenging activities was purified by RP-HPLC while using a linear gradient of acetonitrile (CAN) (Figure 4). All of the chromatographic peaks were collected on their chromatographic peaks and their purities were analyzed. Finally, four APs (RSHP-A, RSHP-B, RSHP-C, and RSHP-D) with retention times of 9.434, 13.435, 14.137, and 18.610 min. had high purities and met the requirement of sequence determination (Table 1). Therefore, RSHP-A, RSHP-B, RSHP-C, and RSHP-D were pooled and lyophilized, and their sequences and activities were determined in the following experiment.

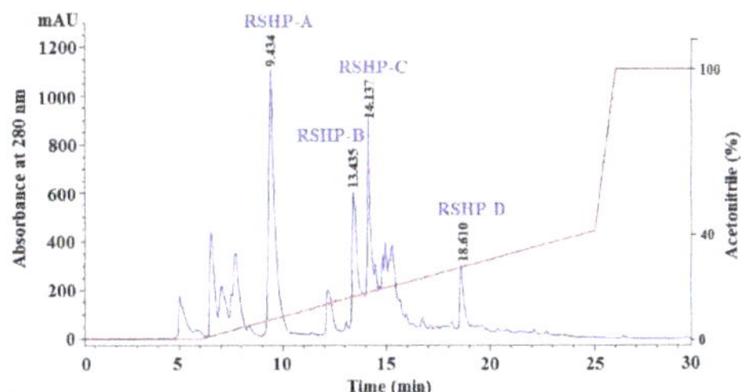

Figure 4. Elution profile of Frac.1 separated by RP-HPLC on a Thermo C-18 column (4.6 × 250 mm) from 0 to 30 min.

Table 1. Retention time (min), amino acid sequences, molecular masses, and radical scavenging activities of four antioxidant peptides (APs) (RSHP-A~D) from water-soluble protein hydrolysate of red stingray (*D. akajei*) cartilages.

		RSHP-A	**RSHP-B**	**RSHP-C**	**RSHP-D**
Retention time (min)		9.434	13.435	14.137	18.610
Amino acid sequence		VPR	IEPH	LEEEE	IEEEQ
Theoretical mass/		370.44/	494.52/	647.69/	646.62/
observed mass (Da)		370.46	494.55	647.64	646.66
	DPPH•	4.61	0.77	0.08	0.15
EC$_{50}$	HO•	1.90	0.46	0.17	0.11
(mg/mL)	O$_2^-$•	3.69	0.70	0.15	0.19
	ABTS$^+$•	4.01	1.30	0.16	0.18

2.2. Amino Acid Sequence and Mass Spectrometry Analysis of APs (RSHP-A~D)

Using a Protein/Peptide Sequencer and electrospray ionization mass spectrometer (ESI-MS), the amino acid sequences of RSHP-A, RSHP-B, RSHP-C, and RSHP-D were determined to be Val-Pro-Arg (VPR), Ile-Glu-Pro-His (IEPH), Leu-Glu-Glu–Glu-Glu (LEEEE), and Ile-Glu-Glu-Glu-Gln (IEEEQ), respectively. The measured MWs of RSHP-A, RSHP-B, RSHP-C, and RSHP-D were 370.46 Da ([M + H]$^+$ 371.68 Da), 494.55 Da ([M + H]$^+$ 495.96 Da), 647.64 Da ([M + H]$^+$ 648.72 Da), and 646.66 Da ([M + H]$^+$ 647.29 Da) (Figure 5 and Table 1), which agreed well with their theoretical masses of 370.44 Da, 494.52 Da, 647.69 Da, and 646.62 Da, respectively.

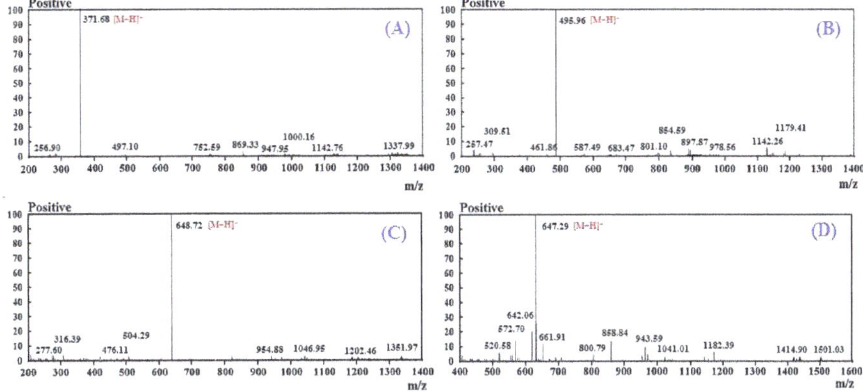

Figure 5. Mass spectrogram of RSHP-A (**A**), RSHP-B (**B**), RSHP-C (**C**), and RSHP-D (**D**).

2.3. Antioxidant Activity

2.3.1. Radical Scavenging Activity

DPPH• Scavenging Activity

The activity curves on DPPH• scavenging of four isolated APs (RSHP-A~D) are shown in Figure 6A, and a positive correlation between the APs concentrations and their radical-scavenging activities was observed. The EC_{50} value of RSHP-B was 1.90 mg/mL, which was significantly ($p < 0.05$) higher than those of RSHP-A (4.61 mg/mL), RSHP-C (3.69 mg/mL), and RSHP-D (4.01 mg/mL), respectively (Table 1). Furthermore, the EC_{50} value of RSHP-B was lower than those of the APs from protein hydrolysates of croceine croaker muscle (YLMSR: 2.74 mg/mL; MILMR: 5.01 mg/mL) [34], loach (PSYV: 17.0 mg/mL) [35], blue mussel (TTANIEDRR: 2.50 mg/mL) [31], salmon (FLNEFLHV: 4.95 mg/mL) [3], bluefin leatherjacket head (WEGPK: 4.44 mg/mL; GVPLT: 4.54 mg/mL) [30], grass carp skin (GFGPL: 2.25 mg/mL; VGGRP: 2.94 mg/mL) [36], skate cartilage (FIMGPY: 2.60 mg/mL; GPAGDY: 3.48 mg/mL; IVAGPQ: 3.93 mg/mL) [21], and scalloped hammerhead muscle (WDR: 3.63 mg/mL; PYFNK: 4.11 mg/mL; LDK: 3.06 mg/mL) [37,38]. However, the EC_{50} value of RSHP-B was higher than those of APs from protein hydrolysates of miiuy croaker swim bladder (FPYLRH: 0.51 mg/mL; GIEWA: 0.78 mg/mL) [39], blue mussel (PIIVYWK: 0.71 mg/mL; FSVVPSPK: 0.94 mg/mL) [31], skipjack tuna bone (GADIVA: 0.57 mg/mL; GAEGFIF: 0.30 mg/mL) [40], grass carp skin (PYSFK: 1.58 mg/mL) [36], and bluefin leatherjacket skin (GSGGL: 0.41 mg/mL; GPGGFI: 0.19 mg/mL; FIGP: 0.12 mg/mL) [41]. The radical-quenching activities of APs are due to their ability to participate in the single electron transfer reaction [1]. The present results illustrated that RSHP-B have a strong ability to act as a free radical scavenger or hydrogen donor to transform DPPH• into harmless compounds and to prevent the single electron transfer reaction.

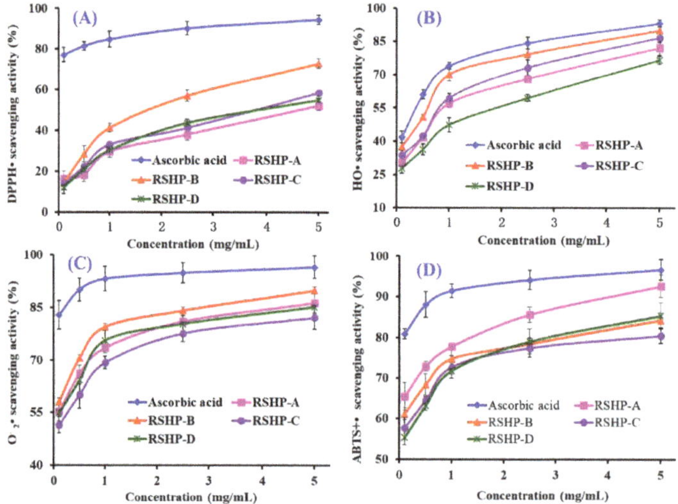

Figure 6. DPPH• (A), HO• (B), O_2^-• (C), and $ABTS^+$• (D) scavenging activities of RSHP-A, RSHP-B, RSHP-C, and RSHP-D from protein hydrolysate of red stingray (*D. akajei*) cartilages. All data are presented as mean ± SD ($n = 3$).

HO• Scavenging Activity

The HO• scavenging activity of four isolated APs (RSHP-A~D) was dose-dependent at the test concentrations (Figure 6B). RSHP-B exhibited the highest HO• scavenging activity, with an EC_{50} value of 0.45 mg/mL, which was significantly ($p < 0.05$) stronger than those of RSHP-A (0.76 mg/mL), RSHP-C

(0.70 mg/mL), and RSHP-D (1.30 mg/mL), respectively (Table 1). The EC_{50} value of RSHP-B was lower than those of APs from the protein hydrolysates of conger eel (LGLNGDDVN: 0.69 mg/mL) [42], weatherfish loach (PSYV: 2.64 mg/mL) [35], miiuy croaker muscle (FWKVV: 0.97 mg/mL; FMPLH: 0.80 mg/mL) [9], mussel sauce (HFGDPFH: 0.50 mg/mL) [43], blue mussel (YPPAK 0.23 mg/mL) [2], hairtail muscle (KA: 1.74 mg/mL; AKG: 2.38 mg/mL; IYG: 2.50 mg/mL) [44], grass carp (PSKYEPFV: 2.86 mg/mL; PYSFK: 2.28 mg/mL; GFGPL: 1.61 mg/mL; VGGRP: 2.06 mg/mL) [36,45], miiuy croaker swim bladder (FPYLRH: 0.68 mg/mL; GIEWA: 0.71 mg/mL) [39], giant squid (NADFGLNGLEGLA: 0.61 mg/mL) [43], and skate cartilage (FIMGPY: 3.04 mg/mL; GPAGDY: 3.92 mg/mL; IVAGPQ: 5.03 mg/mL) [21]. However, the EC_{50} value of RSHP-B was higher than those of the APs from protein hydrolysates of scalloped hammerhead muscle (PYFNK: 0.24 mg/m; LDK: 0.17 mg/mL) [37,38], skipjack tuna bone (GADIVA: 0.25 mg/mL; GAEGFIF: 0.32 mg/mL) [40], blue mussel (YPPAK: 0.23 mg/mL) [3], bluefin leatherjacket skin (GSGGL: 0.18 mg/mL; GPGGFI: 0.09 mg/mL; FIGP: 0.07 mg/mL) [41], and giant squid (NGLEGLK: 0.31 mg/mL) [43]. HO• is one of the most deleterious intracellular ROS and it can cause severe damage to biomacromolecules, including carbohydrates, nucleic acids (mutations), lipids (lipid peroxidation), and amino acids (e.g., conversion of Phe to m-Tyrosine and o-Tyrosine), which further causes a series of chronic diseases [46]. In addition, HO• cannot be eliminated by an enzymatic reaction, and scavenging it from organisms using antioxidants, such as glutathione, sulforaphane, a-tocopherol, curcumin, and ascorbic acid is the only mean to protect important cellular structures from its damage [22,47]. Therefore, four isolated APs, especially RSHP-B, showed a high scavenging ability on HO• and it might have the possibility to serve as a scavenger to weaken the HO• damage in biological systems.

O_2^-• Scavenging Activity

The O_2^-• scavenging activities of four isolated APs (RSHP-A~D) were studied, and the dose-effect relations were observed when their concentrations gradually increased from 0.1 to 5.0 mg/mL (Figure 6C). The EC_{50} values of RSHP-A, RSHP-B, RSHP-C, and RSHP-D were 0.08, 0.17, 0.15, and 0.16 mg/mL, respectively (Table 1). RSHP-A showed stronger O_2^-• scavenging activity than RSHP-B, RSHP-C, and RSHP-D did. The EC_{50} value of RSHP-A was lower than those of APs from protein hydrolysates of skipjack tuna bone (GADIVA: 0.52 mg/mL; GAEGFIF: 0.48 mg/mL) [40], miiuy croaker swim bladder (FPYLRH: 0.34 mg/mL; GIEWA: 0.30 mg/mL) [39], hairtail muscle (KA: 2.08 mg/mL; AKG: 2.54 mg/mL; IYG: 1.36 mg/mL) [44], mussel sauce (HFGDPFH: 0.20 mg/mL) [43], *Mytilus coruscus* (SLPIGLMIAM: 0.32 mg/mL) [48], round scad (HDHPVC: 0.27 mg/mL; HEKVC: 0.24 mg/mL) [49], miiuy croaker muscle (FWKVV: 0.29 mg/mL; FMPLH: 0.15 mg/mL) [9], croceine croaker muscle (YLMR: 0.45 mg/mL; VLYEE: 0.69 mg/mL; MILMR: 0.99 mg/mL) [34,50], and skate cartilage (FIMGPY: 1.61 mg/mL; GPAGDY: 1.66 mg/mL; IVAGPQ: 1.82 mg/mL) [21]. However, the EC_{50} value of RSHP-A was higher than those of the APs from protein hydrolysates of blue mussel (YPPAK: 0.072 mg/mL) [3] and monkfish muscle (LMGQW: 0.042 mg/mL) [51]. O_2^-• is the most common free radical that is generated *in vivo*, and it can promote oxidative reaction to generate hydrogen peroxide and hydroxyl radical [22]. O_2^-• and its derivative radicals seriously threaten the bioactive biomacromolecules and cause different diseases of vital organs of the human body [1,30]. For example, O_2^-• plays crucial roles in cardiovascular and neurodegenerative diseases, as well as in cancer [47]. In organism, SOD can timely catalyze O_2^-• into uninjurious hydrogen peroxide and oxygen under normal conditions. Subsequently, RSHP-A, RSHP-B, RSHP-C, and RSHP-D have high O_2^-• scavenging activity and might be applied to eliminate the O_2^-• damage together with SOD in biological systems.

$ABTS^+$• Scavenging Activity

The abilities of four isolated APs (RSHP-A~D) to scavenge $ABTS^+$• in comparison with ascorbic acid were investigated, and dose-related effects were observed at peptide concentrations that ranged from 0.1 to 5.0 mg/mL (Figure 6D). The EC_{50} values of RSHP-A, RSHP-B, RSHP-C, and RSHP-D were 0.15, 0.11, 0.19, and 0.18 mg/mL, respectively (Table 1). RSHP-B showed the strongest scavenging activity among four isolated APs at all tested concentrations. The EC_{50} value of RSHP-B was lower than those of APs from the protein hydrolysates of croceine croaker muscle (VLYEE: 0.31 mg/mL; YLMSR: 0.42 mg/mL; MILMR: 0.89 mg/mL) [34], hairtail muscle (KA: 1.65 mg/mL; AKG: 0.83 mg/mL; IYG: 0.59 mg/mL) [44], scalloped hammerhead muscle (WDR: 0.34 mg/mL; LDK (0.19 mg/mL) [37,38], salmon (FLNEFLHV: 1.55 mg/mL) [3], grass carp skin (GFGPL: 0.33 mg/mL; VGGRP: 0.47 mg/mL) [36], skipjack tuna bone (GADIVA: 0.41 mg/mL; GAEGFIF: 0.21 mg/mL) [40], bluefin leatherjacket head (WEGPK: 5.41 mg/mL; GPP: 2.47 mg/mL; GVPLT: 3.12 mg/mL) [30], and skate cartilage (FIMGPY: 1.04 mg/mL; GPAGDY: 0.77 mg/mL; IVAGPQ: 1.29 mg/mL) [20]. $ABTS^{+}\bullet$ scavenging assay has been successfully used to evaluate the capacity of dietary antioxidant ingredients, such as polyphenols, polysaccharides, and peptides based on the special chemical properties of formed free radicals [22,38]. In the assay, blue $ABTS^{+}\bullet$ with an absorption maximum of 734 nm is inactivated to its colorless neutral form, when it meets an antioxidant compound. The present results indicated that RSHP-A, RSHP-B, RSHP-C, and RSHP-D could act as electrons or hydrogen atoms donator to inactivate $ABTS^{+}\bullet$.

2.3.2. Fe^{2+}-Chelating Ability

Figure 7 showed that four isolated APs (RSHP-A~D) chelated Fe^{2+} in a dose-effect manner at concentrations that ranged from 0 to 1.0 mg/mL, and RSHP-C showed the highest chelating ability on Fe^{2+} among four isolated APs (RSHP-A~D). Transition metals, such as Fe^{2+} and Cu^{2+}, can serve as catalysts that accelerate the generation of ROS, which further oxidize unsaturated lipids and initiate the oxidative chain reaction [52]. In recent years, some metal chelating peptides have been identified from different food protein resources, such as oyster, anchovy, whey, soybean, and mungbean [53]. Lapsongphon and Yongsawatdigul reported that side chains, including carboxyl (Glu, Asp) and amino groups (Lys, His, Arg), can serve as metal chelators to reduce the available amount of transition metals, causing the inhibition of radical-mediated oxidative chain reactions [42]. Cruz-Huerta et al. have reported that Asp, Glu, and Pro were the most abundant amino acids in the iron-binding peptides from whey protein [54]. GPAGPHGPPGKDGR, AGPHGPPGKDGR, and AGPAGPAGAR from gelatin tryptic hydrolysates of pacific cod skin exhibit high affinity to ferrous ions. Among the groups of the three peptides, the amino and carboxylate terminal groups and peptide bond from peptide backbone, as well as the amino and imine from Arg side chain, are involved in the complexation. In addition, several amino acid side chain groups of GPAGPHGPPGKDGR and AGPHGPPGKDGR, including amino (Lys), imine (His), and carboxylate (Asp), supplied additional iron binding sites [53]. The four isolated APs (RSHP-A~D), especially RSHP-B~D, are rich in acid and basic amino acids. Therefore, they showed higher Fe^{2+}-chelating ability than the positive control of GSH.

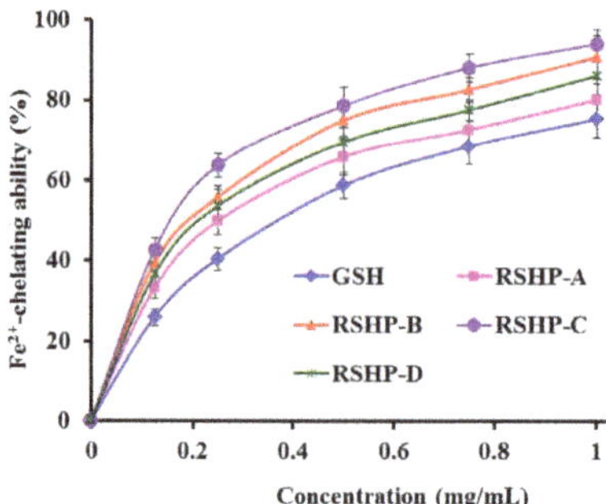

Figure 7. Fe^{2+}-chelating ability of the four isolated APs (RSHP-A~D) from protein hydrolysate of red stingray (*D. akajei*) cartilages. All data are presented as mean ± SD (*n* = 3).

2.3.3. Reducing Power

As shown in Figure 8, four isolated APs (RSHP-A~D) exhibited dose-dependent reducing power at the tested concentrations, and RSHP-B showed the highest capacity to reduce Fe^{3+} to Fe^{2+} among the four isolated APs (RSHP-A~D). In organisms, the antioxidants act as reducing agents to be oxidized for inhibiting the ongoing oxidation reactions. Subsequently, the reducing power is an important indicator to provide information on the potential of samples to serve as antioxidant agents [9,55].

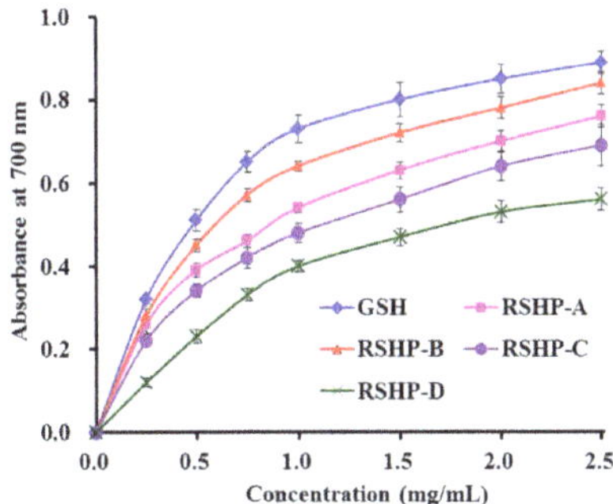

Figure 8. Reducing power of the four isolated APs (RSHP-A~D) from protein hydrolysate of red stingray (*D. akajei*) cartilages. All data are presented as mean ± SD (*n* = 3).

Yang et al. reported that QNDER, KS, KA, AKG, TKA, VK, MK, and IYG from protein hydrolysate of hairtail muscle showed high reducing power, and the absorbance at 700 nm of these eight antioxidant peptides was less than 0.7 at the concentration of 2.5 mg/mL [44]. The data indicated that the reducing power of RSHP-B was stronger than those of the APs from the protein hydrolysate of hairtail muscle at the same concentration. He et al. reported that the absorbance of pentapeptide (FMPLH) from the protein hydrolysate of miiuy croaker muscle at 700 nm was approximate 0.9 at the concentration of 2.5 mg/mL, which was slightly higher than those of RSHP-A~D [9]. The present data indicated that RSHP-B might be applied as a reducing power agent to terminate the *in vivo* oxidation reactions that are in progress.

2.3.4. Lipid Peroxidation Inhibition Ability

Lipid peroxidation is a complex process that involves the formation and propagation of lipid radicals and lipid hydroperoxides in the presence of oxygen [2]. Each assay, such as DPPH, hydroxyl, superoxide anion, and ABTS cation radical scavenging assays measured an antioxidant property that represents a different mechanism, which could not reflect the multiple mechanisms, by which the sample acted as an antioxidant to inhibit lipid oxidation in a food system and living organisms [51]. Therefore, the activities of RSHP-A, RSHP-B, RSHP-C, and RSHP-D against the peroxidation of linoleic acid were investigated and compared to that of the positive control (GSH) and negative control (without antioxidant) in this section (Figure 9). The inhibiting ability of RSHP-B was higher than those of RSHP-A, RSHP-C, and RSHP-D. However, the inhibiting ability of four isolated APs (RSHP-A~D) was lower than that of the positive control of GSH. The result of lipid peroxidation inhibition assay suggested that RSHP-B might inhibit the lipid peroxidation propagation cycle through reacting with peroxyl radicals. In addition, the activity of RSHP-B on lipid peroxidation inhibition slightly less than GSH can be compensable by increasing the dose.

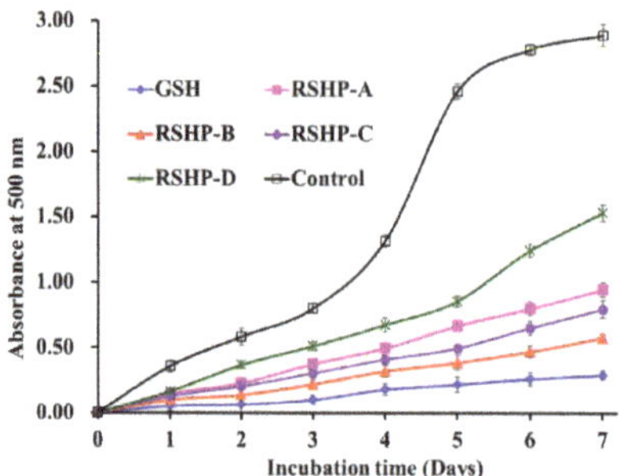

Figure 9. Lipid peroxidation inhibition abilities of the four isolated APs (RSHP-A~D) from protein hydrolysate of red stingray (*D. akajei*) cartilages. All data are presented as mean ± SD (*n* = 3).

3. Discussion

The structural characteristics of APs provide guides for the evaluation of food-derived proteins as potential precursors of APs and predict the possible release of APs from various proteins while using appropriate proteases [1].

Acidic and basic amino acid residues play critical roles in the metal ion chelating activity, which is related to the carboxyl and amino groups in their side chains [56]. The carboxyl residues of acidic amino acids are deprotonated for rendering the complex formation with metal ions, and the amino nitrogen loses its proton nitrogen to allow for unshared pairs of electrons of nitrogen to bind with metal ions [53]. Memarpoor-Yazdi et al. reported similar results, who found that the basic (Arg) and acidic (Asp and Glu) amino acid residues in the sequences of NTDGSTDYGILQINSR and LDEPDPLI were critical to their antioxidant activities [57]. Díaz et al. found that Glu residue is an effective cation chelator that forms complexes with calcium and zinc ions and it may contribute to the antioxidant activity [58]. Therefore, the Glu residue in RSHP-B, RSHP-C and RSHP-D, and Arg residue in RSHP-A might be favorable for their antioxidant activities.

Mirzaei et al. reported that the pyrrolidine ring of Pro residue can interact with the secondary structure of the peptide, thereby increasing the flexibility, and it is also capable of quenching singlet oxygen due to its low ionization potential [59]. Samaranayaka and Li-Chan reported that Pro residue played an important role in the activity of AP that was purified from *Saccharomyces cerevisiae* protein hydrolysate [60]. Therefore, the Pro residue in the amino acid sequences of RSHP-A and RSHP-B should enhance their radical-scavenging activities.

Aromatic amino acid residues, such as Phe, Tyr, His, and Trp, and hydrophobic amino acid residues, including Ala, Val, and Leu, have been reported as critical to the activities of APs [1]. The results from Guo et al. indicated that hydrophobic amino acid residues (e.g., Val, Ile, and Leu) and aromatic amino acid residues (Phe, His, Tyr, and Trp) can enhance the radical-scavenging abilities of APs from Chinese cherry seeds [61]. Therefore, the presence of aromatic and hydrophobic amino acid residues in RSHP-A (Val), RSHP-B (Ile and His), RSHP-C (Leu), and RSHP-D (Ile) should have a positive impact on their radical-scavenging and lipid peroxidation inhibitory activities.

In addition, the activities of APs are dependent on their molecular size. Short peptides with 2–10 amino acid residues have stronger radical-scavenging and lipid peroxidation inhibition activities than their parent native proteins or long-chain peptides [1]. RSHP-A~D exhibited good antioxidant activities, which suggested that they could more effectively and easily interact with free radicals and inhibit the propagation cycles of lipid peroxidation [33]. However, RSHP-B showed the strongest radical scavenging activity (DPPH•, HO•, and ABTS$^+$•), reducing power, and lipid peroxidation inhibition activities, RSHP-A showed the strongest O_2^-• scavenging activity and RSHP-C showed the highest Fe^{2+}-chelating ability among the four isolated APs. There was no consistent trend in the different tested assays. Therefore, more detailed study should be performed to clarify the relationship between the activity and structure of the four isolated APs (RSHP-A~D).

4. Experimental Section

4.1. Materials

Red stingray (*D. akajei*) was purchased from Nanzhen Market in Zhoushan city of China and its cartilages were manually removed with a filleting knife, washed with cold DW, cut into small pieces (about 0.5 cm^2), and then stored at −20 °C. Bovine serum albumin (BSA), DEAE-52 cellulose, D101 macroporous resin, and Sephadex G-15 were purchased from Shanghai Source Poly Biological Technology Co., Ltd. (Shanghai, China). CAN of LC grade and trifluoroacetic acid (TFA) were purchased from Thermo Fisher Scientific Co., Ltd. (Shanghai, China). Four APs (VPR, IEPH, LEEEE, and IEEEQ) with a purity higher than 98% were synthesized in Shanghai Apeptide Co., Ltd. (Shanghai, China).

4.2. Preparation of Water-Soluble Proteins and Hydrolysate from Red Stingray Cartilages

The preparation of water-soluble proteins: The small pieces of cartilages (about 0.5 cm^2) were broken using a hammer, minced to homogenate, and then soaked in 1.0 M guanidine hydrochloride with a solid to solvent ratio of 1:5 (*w/v*) for 48 h, and the liquid supernatant was collected by centrifugation at 4 °C, 12,000× *g* for 10 min and then dialyzed (MW 1 kDa) against 25 volumes of DW for 12 h with solution change every 4 h. Finally, the resulted dialysate (water-soluble proteins) was collected and freeze-dried.

Enzymatic hydrolysis of water-soluble proteins: The freeze-dried water-soluble proteins was dissolved (5 % *w/v*) in 0.2 M phosphate buffer solution (PBS, pH 8.0) and hydrolyzed while using trypsin with a total enzyme dose of 2.5% at 40 °C for 4 h. The process was terminated by heating the hydrolysate to 95 °C for 10 min. Finally, the hydrolysate was centrifuged at 9000× *g* for 15 min, and the supernatant (RSH) was desalted using D101 macroporous resin, freeze-dried, and stored at −20 °C. The concentrations of RSH and its fractions were expressed as mg protein/mL and determined by the dye binding method of Bradford (1976), with BSA as the standard protein

4.3. Isolation of Peptides from RSH

RSH was fractionated into three fractions, termed as RSH-I (MW < 3 kDa), RSH-II (3 < MW < 10 kDa), and RSH-III (MW > 10 kDa), by 3 and 10 kDa ultrafiltration membranes in the Labscale TFF System of Millipor Ltd. (Billerica, MA, USA).

The RSH-I solution (5 mL, 40.0 mg protein/mL) was injected into a DEAE-52 cellulose column (1.6 × 80 cm) pre-equilibrated with DW, and then stepwise eluted with 150 mL DW, 0.1 M, 0.5 M, and 1.0 M NaCl solution at a flow rate of 1.0 mL/min, respectively. Each eluted fraction (5 mL) was collected and measured at 280 nm. Finally, four fractions (RSH-I-1 to RSH-I-4) were pooled on their chromatographic peaks and lyophilized.

RSH-I-4 solution (5 mL, 10.0 mg protein/mL) was fractionated on a Sephadex G-15 column (2.6 × 160 cm) eluted with DW at a flow rate of 0.6 mL/min. Each elate (3 mL) was monitored at 280 nm and then collected on the chromatographic peaks, and two fractions (Frac.1 and Frac.2) were prepared and lyophilized.

A Thermo C-18 column (4.6 × 250 mm, 5μm) (Thermo Co., Ltd., Yokohama, Japan) attached to an Agilent 1260 HPLC system (Agilent Ltd., Santa Rosa, CA, USA) further purified Frac.1. The sample (20 μL) was injected into the HPLC column, eluted with a linear gradient of ACN (0–40% in 0–25 min) in 0.1% TFA at a flow rate of 0.8 mL/min, and monitored at 280 nm. Finally, four APs (RSHP-A~D) were isolated and lyophilized.

4.4. Amino Acid Sequence and Molecular Mass Analysis

Amino acid sequences and molecular masses of RSHP-A, RSHP-B, RSHP-C, and RSHP-D were measured on the previous method [51]. The amino acid sequences of RSHP-A, RSHP-B, RSHP-C, and RSHP-D were determined on an Applied Biosystems 494 protein sequencer (Perkin Elmer/Applied Biosystems Inc., Foster City, CA, USA). Edman degradation was performed according to the standard program that was supplied by Applied Biosystems.

Molecular masses of RSHP-A, RSHP-B, RSHP-C, and RSHP-D were determined using a Q-TOF mass spectrometer (Micromass, Waters, Milford, MA, USA), coupled with an electrospray ionization (ESI) source. Ionization was carried out in a positive mode with a capillary voltage of 3500 V. Nitrogen was maintained at 40 psi for nebulization and 9 L/min at 350 °C for the evaporation temperature. Data were collected in centroid mode from *m/z* 100 to 2000.

4.5. Antioxidant Activity

4.5.1. Radical Scavenging Activity

The radical scavenging and lipid peroxidation inhibition assays were measured on the previous method [38], and the half elimination ratio (EC_{50}) was defined as the concentration where a sample caused a 50% decrease of the initial concentration of DPPH•, HO•, O_2^-•, and $ABTS^+$•, respectively.

4.5.2. Fe^{2+}-Chelating Assay

The Fe^{2+}-chelating assay was performed in accordance with the previous report [62,63]. In brief, the mixed reaction solution (1815 μL DW, 450 μL sample (0, 0.25, 0.5, 0.75, and 1.0 mg/mL) and 45 μL $FeSO_4$ (2 mM)) was energetically shaken and maintained for 0.5 h. After that, 90 μL of ferrozine (5 mM) were added into reaction solution, mixed well, and determined the absorbance at 562 nm. DW was given as the blank control. The Fe^{2+}-chelating ability was calculated using the following formula: Fe^{2+}-chelating ability (%) = [(Absorbance of blank control − Absorbance of sample)/Absorbance of blank control] × 100.

4.5.3. Reducing Power Assay

The reducing power assay was carried out following the previous method [64]. 2.5 mL of 1% $K_3Fe(CN)_6$ solution was blend with 2.0 mL of peptide solution and incubated at 50 °C for 0.5 h. After that, 1.5 mL of 10% trichloroacetic acid was added into the mixed solution. Finally, 2.0 mL of the upper layer, 0.5 mL of 0.1% aqueous $FeCl_3$, and 2.0 mL of DW were mixed, and the absorbance at 700 nm was applied to record the reaction mixture.

4.5.4. Lipid Peroxidation Inhibition Assay

The lipid peroxidation inhibition capacities of APs were measured in a linoleic acid model system, according to the previous methods [65,66]. Briefly, a sample (5.0 mg) was dissolved in 10 mL of PBS (50 mM, pH 7.0) and added to 0.13 mL of a linoleic acid solution and 10 mL of 99.5% ethanol. Subsequently, the total volume was adjusted to 25 mL with DW. The mixture was incubated in a conical flask with a screw cap at 40 °C in a dark room, and the degree of oxidation was evaluated by measuring the $Fe(SCN)_3$ values. The reaction solution (100 μL), incubated in the linoleic acid model system, was mixed with 4.7 mL of 75% ethanol, 0.1 mL of 30% NH_4SCN, and 0.1 mL of 20 mM $FeCl_2$ solution in 3.5% HCl. After 3 min, the thiocyanate value was measured at 500 nm following color development with $FeCl_2$ and thiocyanate at different intervals during the incubation period at 40 °C.

4.6. Statistical Analysis

All data are expressed as the mean ± standard deviation (SD, $n = 3$). An ANOVA test using SPSS 19.0 (Statistical Program for Social Sciences, SPSS Corporation, Chicago, IL, USA) was used to analyze the experimental data. A p-value of less than 0.05 was considered to be statistically significant.

5. Conclusions

In the experiment, four APs (RSHP-A~D) were isolated from water-soluble protein hydrolysate of red stingray (*D. akajei*) cartilages using ultrafiltration and chromatographic methods, and their amino acid sequences were identified as VPR, IEPH, LEEEE, and IEEEQ, respectively. Among the four isolated APs, IEPH showed the strongest scavenging activity on DPPH•, HO•, and ABTS$^+$•, reducing power, and lipid peroxidation inhibition activities; VPR showed the strongest O_2^-• scavenging activity; and, LEEEE showed the highest Fe^{2+}-chelating ability. Subsequently, APs (RSHP-A~D) from water-soluble protein hydrolysate of red stingray cartilages may be possible to serve as antioxidant candidates for new functional foods. Further studies will be done to purify and identify more APs from Frac.1, and more detailed studies on the safety, stability, and structure-activity relationship of the purified APs will also be needed.

Author Contributions: C.-F.C. and B.W. conceived and designed the experiments. X.-Y.P., L.L. and Y.-M.W. performed the experiments and analyzed the data. C.-F.C. and B.W. contributed the reagents, materials, and analytical tools and wrote the paper.

Funding: This work was funded by the National Natural Science Foundation of China (No. 31872547), International S&T Cooperation Program of China (No. 2012DFA30600), and Zhejiang Province Public Technology Research Project (No. LGN18D060002).

Acknowledgments: The authors thank Zhao-Hui Li at Beijing agricultural biological testing center for his technical support on the isolation and amino acid sequence identification of peptides from protein hydrolysate of red stingray (*D. akajei*) cartilages.

Conflicts of Interest: The authors declare no conflicts of interest.

References

1. Sila, A.; Bougatef, A. Antioxidant peptides from marine by-products: Isolation, identification and application in food systems. A review. *J. Funct. Foods* **2016**, *21*, 10–26. [CrossRef]
2. Wang, B.; Li, L.; Chi, C.F.; Ma, J.H.; Luo, H.Y.; Xu, Y.F. Purification and characterisation of a novel antioxidant peptide derived from blue mussel (*Mytilus edulis*) protein hydrolysate. *Food Chem.* **2013**, *138*, 1713–1719. [CrossRef] [PubMed]
3. Ahn, C.B.; Kim, J.G.; Je, J.Y. Purification and antioxidant properties of octapeptide from salmon byproduct protein hydrolysate by gastrointestinal digestion. *Food Chem.* **2014**, *147*, 78–83. [CrossRef] [PubMed]
4. Wang, B.; Wang, Y.M.; Chi, C.F.; Hu, F.Y.; Deng, S.G.; Ma, J.Y. Isolation and characterization of collagen and antioxidant collagen peptides from scales of croceine croaker (*Pseudosciaena crocea*). *Mar. Drugs* **2013**, *11*, 4641–4661. [CrossRef] [PubMed]
5. Lorenzo, J.M.; Vargas, F.C.; Strozzi, I.; Pateiro, M.; Furtado, M.M.; Sant'Ana, A.S.; Rocchetti, G.; Barba, F.J.; Dominguez, R.; Lucini, L.; do Amaral Sobral, P.J. Influence of pitanga leaf extracts on lipid and protein oxidation of pork burger during shelf-life. *Food Res. Int.* **2018**, *114*, 47–54. [CrossRef]
6. Agregán, R.; Franco, D.; Carballo, J.; Tomasevic, I.; Barba, F.J.; Gómez, B.; Muchenje, V.; Lorenzoa, J.M. Shelf life study of healthy pork liver pâté with added seaweed extracts from Ascophyllum nodosum, Fucus vesiculosus and Bifurcaria bifurcate. *Food Res. Int.* **2018**, *112*, 400–411. [CrossRef]
7. Fernandes, R.P.P.; Trindade, M.A.; Tonin, F.G.; Pugine, S.M.P.; Lima, C.G.; Lorenzo, J.M.; de Melo, M.P. Evaluation of oxidative stability of lamb burger with Origanum vulgare extract. *Food Chem.* **2017**, *233*, 101–109. [CrossRef]

8. Najafian, L.; Babji, A.S. A review of fish-derived antioxidant and antimicrobial peptides: Their production, assessment, and applications. *Peptides* **2012**, *33*, 178–185. [CrossRef] [PubMed]
9. He, Y.; Pan, X.; Chi, C.F.; Sun, K.L.; Wang, B. Ten new pentapeptides from protein hydrolysate of miiuy croaker (*Miichthys miiuy*) muscle: Preparation, identification, and antioxidant activity evaluation. *LWT Food Sci. Technol.* **2019**, *105*, 1–8. [CrossRef]
10. Lorenzo, J.M.; Munekata, P.E.S.; Gómez, B.; Barba, F.J.; Mora, L.; Pérez-Santaescolástica, C.; Toldrá, F. Bioactive peptides as natural antioxidants in food products-A review. *Trends Food Sci. Tech.* **2018**, *79*, 136–147. [CrossRef]
11. Wang, L.; Ding, L.; Yu, Z.; Zhang, T.; Ma, S.; Liu, J. Intracellular ROS scavenging and antioxidant enzyme regulating capacities of corn gluten meal-derived antioxidant peptides in HepG2 cells. *Food Res. Int.* **2016**, *90*, 33–41. [CrossRef]
12. Harnedy, P.A.; FitzGerald, R.J. Bioactive peptides from marine processing waste and shellfish: A review. *J. Funct. Foods* **2012**, *4*, 6–24. [CrossRef]
13. Li, X.R.; Chi, C.F.; Li, L.; Wang, B. Purification and identification of antioxidant peptides from protein hydrolysate of scalloped hammerhead (*Sphyrna lewini*) cartilage. *Mar. Drugs* **2017**, *15*, 61. [CrossRef] [PubMed]
14. Bagi, C.M.; Berryman, E.R.; Teo, S.; Lane, N.E. Oral administration of undenatured native chicken type II collagen (UC-II) diminished deterioration of articular cartilage in a rat model of osteoarthritis (OA). *Osteoarthr. Cartilage* **2017**, *25*, 2080–2090. [CrossRef]
15. Luo, Q.B.; Chi, C.F.; Yang, F.; Zhao, Y.Q.; Wang, B. Physicochemical properties of acid- and pepsin-soluble collagens from the cartilage of *Siberian sturgeon*. *Environ. Sci. Pollut. Res.* **2018**, *25*, 31427–31438. [CrossRef]
16. Chi, C.F.; Wang, B.; Li, Z.R.; Luo, H.Y.; Ding, G.F. Characterization of acid-soluble collagens from the cartilages of scalloped hammerhead (*Sphyrna lewini*), red stingray (*Dasyatis akajei*), and skate (*Raja porosa*). *Food Sci. Biotechnol.* **2013**, *22*, 909–916. [CrossRef]
17. Zheng, L.; Ling, P.; Wang, Z.; Niu, R.; Hu, C.; Zhang, T.; Lin, X. A novel polypeptide from shark cartilage with potent anti-angiogenic activity. *Cancer Biol. Ther.* **2007**, *6*, 775–780. [CrossRef]
18. Chen, Q.; Gao, X.; Zhang, H.; Li, B.; Yu, G.; Li, B. Collagen peptides administration in early enteral nutrition intervention attenuates burn-induced intestinal barrier disruption: Effects on tight junction structure. *J. Funct. Foods* **2019**, *55*, 167–174. [CrossRef]
19. Bu, Y.; Elango, J.; Zhang, J.; Bao, B.; Guo, R.; Palaniyandi, K.; Robinson, J.S.; Geevaretnam, J.; Regenstein, J.M.; Wu, W. Immunological effects of collagen and collagen peptide from blue shark cartilage on 6T-CEM cells. *Process. Biochem.* **2017**, *57*, 219–227. [CrossRef]
20. Pan, X.; Zhao, Y.Q.; Hu, F.Y.; Wang, B. Preparation and identification of antioxidant peptides from protein hydrolysate of skate (*Raja porosa*) cartilage. *J. Funct. Foods* **2016**, *25*, 220–230. [CrossRef]
21. Pan, X.; Zhao, Y.Q.; Hu, F.Y.; Chi, C.F.; Wang, B. Anticancer activity of a hexapeptide from skate (*Raja porosa*) cartilage protein hydrolysate in HeLa cells. *Mar. Drugs* **2016**, *14*, 153. [CrossRef]
22. Tao, J.; Zhao, Y.Q.; Chi, C.F.; Wang, B. Bioactive peptides from cartilage protein hydrolysate of spotless smoothhound and their antioxidant activity In vitro. *Mar. Drugs* **2018**, *16*, 100. [CrossRef]
23. Kittiphattanabawon, P.; Benjakul, S.; Visessanguan, W.; Shahidi, F. Isolation and characterization of collagen from the cartilages of brownbanded bamboo shark (*Chiloscyllium punctatum*) and blacktip shark (*Carcharhinus limbatus*). *LWT Food Sci. Technol.* **2010**, *43*, 792–800. [CrossRef]
24. Luo, H.; Xu, J.; Yu, X. Isolation and bioactivity of an angiogenesis inhibitor extracted from the cartilage of Dasyatis akajei. *Asia Pac. J. Clin. Nutr.* **2007**, *16*, 286–289. [PubMed]
25. Li, Z.; Wang, B.; Chi, C.; Gong, Y.; Luo, H.; Ding, G. Influence of average molecular weight on antioxidant and functional properties of cartilage collagen hydrolysates from *Sphyrna lewini, Dasyatis akjei* and *Raja porosa*. *Food Res. Int.* **2013**, *51*, 283–293. [CrossRef]
26. Luo, H.Y.; Yu, X.W.; Qian, X. Purification and bioactivity of a novel angiogenesis inhibitor extracted from the cartilage of Dasyatis akajei. *J. Fish. China* **2007**, *31*, 813–817.
27. Luo, H.; Wu, H.; Yu, X. Effects of mucopolysaccharide from Dasyatis akajei cartilage on plasma lipid and cruor time in rabbits. *Chin. J. Biochem. Pharm.* **2008**, *29*, 23–25.

28. Chi, C.F.; Cao, Z.H.; Wang, B.; Hu, F.Y.; Li, Z.R.; Zhang, B. Antioxidant and functional properties of collagen hydrolysates from spanish mackerel skin as influenced by average molecular weight. *Molecules* **2014**, *19*, 11211–11230. [CrossRef]

29. Sabeena Farvin, K.H.; Andersen, L.L.; Otte, J.; Nielsen, H.H.; Jessen, F.; Jacobsen, C. Antioxidant activity of cod (*Gadus morhua*) protein hydrolysates: Fractionation and characterisation of peptide fractions. *Food Chem.* **2016**, *204*, 409–419. [CrossRef]

30. Chi, C.F.; Wang, B.; Wang, Y.M.; Zhang, B.; Deng, S.G. Isolation and characterization of three antioxidant peptides from protein hydrolysate of bluefin leatherjacket (*Navodon septentrionalis*) heads. *J. Funct. Foods* **2015**, *12*, 1–10. [CrossRef]

31. Park, S.Y.; Kim, Y.S.; Ahn, C.B.; Je, J.Y. Partial purification and identification of three antioxidant peptides with hepatoprotective effects from blue mussel (*Mytilus edulis*) hydrolysate by peptic hydrolysis. *J. Funct. Foods* **2016**, *20*, 88–95. [CrossRef]

32. Wiriyaphan, C.; Chitsomboon, B.; Roytrakul, S.; Yongsawadigul, J. Isolation and identification of antioxidative peptides from hydrolysate of threadfin bream surimi processing byproduct. *J. Funct. Foods* **2013**, *5*, 1654–1664. [CrossRef]

33. Wiriyaphan, C.; Xiao, H.; Decker, E.A.; Yongsawatdigul, J. Chemical and cellular antioxidative properties of threadfin bream (*Nemipterus spp.*) surimi byproduct hydrolysates fractionated by ultrafiltration. *Food Chem.* **2015**, *167*, 7–15. [CrossRef]

34. Chi, C.F.; Hu, F.Y.; Wang, B.; Ren, X.J.; Deng, S.G.; Wu, C.W. Purification and characterization of three antioxidant peptides from protein hydrolyzate of croceine croaker (*Pseudosciaena crocea*) muscle. *Food Chem.* **2015**, *168*, 662–667. [CrossRef]

35. You, S.J.; Wu, J.P. Angiotensin-I converting enzyme inhibitory and antioxidant activities of egg protein hydrolysates produced with gastrointestinal and nongastrointestinal enzymes. *J. Food Sci.* **2011**, *76*, 801–807. [CrossRef] [PubMed]

36. Cai, L.; Wu, X.; Zhang, Y.; Li, X.; Ma, S.; Li, J. Purification and characterization of three antioxidant peptides from protein hydrolysate of grass carp (*Ctenopharyngodon idella*) skin. *J. Funct. Foods* **2015**, *16*, 234–242. [CrossRef]

37. Luo, H.Y.; Wang, B.; Li, Z.R.; Chi, C.F.; Zhang, Q.H.; He, G.Y. Preparation and evaluation of antioxidant peptide from papain hydrolysate of Sphyrna lewini muscle protein. *LWT Food Sci. Technol.* **2013**, *51*, 281–288. [CrossRef]

38. Wang, B.; Li, Z.R.; Chi, C.F.; Zhang, Q.H.; Luo, H.Y. Preparation and evaluation of antioxidant peptides from ethanol-soluble proteins hydrolysate of *Sphyrna lewini* muscle. *Peptides* **2012**, *36*, 240–250. [CrossRef] [PubMed]

39. Zhao, W.H.; Luo, Q.B.; Pan, X.; Chi, C.F.; Sun, K.L.; Wang, B. Preparation, identification, and activity evaluation of ten antioxidant peptides from protein hydrolysate of swim bladders of miiuy croaker (*Miichthys miiuy*). *J. Funct. Foods* **2018**, *47*, 503–511. [CrossRef]

40. Yang, X.R.; Zhao, Y.Q.; Qiu, Y.T.; Chi, C.F.; Wang, B. Preparation and characterization of gelatin and antioxidant peptides from gelatin hydrolysate of Skipjack tuna (*Katsuwonus pelamis*) bone stimulated by in vitro gastrointestinal digestion. *Mar. Drugs* **2019**, *17*, 78. [CrossRef] [PubMed]

41. Chi, C.F.; Wang, B.; Hu, F.Y.; Wang, Y.M.; Zhang, B.; Deng, S.G.; Wu, C.W. Purification and identification of three novel antioxidant peptides from protein hydrolysate of bluefin leatherjacket (*Navodon septentrionalis*) skin. *Food Res. Int.* **2015**, *73*, 124–139. [CrossRef]

42. Ranathunga, S.; Rajapakse, N.; Kim, S.K. Purification and characterization of antioxidative peptide derived from muscle of conger eel (*Conger myriaster*). *Eur. Food Res. Technol.* **2006**, *222*, 310–315. [CrossRef]

43. Rajapakse, N.; Mendis, E.; Jung, W.K.; Je, J.Y.; Kim, S.K. Purification of a radical scavenging peptide from fermented mussel sauce and its antioxidant properties. *Food Res. Int.* **2005**, *38*, 175–182. [CrossRef]

44. Yang, X.R.; Zhang, L.; Ding, D.G.; Chi, C.F.; Wang, B.; Huo, J.C. Preparation, identification, and activity evaluation of eight antioxidant peptides from protein hydrolysate of hairtail (*Trichiurus japonicas*) muscle. *Mar. Drugs* **2019**, *17*, 23. [CrossRef]

45. Ren, J.; Zhao, M.; Shi, J.; Wang, J.; Jiang, Y.; Cui, C.; Kakuda, Y.; Xue, S.J. Purification and identification of antioxidant peptides from grass carp muscle hydrolysates by consecutive chromatography and electrospray ionization-mass spectrometry. *Food Chem.* **2008**, *108*, 727–736. [CrossRef]

46. Chi, C.F.; Hu, F.Y.; Wang, B.; Li, Z.R.; Luo, H.Y. Influence of amino acid compositions and peptide profiles on antioxidant capacities of two protein hydrolysates from skipjack tuna (*Katsuwonus pelamis*) dark muscle. *Mar. Drugs* **2015**, *13*, 2580–2601. [CrossRef]

47. Prasad, A.K.; Mishra, P.C. Scavenging of superoxide radical anion and hydroxyl radical by urea, thiourea, selenourea and their derivatives without any catalyst: A theoretical study. *Chem. Phys. Lett.* **2017**, *684*, 197–204. [CrossRef]

48. Kim, E.K.; Oh, H.J.; Kim, Y.S.; Hwang, J.W.; Ahn, C.B.; Lee, J.S.; Jeon, Y.J.; Moon, S.H.; Sung, S.H.; Jeon, B.T.; Park, P.J. Purification of a novel peptide derived from *Mytilus coruscus* and in vitro/in vivo evaluation of its bioactive properties. *Fish Shellfish Immun.* **2013**, *34*, 1078–1084. [CrossRef]

49. Jiang, H.; Tong, T.; Sun, J.; Xu, Y.; Zhao, Z.; Liao, D. Purification and characterization of antioxidative peptides from round scad (*Decapterus maruadsi*) muscle protein hydrolysate. *Food Chem.* **2014**, *154*, 158–163. [CrossRef] [PubMed]

50. Zhao, Y.Q.; Zeng, L.; Yang, Z.S.; Huang, F.F.; Ding, G.F.; Wang, B. Anti-fatigue effect by peptide fraction from protein hydrolysate of croceine croaker (*Pseudosciaena crocea*) swim bladder through inhibiting the oxidative reactions including DNA damage. *Mar. Drugs* **2016**, *14*, 221. [CrossRef]

51. Chi, C.F.; Wang, B.; Deng, Y.Y.; Wang, Y.M.; Deng, S.G.; Ma, J.Y. Isolation and characterization of three antioxidant pentapeptides from protein hydrolysate of monkfish (*Lophius litulon*) muscle. *Food Res. Int.* **2014**, *55*, 222–228. [CrossRef]

52. Lapsongphon, N.; Yongsawatdigul, J. Production and purification of antioxidant peptides from a mungbean meal hydrolysate by *Virgibacillus sp.* SK37 proteinase. *Food Chem.* **2013**, *141*, 992–999. [CrossRef] [PubMed]

53. Wu, W.; Li, B.; Hou, H.; Zhang, H.; Zhao, X. Identification of iron-chelating peptides from Pacific cod skin gelatin and the possible binding mode. *J. Funct. Foods* **2017**, *35*, 418–427. [CrossRef]

54. Cruz-Huerta, E.; Maqueda, D.M.; de la Hoz, L.; Nunes da Silva, V.S.; Bertoldo Pacheco, M.T.; Amigo, L.; Recio, I. Short communication: Identification of iron-binding peptides from whey protein hydrolysates using iron (III)- immobilized metal ion affinity chromatography and reversed phase-HPLCtandem mass spectrometry. *J. Dairy Sci.* **2016**, *99*, 77–82. [CrossRef]

55. Christodouleas, D.; Fotakis, C.; Papadopoulos, K.; Calokerinos, A.C. Evaluation of total reducing power of edible oils. *Talanta* **2014**, *130*, 233–240. [CrossRef]

56. Gimenez, B.; Aleman, A.; Montero, P.; Gomez-Guillen, M.C. Antioxidant and functional properties of gelatin hydrolysates obtained from skin of sole and squid. *Food Chem.* **2009**, *114*, 976–983. [CrossRef]

57. Memarpoor-Yazdi, M.; Asoodeh, A.; Chamani, J. A novel antioxidant and antimicrobial peptide from hen egg white lysozyme hydrolysates. *J. Funct. Foods* **2012**, *4*, 278–286. [CrossRef]

58. Díaz, M.; Dunn, C.M.; McClements, D.J.; Decker, E.A. Use of caseinophosphopeptides as natural antioxidants in oil-in-water emulsions. *J. Agric. Food Chem.* **2003**, *51*, 2365–2370. [CrossRef] [PubMed]

59. Mirzaei, M.; Mirdamadi, S.; Ehsani, M.R.; Aminlari, M.; Hosseini, E. Purification and identification of antioxidant and ACE-inhibitory peptide from *Saccharomyces cerevisiae* protein hydrolysate. *J. Funct. Foods* **2015**, *19*, 259–268. [CrossRef]

60. Samaranayaka, A.G.P.; Li-Chan, E.C.Y. Food-derived peptidic antioxidants: A review of their production, assessment, and potential applications. *J. Funct. Foods* **2011**, *3*, 229–254. [CrossRef]

61. Guo, P.; Qi, Y.; Zhu, C.; Wang, Q. Purification and identification of antioxidant peptides from Chinese cherry (*Prunus pseudocerasus* Lindl.) seeds. *J. Funct. Foods* **2015**, *19*, 394–403. [CrossRef]

62. Zhang, M.; Mu, T.H. Identification and characterization of antioxidant peptides from sweet potato protein hydrolysates by Alcalase under high hydrostatic pressure. *Innov. Food Sci. Emerg.* **2017**, *43*, 92–101. [CrossRef]

63. Dong, S.; Zeng, M.; Wang, D.; Liu, Z.; Zhao, Y.; Yang, H. Antioxidant and biochemical properties of protein hydrolysates prepared from Silver carp (*Hypophthalmichthys molitrix*). *Food Chem.* **2008**, *107*, 1485–1493. [CrossRef]

64. Li, Z.; Wang, B.; Zhang, Q.; Qu, Y.; Xu, H.; Li, G. Preparation and antioxidant property of extract and semipurified fractions of *Caulerpa racemosa*. *J. Appl. Phycol.* **2012**, *24*, 1527–1536. [CrossRef]

65. Wang, B.; Li, Z.R.; Yu, D.; Chi, C.F.; Luo, H.Y.; Ma, J.Y. Isolation and characterisation of five novel antioxidant peptides from ethanol-soluble proteins hydrolysate of spotless smoothhound (*Mustelus griseus*) muscle. *J. Funct. Foods* **2014**, *6*, 176–185. [CrossRef]

66. Hu, F.Y.; Chi, C.F.; Wang, B.; Deng, S.G. Two novel antioxidant nonapeptides from protein hydrolysate of skate (*Raja porosa*) muscle. *Mar. Drugs* **2015**, *13*, 1993–2009. [CrossRef] [PubMed]

Article

Naturally Drug-Loaded Chitin: Isolation and Applications

Valentine Kovalchuk [1], **Alona Voronkina** [2], **Björn Binnewerg** [3], **Mario Schubert** [3],
Liubov Muzychka [4], **Marcin Wysokowski** [5,6,*], **Mikhail V. Tsurkan** [7], **Nicole Bechmann** [8],
Iaroslav Petrenko [6], **Andriy Fursov** [6], **Rajko Martinovic** [9], **Viatcheslav N. Ivanenko** [10],
Jane Fromont [11], **Oleg B. Smolii** [4], **Yvonne Joseph** [6], **Marco Giovine** [12], **Dirk Erpenbeck** [13],
Michael Gelinsky [14], **Armin Springer** [14,15], **Kaomei Guan** [3], **Stefan R. Bornstein** [16,17] and
Hermann Ehrlich [6,*]

1   Department of Microbiology, National Pirogov Memorial Medical University, Vinnytsia 21018, Ukraine;
    valentinkovalchuk2015@gmail.com
2   Department of Pharmacy, National Pirogov Memorial Medical University, Vinnytsia 21018, Ukraine;
    algol2808@gmail.com
3   Institute of Pharmacology and Toxicology, TU Dresden, Dresden 01307, Germany;
    Bjoern.Binnewerg@tu-dresden.de (B.B.); mario.schubert1@tu-dresden.de (M.S.);
    Kaomei.Guan@tu-dresden.de (K.G.)
4   V.P. Kukhar Institute of Bioorganic Chemistry and Petrochemistry, National Academy of Science of Ukraine,
    Murmanska Str. 1, Kyiv 02094, Ukraine; lmuzychka@rambler.ru (L.M.); smolii@bpci.kiev.ua (O.B.S.)
5   Institute of Chemical Technology and Engineering, Faculty of Chemical Technology, Poznan University of
    Technology, Berdychowo 4, Poznan 60965, Poland
6   Institute of Electronic and Sensor Materials, TU Bergakademie Freiberg, Gustav-Zeuner Str. 3,
    Freiberg 09599, Germany; iaroslavpetrenko@gmail.com (I.P.); andriyfur@gmail.com (A.F.);
    Yvonne.Joseph@esm.tu-freiberg.de (Y.J.)
7   Leibniz Institute for Polymer Research Dresden, Dresden 01069, Germany; tsurkan@ipfdd.de
8   Institute of Clinical Chemistry and Laboratory Medicine, University Hospital Carl Gustav Carus,
    Faculty of Medicine Carl Gustav Carus, TU Dresden, Dresden 01307, Germany;
    Nicole.Bechmann@uniklinikum-dresden.de
9   Institute of Marine Biology, University of Montenegro, Kotor 85330, Montenegro; rajko.mar@ucg.ac.me
10 Department of Invertebrate Zoology, Biological Faculty, Lomonosov Moscow State University,
    Moscow 119992, Russia; ivanenko.slava@gmail.com
11 Aquatic Zoology Department, Western Australian Museum, Locked Bag 49, Welshpool DC,
    Western Australia WA6986, Australia; Jane.Fromont@museum.wa.gov.au
12 Department of Sciences of Earth, Environment and Life, University of Genoa, Corso Europa 26,
    16132 Genova, Italy; mgiovine@unige.it
13 Department of Earth and Environmental Sciences & GeoBio-Center, Ludwig-Maximilians-Universität
    München, Richard-Wagner-Str. 10, Munich 80333, Germany; erpenbeck@lmu.de
14 Centre for Translational Bone, Joint and Soft Tissue Research, Faculty of Medicine and University Hospital
    Carl Gustav Carus of Technische Universität Dresden, Fetscherstraße 74, Dresden 01307, Germany;
    michael.gelinsky@tu-dresden.de (M.G.); Armin.Springer@med.uni-rostock.de (A.S.)
15 Medizinische Biologie und Elektronenmikroskopisches Zentrum (EMZ), Universitätsmedizin Rostock,
    Rostock 18055, Germany
16 Department of Internal Medicine III, University Hospital Carl Gustav Carus, Technische Universität
    Dresden, Dresden 01307, Germany; Stefan.Bornstein@uniklinikum-dresden.de
17 Diabetes and Nutritional Sciences Division, King's College London, London WC2R 2LS, UK
*   Correspondence: marcin.wysokowski@put.poznan.pl (M.W.); hermann.ehrlich@esm.tu-freiberg.de (H.E.);
    Tel.: +0048-61-665-3748 (M.W.); +49-3731-39-2867 (H.E.)

Received: 12 September 2019; Accepted: 8 October 2019; Published: 10 October 2019

Abstract: Naturally occurring three-dimensional (3D) biopolymer-based matrices that can be used in different biomedical applications are sustainable alternatives to various artificial 3D materials. For this purpose, chitin-based structures from marine sponges are very promising substitutes. Marine sponges from the order Verongiida (class Demospongiae) are typical examples of demosponges with well-developed chitinous skeletons. In particular, species belonging to the family Ianthellidae possess chitinous, flat, fan-like fibrous skeletons with a unique, microporous 3D architecture that makes them particularly interesting for applications. In this work, we focus our attention on the demosponge *Ianthella flabelliformis* (Linnaeus, 1759) for simultaneous extraction of both naturally occurring ("ready-to-use") chitin scaffolds, and biologically active bromotyrosines which are recognized as potential antibiotic, antitumor, and marine antifouling substances. We show that selected bromotyrosines are located within pigmental cells which, however, are localized within chitinous skeletal fibers of *I. flabelliformis*. A two-step reaction provides two products: treatment with methanol extracts the bromotyrosine compounds bastadin 25 and araplysillin-I N20 sulfamate, and a subsequent treatment with acetic acid and sodium hydroxide exposes the 3D chitinous scaffold. This scaffold is a mesh-like structure, which retains its capillary network, and its use as a potential drug delivery biomaterial was examined for the first time. The results demonstrate that sponge-derived chitin scaffolds, impregnated with decamethoxine, effectively inhibit growth of the human pathogen *Staphylococcus aureus* in an agar diffusion assay.

Keywords: chitin; scaffolds; pigmental cells; demosponges; Ianthella; bromotyrosines; decamethoxine; drug delivery

---

1. Introduction

Development of three-dimensional (3D) scaffolds based on natural biopolymers is a recent trend in materials and biomaterials science. The scientific community now focuses on development of fabrication methods which will allow for precise control of the architecture and pore structures in such scaffolds. Evolutionary optimized 3D constructs of natural origin can be found in marine demosponges (phylum Porifera, class Demospongiae), which are recognized among the first multicellular organisms on our planet. These organisms evolved and survived for more than 600 million years due to their ability to excellently combine the mechanical stability of their voluminous, fiber-based, water-filtering skeletons [1] and their diverse chemical defense strategies [2,3] due to biosynthesis of secondary metabolites with anti-predatory and antibiotic properties [4]. The demosponges can produce both proteinaceous (spongin)- or polysaccharide (chitin)-based and up to 2-m-high skeletons (for an overview, see References [5–10]). Representatives of the family Ianthellidae (Hyatt, 1875) possess chitinous, flat, fan-like skeletons with unique 3D architecture [11,12] (Figures 1 and 2A), some of which can reach up to 2 m in diameter [12].

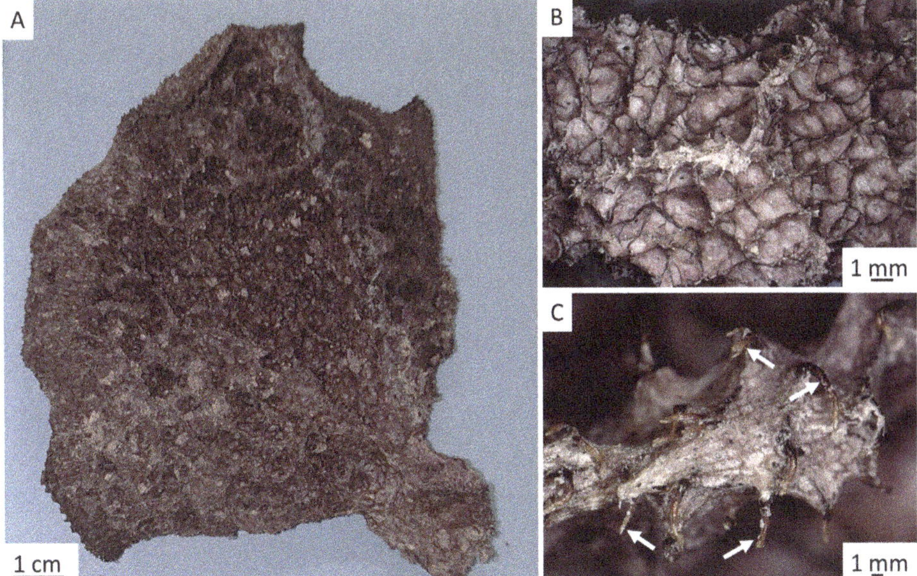

Figure 1. The marine demosponge *Ianthella flabelliformis* (Linnaeus, 1759) (**A**), as collected after air-drying, exhibits a characteristic fan-like and meshwork morphology (**B**). Chitin-based skeletal fibers (arrows) are visible between tissue-like layers (**C**).

One main advantage of skeletons of the ianthellid sponges with respect to their practical application is that chitinous scaffolds can be easily isolated from them (for details, see Reference [11]). These scaffolds exhibit the characteristic shape, size, and meshwork-like structural motif of demosponges, throughout which mesophyll cells are perfectly distributed (Figure 1B,C). Recently, 3D chitin scaffolds isolated from the demosponge *Ianthella basta* (Pallas, 1766) were successfully used in tissue engineering of human mesenchymal stromal cells (hMSC), as well as human dermal MSCs [13]. Additional advantages of these sponges for use in technological [14,15] and biomedical applications [16] is their exceptional ability to regenerate their skeletons in situ, with a growth rate of up to 12 cm/year [12]. Consequently, marine farming of these sponges in coastal areas in Australia and Guam (United States of America) [17] is planned to be developed in the near future.

Currently, our strategy regarding the ianthellid sponges is focused on development of methods for simultaneous extraction of both naturally pre-designed chitin (as "ready-to-use" scaffolds) and biologically active bromotyrosines which are recognized as potential antibiotic, antitumor (for a review, see References [18,19]), and marine antifouling [20,21] substances. The pharmacological potential of these marine demosponges as producers of bromotyrosine-related bastadins was very positively approved [22–27].

In this work, we focus our attention on *Ianthella flabelliformis* (Linnaeus, 1759) (Demospongiae, Verongiida, Ianthellidae) [28], originally designated by Linneaus as *Spongia flabelliformis* [29] and later transferred to the genus *Ianthella* (Gray, 1869), for simultaneous extraction of both naturally occurring chitin scaffolds, and biologically active bromotyrosines which are recognized as potential antibiotic, antitumor, and marine antifouling substances. In this species, bromotyrosines are located within specialized, pigmental cells that are tightly associated with skeletal chitin. Special attention was paid to investigations on the applicability of 3D chitinous spacer fabric-like scaffolds as potential drug delivery biomaterials, filled with a quaternary ammonium compound, decamethoxine [30], which is a well-recognized antiseptic against diverse human diseases [31]. We used a clinical strain of the human pathogen *Staphylococcus aureus* as the test microorganism in this study.

2. Results

The methodology of how to obtain chitin in the size, shape, and unique architecture of the *I. flabelliformis* fan-like skeleton is described in Section 4.2. However, one of the most important steps is obtaining a pigmented skeletal structure that is tissue-free (Figure 2A,B). Additional treatment with 5% NaOH at 37 °C over 4 h exposes chitinous fibers with visible inner channels which are typically located within skeletal fibers of verongiid demosponges (see also References [9,11]). These channels permit verongiid chitin to be saturated with diverse liquids due to the capillary effect (for details, see Reference [32]), and the presence of similar channels in our samples suggests that a similar mechanism of capillary uptake might be possible in the skeleton of *I. flabelliformis* (Figure 2C).

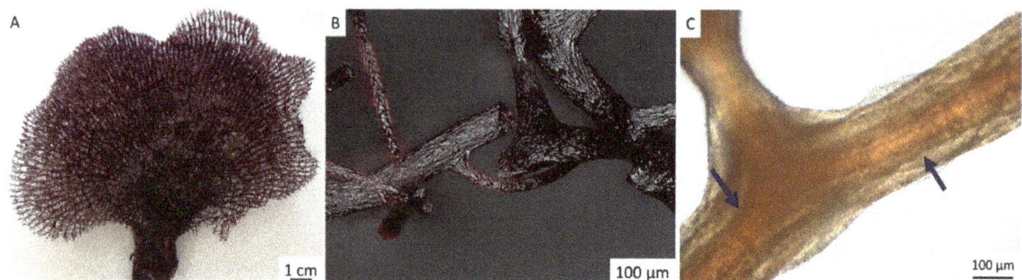

Figure 2. The cell-free macerated skeleton of *I. flabelliformis* (**A,B**) is made of anastomosing, interconnected tube-based chitinous fibers (**B**). Partial depigmentation using alkali treatment leads to visualization of the inner channel (arrows, **C**) which is located within each fiber.

Importantly, using this method, isolation of pigmented chitinous fibers (Figure 3) is possible, in which the presence of reddish-violet cells was visualized using light microscopy (Figure 3C). These so-called "fiber cells" [33], *"pigmental cells"* [34–37], or *"spherulous cells"* [38,39] are presumable characteristic features for representatives of the Ianthellidae family [37,40]. The size of these cells in *I. flabelliformis* is 11.8 µm ± 1.15 µm.

Single-spot energy-dispersive X-ray spectroscopy (EDX) analysis using SEM strongly confirmed the presence of bromine within individual pigmental cells (Figure 4). These results correlate well with those reported previously for similar cells observed in the verongiid demosponge *Aplysina aerophoba* (Nardo, 1833) [39].

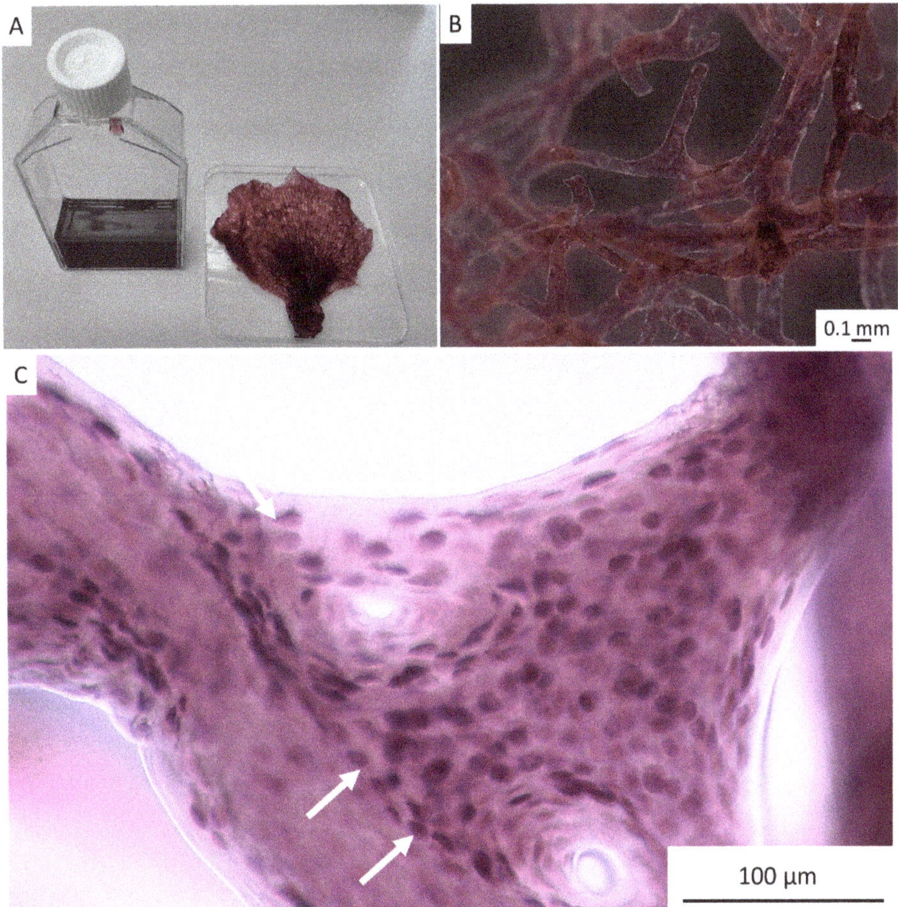

Figure 3. The bromotyrosine-containing extract isolated from skeleton of *I. flabelliformis* (**A**) is one of the sources of pharmacologically relevant reagents. The dark-reddish color of chitinous skeletal fibers (**B**) is determined by the presence of pigmental cells, or spherulocites (**C**), special chitin-associated bromotyrosine-producing cells (arrows).

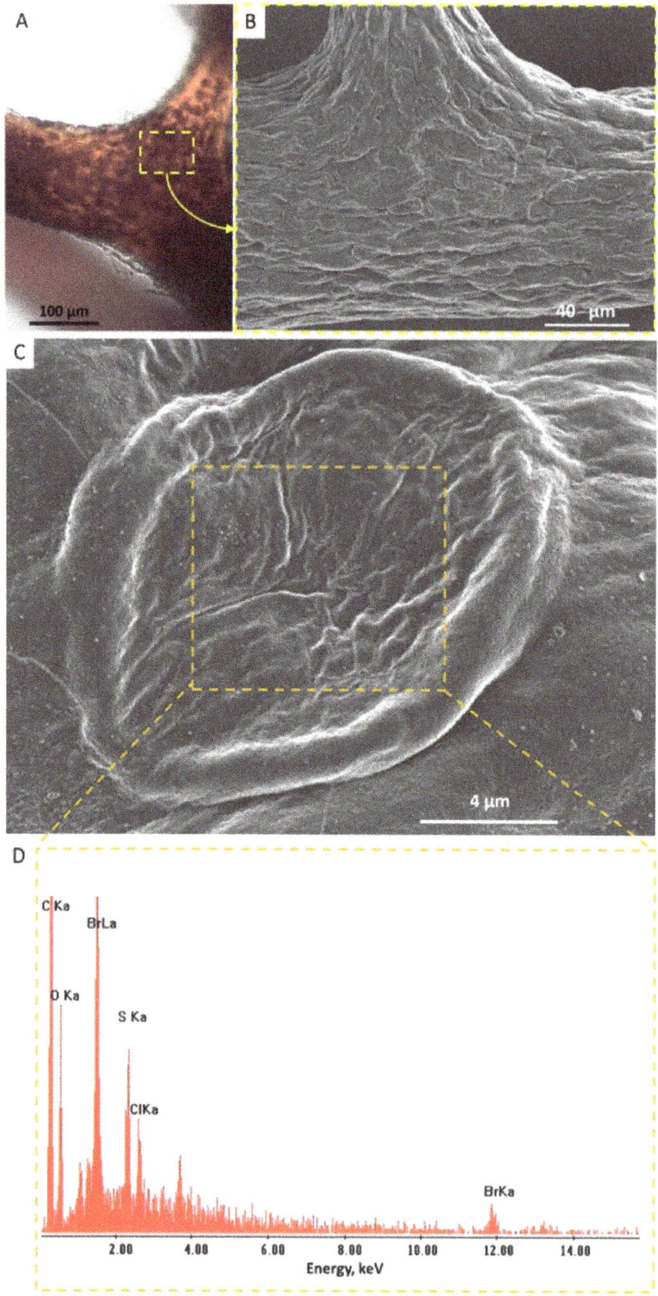

Figure 4. Pigmental cells located within fibers of *I. flabelliformis* chitin are clearly visible using light microscopy (**A**) (see also Figure 3C). These cells are observable using SEM (**B,C**). Single-spot energy-dispersive X-ray spectroscopy (EDX) analysis shows strong evidence of the presence of bromine within individual pigmental cells (**D**).

TEM analysis of the cross-sectioned chitinous fiber of *I. flabelliformis* studied during the isolation step, as represented in Figures 3C and 4A, showed the location of such cells between the alternating chitinous layers (Figure 5).

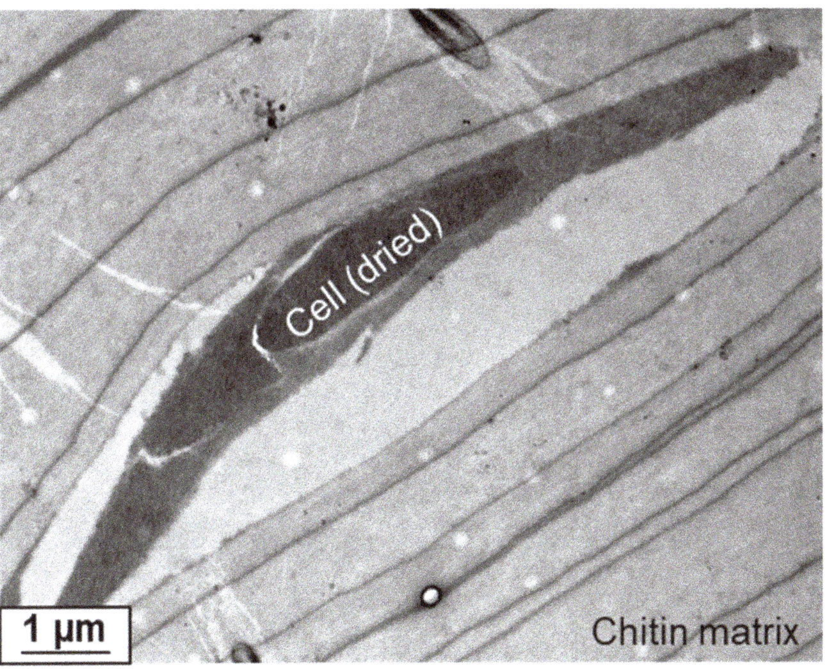

Figure 5. TEM image of the cross-section through the chitinous sponge matrix of *I. flabelliformis* showing the interlayer location of the pigmental cell that lost its oval morphology due to the drying procedure. This cell is definitively of eukaryotic and not bacterial origin.

Due to the evident presence of bromine within pigmented cells of *I. flabelliformis*, we suggest that bromotyrosines are responsible for this naturally occurring event. Thus, the corresponding methanol extract obtained after treatment of the pigmented skeletal fibers (Figure 3) was subjected to HPLC for isolation, analysis, and identification of possible bromotyrosine derivatives. Compounds were identified by analyzing the mass spectra and comparing with the data obtained for the same compounds previously isolated from *I. flabelliformis* sponge [41,42]. Our data (see Supplementary Materials) indicate that the isolated compounds correspond to bastadin 25 (MS (electrospray ionization, ESI): m/z 1032 [M − H]$^-$) [41] and araplysillin-I N^{20}-sulfamate (MS (ESI): m/z 794 [M − Na]$^-$) [42,43].

Further treatment of pigmented chitinous fibers using acetic acid and alkali (see Section 4.2) (Figure 6) led to the isolation of ready-to-use 3D chitinous scaffolds which visually resemble the well-known architecture of fabric-based bandage materials (Figure 7).

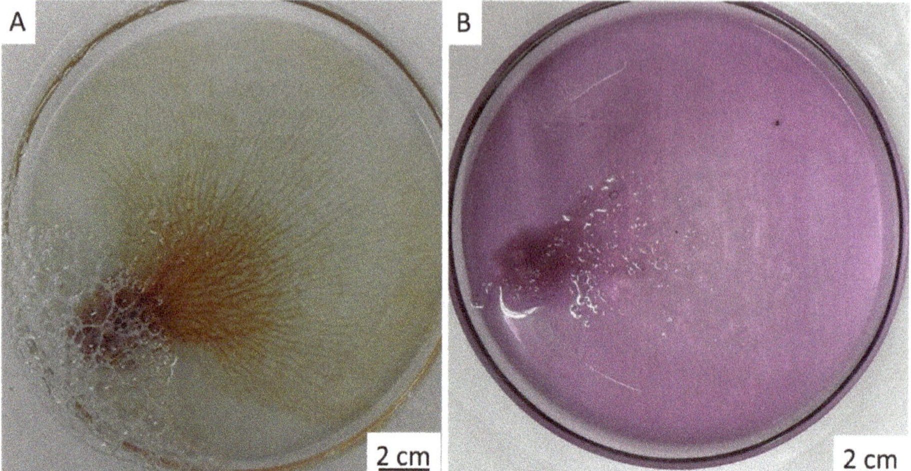

Figure 6. The spherulocite-free chitinous skeleton of *I. flabelliformis* can be isolated after alternating treatment of the construct with acetic acid (**A**) and NaOH (**B**) (see also Figure 7).

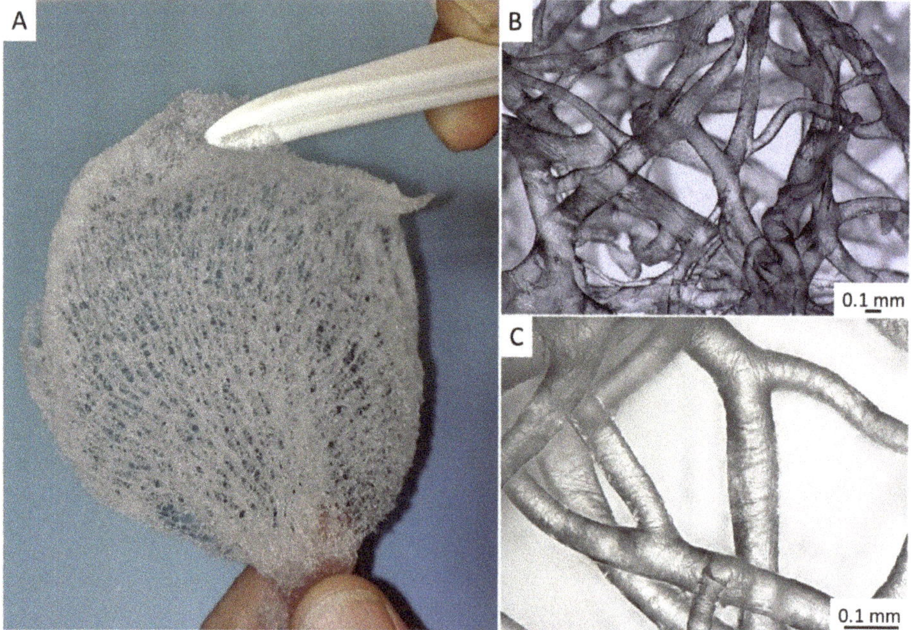

Figure 7. The chitinous skeleton isolated from *I. flabelliformis* (**A**) represents a mechanically elastic, flat, but still three-dimensional (3D)-based construct made of interconnected tubular fibers (**B**). These fibers show excellent capacity for saturation with diverse liquids including water (**C**).

As reported by us previously [13], similar tubular, flat, 3D scaffolds isolated from the marine demosponge *Ianthella basta* closely related to *I. flabelliformis* [11] were applied in the tissue engineering of hMSCs. However, in this study, we decided to examine the possibility of such constructs as drug delivery matrices for the future development of alternatives to well-recognized antimicrobial textiles (for an overview, see References [44–47]).

We used here the chemical compound 10-[dimethyl-[2-[(1*R*,2*S*,5*R*)-5-methyl-2-propan-2-ylcyclohexyl]oxy-2-oxoethyl]azaniumyl]decyl-dimethyl-[2-[(1*R*,2*S*,5*R*)-5-methyl-2-propan-2-ylcyclohexyl]oxy-2-oxoethyl]azanium;dichloride (Figure 8A), known also as decamethoxine (molecular formula $C_{38}H_{74}Cl_2N_2O_4$; molecular weight 693.9 g/mol). It is highly soluble both in water and ethanol. Due to the characteristic nanoporous structure of tubular chitin in verongiid sponges [9,11] (Figure 8B), we suggest that this reagent can easily move through membrane-like walls of chitin fiber both via absorption from the solution and via diffusion (after drying and fixation) on and within the chitinous matrix (Figure 8C). The principal scheme is shown in Figure 8.

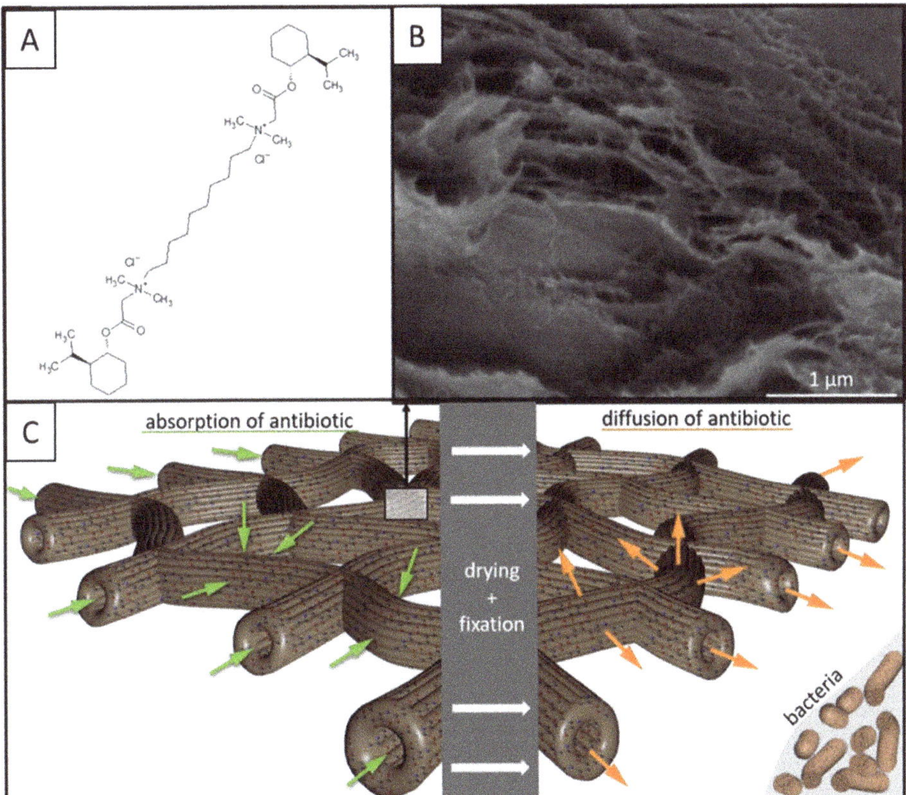

Figure 8. Three-dimensional chitin scaffold as a potential drug delivery matrix. (**A**) Chemical formula of decamethoxine which was used in our study as a model substance with antibiotic activity. The nanoporous structure of *I. flabelliformis* chitin-based tube walls (SEM image, **B**) may be responsible for both the initial absorption of the substance into the organic matrix and for the following diffusion of this substance from the chitinous construct being transferred to the agar plate contaminated with corresponding bacterial cultures (**C**).

The 3D chitin scaffold of *I. flabelliformis* impregnated with a 0.1% water solution of decamethoxine and dried with sterile filter paper inhibited the growth of *S. aureus* in an agar diffusion assay (6 mm from the edge of the matrix) (Figure 9(3)). The chitin scaffold impregnated with 0.1% ethanol solution of decamethoxine and then dried for ethanol evaporation showed less activity (4-mm growth inhibition from the edge of the matrix) (Figure 9(4)). This phenomenon is based, probably, on the difference between water and ethanol with respect to their hydrophilicity. Both samples retained properties to release antiseptic on their second and third use, after moving to a fresh microbe culture. Control samples of chitinous scaffolds used without any treatment, with the exception of sterilization at 121 °C (Figure 9(1),(2)), did not show any antibacterial activity under the same experimental conditions.

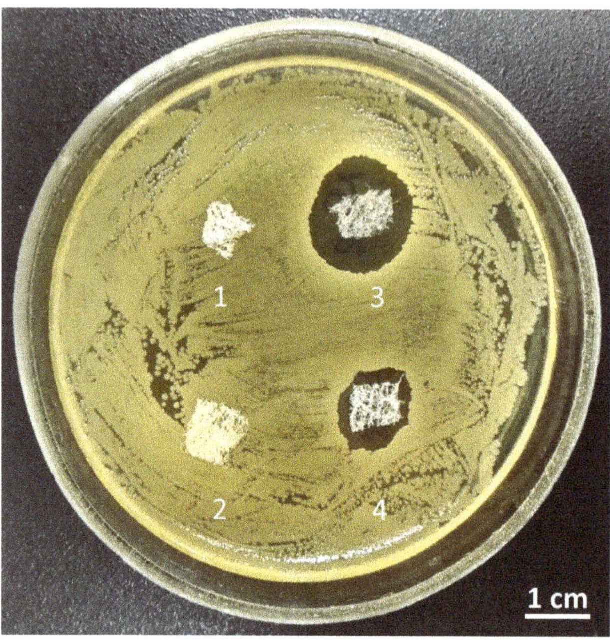

Figure 9. Sponge chitin filled with decamethoxine as an antimicrobial fabric against *Staphylococcus aureus*. 1—Sterile *I. flabelliformis* chitin scaffold (see enlarged image in Figure 7) washed with distilled water as a control sample. 2—Chitin scaffold washed with 70% ethanol and dried as a control sample. 3—Chitin scaffold impregnated with 0.1% water solution of decamethoxine. 4—Chitin scaffold impregnated with 0.1% ethanolic solution of decamethoxin. All observations were carried out after 24 h.

3. Discussion

After the discovery of pigmental cells in the fibrous skeletons of Ianthellids by Flemming in 1872, there were numerous reports discussed by de Laubenfels in 1948 where the possible origin, location, and function of these cells was proposed. Carter, for example, described *Ianthella* sponge in 1881 as follows: *"Sarcode charged with dark purple pigmental cells, especially numerous on the surface and in the horny laminae of the fiber, which appear to be secreted by them."* [34]. At the same time, Lendenfeld (1882) [33] stated the following" *"The most characteristic whereby* Ianthella *is set off is its unique content of fiber cells. These are not scattered through the pith, but are either actually embedded in the fiber walls, or adherent to the inner surface of the hollow cylinder of the fiber which surrounds the pith. These cells must have reached their buried location alive, but how they obtained food and oxygen seems mysterious. Perhaps the pith is quite permeable and acts as supply tube. Another hypothesis is that cells get caught in the forming fiber and die, but leave a pattern of their shape behind, like footprints, or their original substance replaced by bacteria-resistant chemicals."* [33]. Indeed, until the discovery of chitin in fibrous skeletons of verongiid demosponges by our group in 2007 [8], as well as in the verongiid species of the genus *Ianthella* [11], the most accepted opinion was that skeletal fibers are made of a "horny", proteinaceous, biomaterial called spongin [6]. It is also well known that pigment oxidation is responsible for the rapid *Ianthella* sponge tissue changes in color (mostly from yellow to purple, or even blackish purple) [33,35,48]. Preserved specimens usually remain in the dark violet condition (Figures 1, 3 and 4A).

In this study, we showed that the pigmental cells of *I. flabelliformis* are located within skeletal fibers between microlayers of chitin (Figure 5) and contain bromotyrosines. Previously, the biosynthesis of several bromotyrosine-related compounds, i.e., bastadins, by *I. flabelliformis* demosponge was reported [41]. Compounds such as bastadin 25, 15-*O*-sulfonatobastadin 11, and bastadin 26 were identified in *I. flabelliformis*; however, their origin was heretofore elusive, with their production in special cells discovered here. Interestingly, there are still no reports of bastadin or araplysillin as typical bromotyrosines of *Ianthella* species being microbially derived [48]. The intriguing question about the possible ancient microbial origin of the pigmental cells remains open.

The biological function of bromotyrosine-producing cells could be based on previously reported results [49] concerning the inhibition of microbial chitinases using bromotyrosines. In this case, verongiids developed a unique chemical defense strategy to protect their skeletal fibrous chitin from bacterial and fungal invasion. Taking into account our discovery of exceptionally preserved chitin in 505-million-years-old fossil remains of the verongiid sponge *Vauxia gracilenta* [1], we believe that the appearance of this strategy was crucial in the evolution of the sponges belonging to the order Verongiida. Previously, we also showed that partially depigmented chitinous skeletons of selected verongiids are still resistant to diverse bacterial chitinases under experimental conditions in vitro [11]. Only completely purified sponge chitin becomes soluble in chitinase-containing solutions [50,51].

In the near future, we plan to use such a "naturally loaded" bromotyrosine chitin (Figure 3C) to study the possible diffusion of corresponding bromotyrosines using model systems with sea water, physiological solutions, and artificial body fluid.

To our best knowledge, there are no reports on the application of pure chitinous scaffolds for drug delivery. Most papers are dedicated to chitosan or ionically cross-linked chitin microspheres [52]. In one case, a chitin–amphipathic anion/quaternary ammonium salt dressing was prepared [53]. In our study, however, we utilized a recognized antibacterial compound—decamethoxine.

Decamethoxine (its structural formula is shown in Figure 8A) is a cationic gemini surfactant [54], which exhibits strong bactericidal and fungicidal effects. It modifies the permeability of the microbial cell membrane, resulting in the destruction and death of diverse microorganisms [55]. For example, it has a wide spectrum of antimicrobial action on Gram-positive bacteria (*Staphylococcus, Streptococcus, Pneumococcus*), Gram-negative bacteria (*Pseudomonoas, Neisseria gonorrhoeae, Chlamydia trachomatis*) [56], protozoa, dermatophyte, yeast-like fungi of *Candida* genus, and viruses [57]. It was also proven that decamethoxine at a concentration of 10 µg/ml drastically reduces the adhesion of coryneform bacteria, *Salmonella, Staphylococcus*, and *Escherichia* [56]. Its method of action may be achieved via adhesion or competitive binding to bacterial adhesins, or to the surface receptors of host cells. Due to its high bacteriostatic effect, decamethoxine is used for the disinfection of surfaces of diverse surgical tools [58], as well as of contact lenses [59].

Our first results (Figure 9) confirmed that decamethoxine can be successfully absorbed from corresponding water and ethanol-containing solutions by chitinous scaffolds isolated from *I. flabelliformis*. Furthermore, this compound can subsequently diffuse from the chitinous matrix surface, as well as, probably, from the inner space of microtubular and nanoporous structures. The appearance of death zones around colonies of *S. aureus* during 24 h of incubation confirms the antibiotic activity of decamethoxine through diffusion from the chitinous scaffold. Now, we need a longer assay including studies on a Fickian diffusion, as well as on possible non-Fickian behavior [60,61], of this previously non-investigated microtubular chitin matrix. On this first stage, we did not differentiate between the release of substance adsorbed to the outside of the matrix, substance absorbed via nanopores, substance sucked up and released via capillary action, etc. Consequently, it is also not clear with what kind of diffusion-controlled system (matrix-type system or reservoir-type system) [62] is used here. Understanding the structure–function relationship of the sponge biomaterial system represented in this study as a new antimicrobial drug release scaffold [63] could be the key to the successful application of this special delivery system. The drug release kinetics [64] with respect to decamethoxine and other antimicrobial compounds which can be used in naturally pre-structured sponge chitin will be studied in detail in the near future.

4. Materials and Methods

4.1. Location and Collection

The sponge *Ianthella flabelliformis* (WAM Z87073) was collected by J. Fromont and L. Kirkendale at station SOL47/W/A042 (15°36′46.10″ south (S), 124°04′22.92″ east (E) to 15°36′44.77″ S, 124°04′22.38″ E), Kimberley, Western Australia in March 2015 at a depth of 35.3–35.5 m. Morphological identification was supported by molecular barcoding and comparison against reference materials of *I. flabelliformis* and other *Ianthella* spp. from the Western Australian Museum using the 28S ribosomal RNA (rRNA) C-region barcoding region for sponges (see Reference [65] for methodological details).

4.2. Isolation of Chitinous Skeleton from the Sponge and Identification of Selected Bromotyrosines

The isolation of chitinous scaffolds from the ianthellid sponges was conducted as described by us previously [11]. In brief, it was performed in three main steps: (i) sponge skeletons were washed three times with distilled water for the removal of water-soluble compounds, and then bromotyrosines were extracted with methanol; (ii) residual fragments were treated with NaOH (2.5 M, Merck) at 37 °C for 72 h for deproteinization; (iii) lastly, the isolated scaffolds were treated with acetic acid (20%, Roth) at 37 °C for a period of 6 h to remove residual calcium and magnesium carbonates, and then washed in distilled water up to pH 6.8. This isolation procedure was repeated three times to obtain colorless tubular scaffolds. The purity of isolated chitinous scaffolds was proven using standard analytical procedures as described previously [11].

The methanolic extracts of sponge fragments shown in Figure 3C were analyzed using a Shimadzu HPLC system, coupled to an ultraviolet–visible light (UV–Vis) detector (Shimadzu, Kyoto, Japan; Waters SunFire Prep OBD C18 column (30 × 75 mm)). Routine detection was at 215 and 241 nm. A solvent system consisting of MeCN (A) and H_2O (B) at a gradient increasing linearly from 0% to 100% was used for compound separation. LCMS analyses were carried out on an Agilent 1100 (Agilent, Santa Clara, California, USA) LC system equipped with a G1956 MSD detector. A Zorbax C18 RR column was used, and gradient elution with 0.1% HCOOH in H_2O–MeCN was applied.

4.3. Antimicrobial Activity of Chitin Matrix

The prepared chitin scaffold of *I. flabelliformis* was cut into 1-cm^2 squares, washed twice for 15 min in sterile distilled water, and put into 0.1% (*w/v*) water or ethanol solution of decamethoxine (Yuria-pharm, Kyiv, Ukraine). Samples used for control were put into sterile distilled water or 70% ethanol. After 2 h of incubation, the samples were dried with sterile filter paper (for the water solution) or in a thermostat (for ethanol-based solutions). Dry samples were placed on a Petri dish with fresh culture of a clinical strain of *Staphylococcus aureus* ATCC 6538P (FDA 209P) on meat–peptone agar (MPA) and cultivated for 24 h at 37 °C. In 24 h, the zones of growth inhibition for first use were measured, and samples were moved with sterile forceps to a Petri dish with a fresh daily culture of *S. aureus*. Cultivation and measuring were repeated three times with the same samples of chitin scaffold and fresh cultures (first use, second use, and third use). All tests were provided with proper control (sterility control of nutritive environment (MPA), and control of microorganism growth without compound).

4.4. Stereomicroscopy and Light Microscopy Imaging

I. flabelliformis sponge samples in different stages of chemical treatment and isolated chitinous scaffolds were observed with a Keyence VHX-6000 (Keyence, Osaka, Japan) digital optical stereomicroscope, and using a BZ-9000 microscope (Keyence, Osaka, Japan) in the light microscopy mode (Machalowski et al., 2019).

4.5. Transmission Electron Microscopy (TEM), Scanning Electron Microscopy (SEM), and EDX

For TEM investigations, samples of *I. flabelliformis* chitin as represented in Figure 3C were fixed with 2.5% glutaraldehyde in phosphate-buffered saline (PBS) at room temperature, post-fixed with 1.5% osmium tetroxide, dehydrated in a graduated series of acetone (including a staining step with 1% uranylacetate), and embedded in Epoxy resin according to Spurr (1969) [66]. Ultra-thin sections (about 70 nm) of samples were prepared on a Leica EM UC6 ultramicrotome (Leica, Wetzlar, Germany) equipped with a Diatome diamond knife, mounted on Pioloform-coated copper grids, post-stained with uranylacetate and lead citrate (according to Reynolds, 1963 [67]), and analyzed using a Zeiss CTEM 902 (Carl Zeiss, Wetzlar, Germany) at 80 kV (University of Bayreuth).

For SEM analysis, samples were prepared as described for TEM analyses without osmium tetroxide and uranylacetate. For elemental analyses (EDX), the block faces of samples were cut on a Leica EM UC6 ultramicrotome equipped with a Diatome diamond knife, carbon-coated, mounted on an SEM sample holder, and analyzed on a Philips ESEM XL 30 (FEI Company, Peabody, MA, USA) at suitable accelerating voltages. For EDX spectra, accelerating voltages between 15 kV and 30 kV were used.

5. Conclusions

The demand of biomaterials of natural origin is increasing for different reasons, including the major sustainability in their production pipeline leading to the reduction of environmental impact of microplastics and of CO_2 emissions. In this context, the chitin-based materials of marine origin are considered of recent interest, and the results of this work clearly underline that sponges of the order Verongiida can now be considered as a relevant resource in this perspective. The methodology here developed for the first time on *I. flabelliformis* allows a double exploitation of this sponge species: (i) as a source of bromotyrosines of well-known pharmaceutical properties, and (ii) as a source of unique biomaterial that could potentially substitute artificial fabric-based bandages, for its peculiar structural organization and for its capacity to be filled with anti-bacterial compounds. The purified 3D chitin matrix isolated from *I. flabelliformis* does not have innate antibacterial activity, but is prospective as a naturally prefabricated dressing material due to its ability for incorporation of antiseptic solutions using the capillary effect or the fixing of dry antiseptic on the surface of its nanoporous, membrane-like chitin microtubes. Consequently, the very promising results here shown are prodromal for further development both in the field of mariculture and of sponge cell biology. The production of chitin-based scaffolds will be in fact realistically sustainable only after a careful analysis of the farming potentialities of this sponge species. Its widespread presence in the Indo-Pacific does not exclude "a priori" the possibility to develop sponge farming facilities in many geographical areas, and the concrete feasibility of this approach is the consequential evolution of this study. The bromotyrosines produced by *I. flabelliformis* are also well-known interesting compounds. On this specific topic, the interest will be to improve their production not only via simple extraction from farmed sponges but also by developing specific bioreactors for culturing the bromotyrosine-producing cells isolated from the sponge chitinous structures (Figure 10).

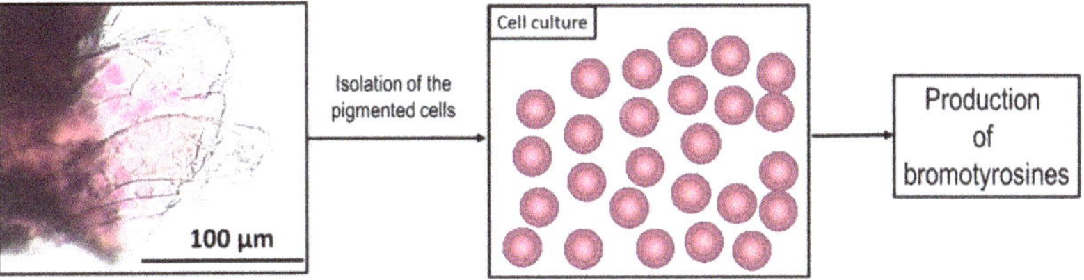

Figure 10. Schematic overview of the challenging tasks: in the near future, we must elaborate an effective method for the isolation of pigmental, bromotyrosine-producing cells from chitinous skeletal fibers of ianthellids with the aim of obtaining cell cultures which should be able to synthetize corresponding bromotyrosines using bioreactors.

Supplementary Materials: The following are available online at http://www.mdpi.com/1660-3397/17/10/574/s1: Figure S1: Structures of bromotyrosines isolated from *I. flabelliformis*, Figure S2: LCMS spectrum of isolated compound (Bastadin 25), Figure S3: LCMS spectrum of isolated compound (Araplysillin-I N20-sulfamate).

Author Contributions: Conceptualization, H.E., K.G., V.K., O.B.S., and S.R.B.; methodology, M.S., A.V., L.M., I.P., D.E., M.V.T., R.M., N.B., and A.F.; validation, H.E., S.R.B., and Y.J.; investigation, B.B., M.S., A.V., L.M., M.W., I.P., V.K., M.V.T., J.F., V.N.I., D.E., A.S., and O.B.S.; resources, J.F., D.E., K.G., and H.E.; writing—original draft preparation, H.E., A.V., M.W., and M.G. (Marco Giovine); writing—review and editing, M.W., J.F., M.G. (Marco Giovine), and V.N.I.; supervision, H.E.; funding acquisition, H.E., K.G., and M.G. (Michael Gelinsky).

Funding: This work was partially supported by DFG Project HE 394/3, SMWK Project no. 02010311 (Germany), and DAAD-Italy Project "Marine Sponges as Sources for Bioinspired Materials Science" (No. 57397326). Work in the Guan group is financially supported by the Free State of Saxony and the European Union EFRE (SAB project "PhänoKard") and by the DFG (GU595/3-1, IRTG2251). Björn Binnewerg was supported by the Else Kröner Fresenius Stiftung (Else-Kröner Promotionskolleg, EKFS Foundation). M.W. is financially supported by the Polish National Agency for Academic Exchange (PPN/BEK/2018/1/00071).

Acknowledgments: The authors acknowledge Oliver Gomez, Western Australian Museum, for technical support. Collection of the sponge was undertaken as part of the Western Australian Marine Science Institution project (Theme 1.1.1, http://www.wamsi.org.au/Kimberley-science-node). D.E. would like to thank Leonard Namuth and Gert Wörheide for various aspects of the barcoding work. The authors are thankful to Sarah Tsurkan for English proof reading.

Conflicts of Interest: The authors declare no conflicts of interest.

References

1. Ehrlich, H.; Rigby, J.K.; Botting, J.P.; Tsurkan, M.V.; Werner, C.; Schwille, P.; Petrášek, Z.; Pisera, A.; Simon, P.; Sivkov, V.N.; et al. Discovery of 505-million-year old chitin in the basal demosponge *Vauxia gracilenta*. *Sci. Rep.* **2013**, *3*, 3497. [CrossRef] [PubMed]

2. Mioso, R.; Marante, F.J.T.; Bezerra, R.D.S.; Borges, F.V.P.; Santos, B.V.D.O.; De Laguna, I.H.B. Cytotoxic compounds derived from marine sponges: A review (2010–2012). *Molecules* **2017**, *22*, 208. [CrossRef] [PubMed]

3. Helber, S.B.; Hoeijmakers, D.J.J.; Muhando, C.A.; Rohde, S.; Schupp, P.J. Sponge chemical defenses are a possible mechanism for increasing sponge abundance on reefs in Zanzibar. *PLoS ONE* **2018**, *13*, e0197617. [CrossRef] [PubMed]

4. Ebel, R.; Brenzinger, M.; Kunze, A.; Gross, H.J.; Proksch, P. Wound activation of protoxins in marine sponge *Aplysina aerophoba*. *J. Chem. Ecol.* **1997**, *23*, 1451–1462. [CrossRef]

5. Green, D.W.; Howard, D.; Yang, X.; Kelly, M.; Oreffo, R.O.C. Natural marine sponge fiber skeleton: A biomimetic scaffold for human osteoprogenitor cell attachment, growth, and differentiation. *Tissue Eng.* **2003**, *9*, 1159–1166. [CrossRef] [PubMed]

6. Jesionowski, T.; Norman, M.; Żółtowska-Aksamitowska, S.; Petrenko, I.; Joseph, Y.; Ehrlich, H. Marine spongin: Naturally prefabricated 3D scaffold-based biomaterial. *Mar. Drugs* **2018**, *16*, 88. [CrossRef] [PubMed]

7. Ehrlich, H.; Wysokowski, M.; Żółtowska-Aksamitowska, S.; Petrenko, I.; Jesionowski, T. Collagens of poriferan origin. *Mar. Drugs* **2018**, *16*, 79. [CrossRef] [PubMed]

8. Ehrlich, H.; Maldonado, M.; Spindler, K.-D.; Eckert, S.; Hanke, T.; Born, R.; Goebel, C.; Simon, P.; Heinemann, S.; Worch, H. First evidence of chitin as a component of the skeletal fibres of marine sponges. Part I. Verongidae (Demospongia: Porifera). *J. Exp. Zool. B Mol. Dev. Evol.* **2007**, *308*, 347–356. [CrossRef]

9. Ehrlich, H.; Ilan, M.; Maldonado, M.; Muricy, G.; Bavestrello, G.; Kljajic, Z.; Carballo, J.L.; Schiaparelli, S.; Ereskovsky, A.; Schupp, P.; et al. Three-dimensional chitin-based scaffolds from Verongida sponges (Demospongiae: Porifera). Part I. Isolation and identification of chitin. *Int. J. Biol. Macromol.* **2010**, *47*, 132–140. [CrossRef]

10. Ehrlich, H.; Shaala, L.A.; Youssef, D.T.A.; Zoltowska-Aksamitowska, S.; Tsurkan, M.; Galli, R.; Meissner, H.; Wysokowski, M.; Petrenko, I.; Tabachnick, K.R.; et al. Discovery of chitin in skeletons of non-verongiid Red Sea demosponges. *PLoS ONE* **2018**, *13*, 1–18. [CrossRef]

11. Brunner, E.; Ehrlich, H.; Schupp, P.; Hedrich, R.; Hunoldt, S.; Kammer, M.; Machill, S.; Paasch, S.; Bazhenov, V.V.; Kurek, D.V.; et al. Chitin-based scaffolds are an integral part of the skeleton of the marine demosponge *Ianthella basta*. *J. Struct. Biol.* **2009**, *168*, 539–547. [CrossRef] [PubMed]

12. Rohde, S.; Schupp, P.J. Growth and regeneration of the elephant ear sponge *Ianthella basta* (Porifera). *Hydrobiologia* **2012**, *687*, 219–226. [CrossRef]

13. Mutsenko, V.V.; Gryshkov, O.; Lauterboeck, L.; Rogulska, O.; Tarusin, D.N.; Bazhenov, V.V.; Schütz, K.; Brüggemeier, S.; Gossla, E.; Akkineni, A.R.; et al. Novel chitin scaffolds derived from marine sponge *Ianthella basta* for tissue engineering approaches based on human mesenchymal stromal cells: Biocompatibility and cryopreservation. *Int. J. Biol. Macromol.* **2017**, *104*, 1955–1965. [CrossRef] [PubMed]

14. Stepniak, I.; Galinski, M.; Nowacki, K.; Wysokowski, M.; Jakubowska, P.; Bazhenov, V.V.; Leisegang, T.; Ehrlich, H.; Jesionowski, T. A novel chitosan/sponge chitin origin material as a membrane for supercapacitors-preparation and characterization. *RSC Adv.* **2016**, *6*, 4007–4013. [CrossRef]

15. Wysokowski, M.; Petrenko, I.; Stelling, A.L.; Stawski, D.; Jesionowski, T.; Ehrlich, H. Poriferan chitin as a versatile template for extreme biomimetics. *Polymers (Basel)* **2015**, *7*, 235–265. [CrossRef]

16. Ehrlich, H. Chitin of poriferan origin as a unique biological material. In *Blue Biotechnology: Production and Use of Marine Molecules*; Bates, S.S., La Barre, S., Eds.; Wiley-VCH: Weinheim, Germany, 2019; pp. 821–854.

17. Fromont, J.; Wahab, M.A.A.; Gomez, O.; Ekins, M.; Grol, M.; Hooper, J.N.A. Patterns of sponge biodiversity in the Pilbara, Northwestern Australia. *Diversity* **2016**, *8*, 21. [CrossRef]

18. Bechmann, N.; Ehrlich, H.; Eisenhofer, G.; Ehrlich, A.; Meschke, S.; Ziegler, C.G.; Bornstein, S.R. Anti-tumorigenic and anti-metastatic activity of the sponge-derived marine drugs aeroplysinin-1 and isofistularin-3 against pheochromocytoma in vitro. *Mar. Drugs* **2018**, *16*, 172. [CrossRef]

19. Kumar, M.S.; Adki, K.M. Marine natural products for multi-targeted cancer treatment: A future insight. *Biomed. Pharmacother.* **2018**, *105*, 233–245. [CrossRef]

20. Niemann, H.; Hagenow, J.; Chung, M.-Y.; Hellio, C.; Weber, H.; Proksch, P. SAR of Sponge-inspired hemibastadin congeners inhibiting blue mussel phenoloxidase. *Mar. Drugs* **2015**, *13*, 3061–3071. [CrossRef]

21. Bayer, M.; Hellio, C.; Marechal, J.P.; Frank, W.; Lin, W.; Weber, H.; Proksch, P. Antifouling bastadin congeners target mussel phenoloxidase and complex copper(II) Ions. *Mar. Biotechnol.* **2011**, *13*, 1148–1158. [CrossRef]

22. Calcul, L.; Inman, W.D.; Morris, A.A.; Tenney, K.; Ratnam, J.; McKerrow, J.H.; Valeriote, F.A.; Crews, P. Additional insights on the bastadins: Isolation of analogues from the sponge *Ianthella* cf. reticulata and exploration of the oxime configurations. *J. Nat. Prod.* **2010**, *73*, 365–372. [CrossRef] [PubMed]

23. Franklin, M.A.; Penn, S.G.; Lebrilla, C.B.; Lam, T.H.; Pessah, I.N.; Molinski, T.F. Bastadin 20 and bastadin O-sulfate esters from *Ianthella basta*: Novel modulators of the Ry1R FKBP12 receptor complex. *J. Nat. Prod.* **1996**, *59*, 1121–1127. [CrossRef] [PubMed]

24. Greve, H.; Kehraus, S.; Krick, A.; Kelter, G.; Maier, A.; Fiebig, H.H.; Wright, A.D.; König, G.M. Cytotoxic bastadin 24 from the Australian sponge *Ianthella quadrangulata*. *J. Nat. Prod.* **2008**, *71*, 309–312. [CrossRef] [PubMed]

25. Mathieu, V.; Wauthoz, N.; Lefranc, F.; Niemann, H.; Amighi, K.; Kiss, R.; Proksch, P. Cyclic versus hemi-bastadins. Pleiotropic anti-cancer effects: From apoptosis to anti-angiogenic and anti-migratory effects. *Molecules* **2013**, *18*, 3543–3561. [CrossRef] [PubMed]

26. Le Norcy, T.; Niemann, H.; Proksch, P.; Tait, K.; Linossier, I.; Réhel, K.; Hellio, C.; Faÿ, F. Sponge-inspired dibromohemibastadin prevents and disrupts bacterial biofilms without toxicity. *Mar. Drugs* **2017**, *15*, 222. [CrossRef]

27. Gartshore, C.J.; Salib, M.N.; Renshaw, A.A.; Molinski, T.F. Isolation of Bastadin-6-O-Sulfate and expedient purifications of Bastadins-4, −5 and −6 from extracts of *Ianthella basta*. *Fitoterapia* **2018**, *126*, 16–21. [CrossRef] [PubMed]

28. Van Soest, R.W.M.; Boury-Esnault, N.; Hooper, J.N.A.; Rützler, K.; de Voogd, N.J.; Alvarez, B.; Hajdu, E.; Pisera, A.B.; Manconi, R.; Schönberg, C.; et al. World Porifera Database. *Ianthella Flabelliformis* (Linnaeus, 1759). Available online: http://www.marinespecies.org/aphia.php?p=taxdetails&id=169690 (accessed on 8 October 2019).

29. Linnaeus, C. *Systema Naturæ per Regna Tria Naturæ, Secundum Classes, Ordines, Genera, Species, Cum Characteribus, Differentiis, Synonymis, Locis*; Vindobonae: Vienna, Austria, 1759. [CrossRef]

30. Weuffen, W.; Berencsi, G.; Groschel, D.; Kemter, B.; Kramer, A.K.A.P. *Handbuch der Antiseptik; Band 2: Antiseptika; Teil 3: Antibakterielle, Antifungielle und antivirale Antiseptik-Ausgewählte Wirkstoffe*; VEB Verlag Volk und Gesundheit.: Berlin, Germany, 1987.

31. Paliy, G.K.; Nazarchuk, O.A.; Kulakov, O.I.; Paliy, V.G.; Nazarchuk, S.A.; Paliy, D.V.; Kordon, Y.V.; Gonchar, O.O. Substaniation of antimicrobial dressings use in surgery. *Med. Perspekt.* **2014**, *19*, 152–158. [CrossRef]

32. Klinger, C.; Żółtowska-Aksamitowska, S.; Wysokowski, M.; Tsurkan, M.V.; Galli, R.; Petrenko, I.; Machałowski, T.; Ereskovsky, A.; Martinović, R.; Muzychka, L.; et al. Express method for isolation of ready-to-use 3D chitin scaffolds from *Aplysina archeri* (Aplysineidae: Verongiida) demosponge. *Mar. Drugs* **2019**, *17*, 131. [CrossRef]

33. Lendenfeld, R.L.R. Das hornfaserwachstum der aplysinidae. *Zool. Anz.* **1882**, *5*, 636.

34. Carter, H.J. Contributions to our knowledge of the Spongida. *Ann. Mag. Nat. Hist.* **1881**, *8*, 241–259. [CrossRef]

35. Lendenfeld, R.L.R. *A Monograph of the Horny Sponges*; Triibner and Co.: London, UK, 1889.

36. Polejaeff, N.N. Report on the Keratosa collected by H.M.S. Challenger during the years 1873–1876. In *Report on the Scientific Results of the Voyage of H.M.S. Challenger during the Years 1873–1876*; for Her Majesty's Stationary Office, Zoology: London, UK; Edinburgh, Dublin, Ireland, 1884; pp. 1–88.

37. De Laubenfels, M.W. The order of Keratosa of the phylum Porifera. A monographic study. *Occ. Pap. Allan Hancock Found.* **1948**, *3*, 1–217.

38. Vacelet, J. Les cellules a inclusions de léponge cornée *Verongia cavernicola*. *J. Microsc.* **1967**, *6*, 237–240.

39. Turon, X.; Becerro, M.A.; Uriz, M.J. Distribution of brominated compounds within the sponge Aplysina aerophoba: Coupling of X-ray microanalysis with cryofixation techniques. *Cell Tissue Res.* **2000**, *301*, 311–322. [CrossRef] [PubMed]

40. Bergquist, P.R.; de Cook, S.C. Order Verongida Bergquist, 1978. In *Systema Porifera*; Hooper, J.N.A., Van Soest, R.W.M., Willenz, P., Eds.; Springer: Boston, MA, USA, 2002; p. 1081.

41. Carroll, A.R.; Kaiser, S.M.; Davis, R.A.; Moni, R.W.; Hooper, J.N.A.; Quinn, R.J. A bastadin with potent and selective δ-opioid receptor binding affinity from the Australian sponge *Ianthella flabelliformis*. *J. Nat. Prod.* **2010**, *73*, 1173–1176. [CrossRef] [PubMed]

42. Motti, C.A.; Freckelton, M.L.; Tapiolas, D.M.; Willis, R.H. FTICR-MS and LC-UV/MS-SPE-NMR applications for the rapid dereplication of a crude extract from the sponge *Ianthella flabelliformis*. *J. Nat. Prod.* **2009**, *72*, 290–294. [CrossRef] [PubMed]

43. Rogers, E.W.; Molinski, T.F. Highly polar spiroisoxazolines from the sponge *Aplysina fulva*. *J. Nat. Prod.* **2007**, *70*, 1191–1194. [CrossRef]

44. Höfer, D. Antimicrobial Textiles – Evaluation of their effectiveness and safety. In *Biofunctional Textiles and the Skin*; Hipler, U.-C., Elsner, P., Eds.; Karger: Basel, Switzerland, 2006; pp. 42–50.

45. Kramer, A.; Guggenbichler, P.; Heldt, P.; Jünger, M.; Ladwig, A.; Thierbach, H.; Weber, U.; Daeschlein, G. Hygienic relevance and risk assessment of antimicrobial-impregnated textiles. *Curr. Probl. Dermatol.* **2006**, *33*, 78–109. [PubMed]

46. Elsner, P. Antimicrobials and the skin: Physiological and pathological flora. *Curr. Probl. Dermatol.* **2006**, *33*, 35–41.

47. Gokarneshan, N.; Nagarajan, V.B.; Viswanath, S.R. Developments in antimicrobial textiles—Some insights on current research trends. *Biomed. J. Sci Tech. Res.* **2017**, *1*, 230–233.

48. Freckelton, M.L.; Luter, H.M.; Andreakis, N.; Webster, N.S.; Motti, C.A. Qualitative variation in colour morphotypes of *Ianthella basta* (Porifera: Verongida). In *Ancient Animals, New Challenges. Developments in Hydrobiology*; Springer: Dordrecht, The Netherlands, 2011.

49. Tabudravu, J.N.; Eijsink, V.G.H.; Gooday, G.W.; Jaspars, M.; Komander, D.; Legg, M.; Synstad, B.; Van Aalten, D.M.F. Psammaplin A, a chitinase inhibitor isolated from the Fijian marine sponge *Aplysinella rhax*. *Bioorganic Med. Chem.* **2002**, *10*, 1123–1128. [CrossRef]

50. Ehrlich, H.; Bazhenov, V.V.; Debitus, C.; de Voogd, N.; Galli, R.; Tsurkan, M.V.; Wysokowski, M.; Meissner, H.; Bulut, E.; Kaya, M.; et al. Isolation and identification of chitin from heavy mineralized skeleton of *Suberea clavata* (Verongida: Demospongiae: Porifera) marine demosponge. *Int. J. Biol. Macromol.* **2017**, *104*, 1706–1712. [CrossRef] [PubMed]

51. Shaala, L.A.; Asfour, H.Z.; Youssef, D.T.A.; óltowska-Aksamitowska, S.Z.; Wysokowski, M.; Tsurkan, M.; Galli, R.; Meissner, H.; Petrenko, I.; Tabachnick, K.; et al. New source of 3D chitin scaffolds: The red sea demosponge *Pseudoceratina arabica* (pseudoceratinidae, verongiida). *Mar. Drugs* **2019**, *17*, 92. [CrossRef] [PubMed]

52. Shang, Y.; Ding, F.; Xiao, L.; Deng, H.; Du, Y.; Shi, X. Chitin-based fast responsive pH sensitive microspheres for controlled drug release. *Carbohydr. Polym.* **2014**, *102*, 413–418. [CrossRef]

53. Zhou, D.; Yang, R.; Yang, T.; Xing, M.; Gaoxing, L.G. Preparation of chitin-amphipathic anion/quaternary ammonium salt ecofriendly dressing and its effect on wound healing in mice. *Int. J. Nanomed.* **2018**, *13*, 4157–4169. [CrossRef] [PubMed]

54. Dement'eva, O.V.; Naumova, K.A.; Senchikhin, I.N.; Roumyantseva, T.B.; Rudoy, V.M. Sol–gel synthesis of mesostructured SiO2 containers using vesicles of hydrolyzable bioactive gemini surfactant as a template. *Colloid J.* **2017**, *79*, 451–458. [CrossRef]

55. Shchetina, V.N.; Belanov, E.F.; Starobinets, Z.G.; Volianskiĭ, I. The effect of dexamethoxin on the integrity of cytoplasmic membrane in gram-positive and gram-negative microorganisms. *Mikrobiol. Zh* **1990**, *52*, 24–28. [PubMed]

56. Nazarchuk, O.A.; Chereshniuk, I.L.; Nazarchuk, H.H. The research of antimicrobial efficacy of antiseptics decamethoxin, miramistin and their effect on nuclear DNA fragmentation and epithelial cell cycle. *Wiad. Lek.* **2019**, *72*, 374–380.

57. Fuss, J.; Palii, V.; Voloboyeva, A. evaluating the effectivenes of antiseptic solution decasan in treatment of necrotic soft tisue diseases. *Pol. Przegl. Chir.* **2016**, *88*, 233–237. [CrossRef]

58. Paliy, G.K.; Lukhimel, A.D.; Onofreĭchuk, I.F. Disinfection of surgical silk with decamethoxin, antibiotics and their combinations. *Antibiotiki* **1978**, *88*, 629–633.

59. Galatenko, N.A.; Nechaeva, L.; Bufius, N.N. Possibilities of the chemical sterilization of soft contact lenses made of polyacrylamide. *Gig. Sanit.* **1991**, *7*, 67–68.

60. Ritger, P.L.; Peppas, N.A.A. Fickian and non-Fickian release from non-swellable devices in the form of slabs, spheres, cylinders or discs. *J. Control. Release* **1987**, *5*, 23–36. [CrossRef]

61. Peppas, A.N. Analysis of Fickian and non-Fickian drug release from polymers. *Pharm. Acta Helv.* **1985**, *60*, 110–111.

62. He, C.; Nie, W.; Feng, W. Engineering of biomimetic nanofibrous matrices for drug delivery and tissue engineering. *J. Mater. Chem. B* **2014**, *2*, 7828–7848. [CrossRef]

63. Prabu, P.; Kim, K.W.; Dharmaraj, N.; Park, J.H.; Khil, M.S.; Kim, H.Y. Antimicrobial drug release scaffolds of natural and synthetic biodegradable polymers. *Macromol. Res.* **2008**, *16*, 303–307. [CrossRef]

64. Kamaly, N.; Yameen, B.; Wu, J.; Omid, C.; Farokhzad, O.C. Degradable controlled-release polymers and polymeric nanoparticles: Mechanisms of controlling drug release. *Chem. Rev.* **2016**, *116*, 2602–2663. [CrossRef]

65. Erpenbeck, D.; Voigt, O.; Al-Aidaroos, A.M.; Berumen, M.L.; Büttner, G.; Catania, D.; Guirguis, A.N.; Paulay, G.; Schätzle, S.; Wörheide, G. Molecular biodiversity of Red Sea demosponges. *Mar. Pollut. Bull.* **2016**, *105*, 507–514. [CrossRef]

66. Spurr, A.R. A low-viscosity epoxy resin embedding medium for electron microscopy. *J. Ultrastruct. Res.* **1969**, *26*, 31–43. [CrossRef]

67. Reynolds, E.S. The use of lead citrate at high pH as an electron-opaque stain in electron microscopy. *J. Cell Biol.* **1963**, 208–212. [CrossRef]

Article

Collagen Peptides from Swim Bladders of Giant Croaker (*Nibea japonica*) and Their Protective Effects against H_2O_2-Induced Oxidative Damage toward Human Umbilical Vein Endothelial Cells

Jiawen Zheng [1,†], Xiaoxiao Tian [1,†], Baogui Xu [1], Falei Yuan [1], Jianfang Gong [2] and Zuisu Yang [1,*]

[1]   Zhejiang Provincial Engineering Technology Research Center of Marine Biomedical Products, School of Food and Pharmacy, Zhejiang Ocean University, Zhoushan 316022, China; jwzheng1996@163.com (J.Z.); TIANXIAOXIAO0208@163.com (X.T.); xubaogui96@163.com (B.X.); yuanfalei@zjou.edu.cn (F.Y.)

[2]   Donghai Science and Technology College, Zhejiang Ocean University, Zhoushan 316000, China; gjf200527@163.com

*   Correspondence: abc1967@126.com; Tel.: +86-0580-226-0600; Fax: +86-0580-254-781

†   These authors contributed equally to this work.

Received: 18 July 2020; Accepted: 15 August 2020; Published: 18 August 2020

Abstract: Five different proteases were used to hydrolyze the swim bladders of *Nibea japonica* and the hydrolysate treated by neutrase (collagen peptide named SNNHs) showed the highest DPPH radical scavenging activity. The extraction process of SNNHs was optimized by response surface methodology, and the optimal conditions were as follows: a temperature of 47.2 °C, a pH of 7.3 and an enzyme concentration of 1100 U/g, which resulted in the maximum DPPH clearance rate of 95.44%. Peptides with a Mw of less than 1 kDa (SNNH-1) were obtained by ultrafiltration, and exhibited good scavenging activity for hydroxyl radicals, ABTS radicals and superoxide anion radicals. Furthermore, SNNH-1 significantly promoted the proliferation of HUVECs, and the protective effect of SNNH-1 against oxidative damage of H_2O_2-induced HUVECs was investigated. The results indicated that all groups receiving SNNH-1 pretreatment showed an increase in GSH-Px, SOD, and CAT activities compared with the model group. In addition, SNNH-1 pretreatment reduced the levels of ROS and MDA in HUVECs with H_2O_2-induced oxidative damage. These results indicate that collagen peptides from swim bladders of *Nibea japonica* can significantly reduce the oxidative stress damage caused by H_2O_2 in HUVECs and provides a basis for the application of collagen peptides in the food industry, pharmaceuticals, and cosmetics.

Keywords: *Nibea japonica*; swim bladder; marine collagen peptides; antioxidant activity

1. Introduction

Redox reactions are basic physiological and chemical reactions that are constantly carried out in the human body. During metabolism in the human body, large amounts of reactive oxygen species (ROS) are generated and the antioxidant defense system in the body removes active oxygen free radicals in order to prevent damage to the body [1,2]. Under normal circumstances, a healthy human body has an efficient and dynamic antioxidant defense system where the generation and removal of oxidative radicals in the human body are in a dynamic balance. However, once the human body generates excessive reactive oxygen radicals or the antioxidant defense system becomes inefficient, this dynamic balance is disturbed, which generates a state of oxidative stress, leading to aging and a number of chronic diseases, such as diabetes and coronary arteriosclerosis [1,3]. ROS are the products of redox reactions in the body and are oxygen free radicals. In addition, ROS play direct, critical roles in the body's oxidative stress reaction and are one of the most harmful free radicals to human tissues. Furthermore, ROS can produce excessive oxidation on cell membrane structures, nucleic acids, lipids and proteins as well as cause changes in tissue structure and produce adverse effects, such as physiological dysfunction [4–6]. Under normal circumstances, the body can remove excess ROS in two ways to reduce the damage caused by oxidative stress. One way is through a series of antioxidant mechanisms in the body, which clear excess ROS and maintain the dynamic balance of ROS. These antioxidant mechanisms involve glutathione peroxidase (GSH-Px), superoxide dismutase (SOD), catalase (CAT) and other small-molecule reducing substances. The second way is the intake of external substances, such as vitamin C, carotenoids, glutathione (GSH) and melatonin [7,8]. Therefore, supplementing with antioxidants to eliminate excess ROS may be an effective means for treating chronic diseases [7].

In addition, antioxidant peptides prevent peroxidation in the body and help the body clear active oxygen free radicals [9]. In recent years, an increasing number of studies have found that certain peptides have the ability to scavenge free radicals, which provides a new idea for identifying novel antioxidants. However, there are only a few types of naturally-occurring, antioxidant-active peptides, which has made obtaining novel antioxidant-active peptides by hydrolyzing macromolecular proteins a research hotspot [10,11]. In addition, research related to food health products, cosmetics, and pharmaceutical industries with antioxidants as the mechanism will also move in the direction of researching active antioxidant peptides [12]. For example, Chi et al. [13] obtained three antioxidant peptides from the protein hydrolysate of bluefin leatherjacket (*Navodon septentrionalis*) skin, which all had molecular weights (Mw) of less than 700 Da. In addition, the strength of the antioxidant activities of these peptides increased as their Mw decreased. Another example is a small-molecule antioxidant peptide obtained from horse mackerel (*Magalaspis cordyla*) viscera protein. This antioxidant peptide had a Mw of 518.5 Da and had a good ability to scavenge 2, 2-diphenyl-1-picrylhydrazyl (DPPH) and hydroxyl radicals [14]. You et al. [15–17] hydrolyzed loach, jellyfish and grass carp. After these hydrolyzed polypeptide products were administered to mice over time by oral gavage, the activity of the antioxidant enzyme system, SOD and GSH-Px, was detected to varying degrees. Taken together, these results indicate that marine proteins are a high-quality raw material for preparing antioxidant peptides.

Previously, acid soluble collagen (ASC) and pepsin soluble collagen (PSC) were extracted from the swim bladders and skin of *Nibea japonica*. Both ASC and PSC are type I collagens, which have potential wound healing functions and can be applied to the field of cosmetics [18–20]. However, there have been no reports on the extraction of antioxidant peptides from the swim bladders of *Nibea japonica*. Therefore, our study further utilizes the swim bladders of *Nibea japonica* and used neutrase to obtain products with the highest DPPH free radical scavenging rate (SNNHs). After ultrafiltration, the hydrolysate with Mw of less than 1 kDa (SNNH-1) had the highest scavenging rate and was extracted for functional evaluation. Here, the antioxidant activity of SNNH-1 in vitro was analyzed. Our findings showed that SNNH-1 can be used as a marine antioxidant and provide a basis for its application in the food and pharmaceutical fields.

2. Results and Discussion

2.1. Single Factor Results

2.1.1. Selection of the Optimal Enzyme

To obtain the hydrolysate with ideal activity, research must be carried out to find the best proteolytic enzyme. The defatted swim bladders were used to hydrolyze the protein for 4 h at the optimal temperature and pH of each enzyme (trypsin at pH 8.0, 50 °C; neutrase at pH 7.0, 45 °C; alcalase at pH 9.0, 50 °C; pepsin at pH 2.0, 37 °C; and papain at pH 6.0, 55 °C). Each enzyme was used at 1000 U/g. The swim bladders were mixed with water at the ratio of 1:4. To inactivate the enzyme, the hydrolysate was boiled for 10 min. Subsequently, the hydrolysate was centrifuged at 12,000× g rpm for 10 min at 4 °C. The supernatant was stored at −80 °C overnight, freeze-dried, and stored at −20 °C. The optimal hydrolase of the swim bladders of *Nibea japonica* was screened by the DPPH clearance rate.

As shown in Figure 1, the neutrase hydrolysate exhibited the highest DPPH radical scavenging activity. Therefore, neutrase was selected for the preparation of proteolytic products of the swim bladder from *Nibea japonica*. Neutrase hydrolysate obtained under the optimum conditions was called SNNHs and was stored at −20 °C.

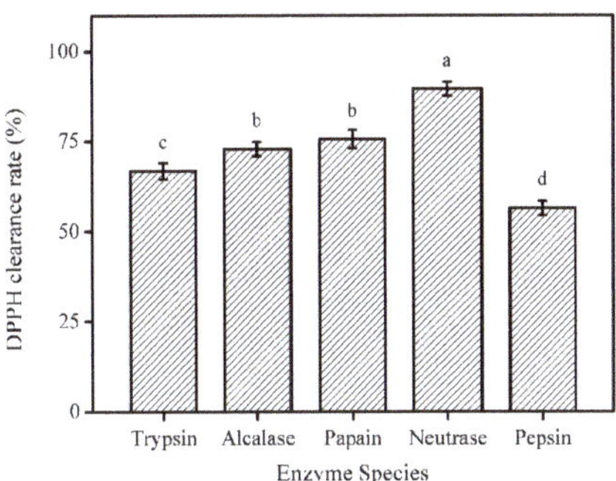

Figure 1. DPPH radical scavenging activity of hydrolysates produced by various proteases. The concentration of hydrolyzed product was 10 mg/mL. All results were triplicates of the mean ± SD. Different letters indicate significant differences between groups ($p < 0.05$).

2.1.2. Single Factor Experiments

Figure 2 shows the influence of five single factors on DPPH radical scavenging activity. Within the range of the five single factors, the DPPH clearance rate was first increased to the maximum value and then decreased. The optimal enzymolysis conditions corresponding to the highest level of DPPH clearance in the range test were as follows: enzyme concentration of 1000 U/g, a solid:liquid ratio of 1:5 (*w/v*), a hydrolysis time of 6.0 h, a pH of 7.0, and a temperature of 45 °C.

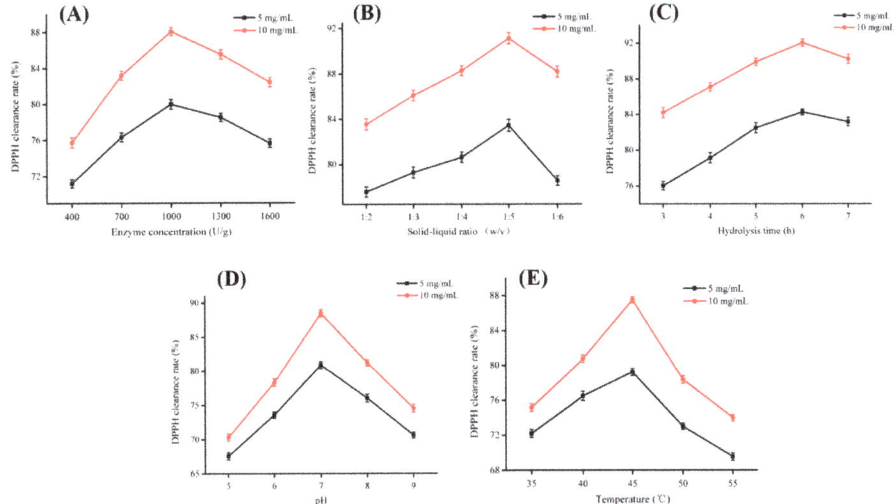

Figure 2. Effects of five single factors on DPPH radical scavenging activity. (**A**) Enzyme concentration (U/g); (**B**) solid:liquid ratio (*w/v*); (**C**) hydrolysis time (h); (**D**) pH; and (**E**) temperature (°C). The concentration of hydrolyzed product was 5 mg/mL and 10 mg/mL, respectively. All results were triplicates of the mean ± SD.

2.2. Optimization of Extraction Parameters by Response Surface Methodology (RSM)

2.2.1. Response Surface Analysis

According to the results of the single factor experiment, the RSM of the Box–Benhnken (BBD) was used to analyze the optimal levels of three independent factors that had significant influences on DPPH radical scavenging rate. The experimental design and results are listed in Table 1. Based on the regression analysis of the data, the second-order polynomial function was used to predict the effect of these three factors on DPPH clearance, as follows: $Y = 94.16 + 3.37X_1 + 1.97X_2 + 1.47X_3 - 0.47X_1X_2 + 0.14X_1X_3 - 1.18X_2X_3 - 3.68X_1^2 - 2.34X_2^2 - 1.79X_3^2$ (where Y was the DPPH clearance rate, and X_1, X_2 and X_3 were the temperature, pH, and enzyme concentration, respectively).

Table 1. The Box–Behnken design and the response for the DPPH clearance rate.

Runs	Temperature (X_1)	pH (X_2)	Enzyme Concentration (X_1)	DPPH Clearance Rate (%) (Y)
1	40	7	700	84.43
2	50	7	1300	93.24
3	45	8	1300	93.09
4	45	7	1000	94.33
5	45	6	1300	90.92
6	50	7	700	91.03
7	45	7	1000	93.78
8	40	8	1000	86.98
9	45	6	700	84.61
10	50	8	1000	92.62
11	40	7	1300	86.06
12	50	6	1000	90.24
13	45	8	700	91.51
14	45	7	1000	94.36
15	40	6	1000	82.71

Table 2 presents the analysis of variance (ANOVA) results of the quadratic model, from which the significance of the quadratic model can be determined. When the value of F was larger and the value of p was smaller, the corresponding variable was more statistically significant. If the p value was greater than 0.05, then the model item was not statistically significant. The F-value and p-value ($p = 0.0004$) demonstrate that the model had high significance. The model's ANOVA decision coefficient ($R^2 = 0.9860$) and adjusted decision coefficient ($R_{Adj}^2 = 0.9615$) also show that the model was highly significant. Therefore, we decided on this model for optimization.

In addition, three-dimensional response surfaces and contour plots generated by the model equation can intuitively explain the interaction between the two factors (Figure 3) and show the optimal levels of each component required for the maximum DPPH clearance rate. The maximum DPPH radical scavenging rate was 95.44% at the following conditions: temperature was 47.2 °C, pH was 7.3, and enzyme concentration was 1100 U/g.

Table 2. Analysis of variance of the regression model.

Source	Sum of Squares	df	Mean Square	F Value	p Value
Model	218.02	9	24.22	39.82	0.0004
X_1	90.79	1	90.79	149.22	<0.0001
X_2	30.89	1	30.89	50.77	0.0008
X_3	17.20	1	17.20	28.27	0.0031
X_1X_2	0.89	1	0.89	1.47	0.2798
X_1X_3	0.084	1	0.084	0.14	0.7253
X_2X_3	5.59	1	5.59	9.19	0.0290
X_1^2	50.03	1	50.03	82.22	0.0003
X_2^2	20.19	1	20.19	33.18	0.0022
X_3^2	11.78	1	11.78	19.35	0.0070
Residual	3.04	5	0.61		
Lack of fit	2.83	3	0.94	8.84	0.1033
Pure Error	0.21	2	0.11		
Cor Total	221.07	14			
R^2					0.9862
R_{adj}^2					0.9615

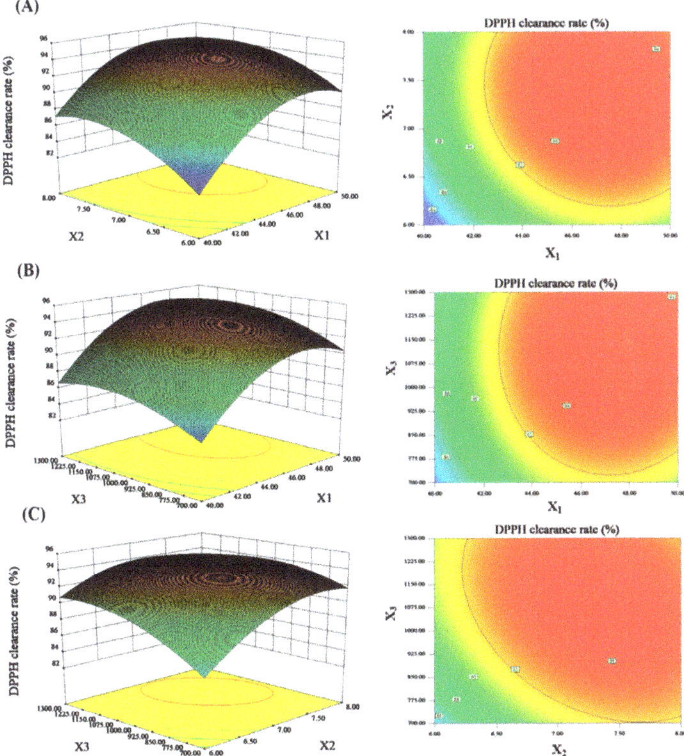

Figure 3. Response surface plots showing the effects of different variables. Images on the left represent three-dimensional response surface plots, whereas images on the right represent two-dimensional contour plots. Images represent the following: (**A**) Effects of temperature (X1) and pH (X2) on the DPPH radical scavenging activity; (**B**) Effects of temperature (X1) and enzyme concentration (X3) on the DPPH radical scavenging activity; and (**C**) Effects of pH (X2) and enzyme concentration (X3) on the DPPH clearance rate.

2.2.2. Validation of the Models

Experiments were performed in triplicate and were carried out under the optimal extraction conditions: temperature was 47.2 °C, pH was 7.3, and enzyme concentration was 1100 U/g. Under similar conditions, the average DPPH radical scavenging rate was 94.85%, which was in good agreement with the predicted value.

2.3. Molecular Weight Distribution of SNNHs

Antioxidant peptides with different Mw have different free radical scavenging capabilities. To determine the Mw distribution of SNNHs, taking the retention time (*Rt*) as the abscissa and log(*Mw*) as the ordinate, the following regression equation was obtained: $\log(Mw) = -0.477Rt + 7.0117$. The value of the measurement coefficient (R^2) was 0.9996, which showed a good linear relationship. This indicated that the relative Mw of the SNNHs could be analyzed based on this linear regression equation (Figure 4A).

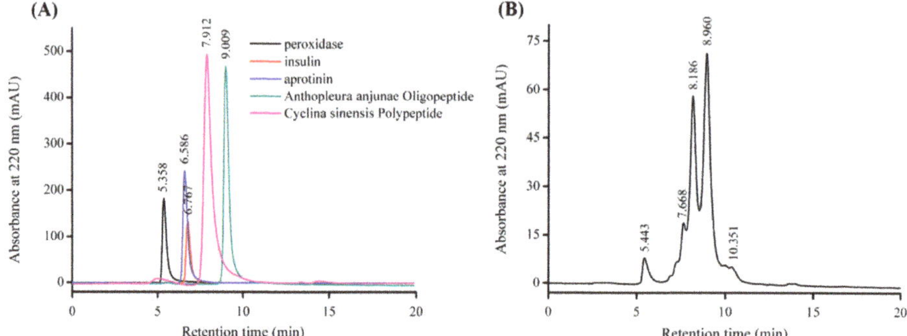

Figure 4. (**A**) The HPLC chromatograms of the standard molecular weight samples; (**B**) the molecular weight distribution of SNNHs from the swim bladders of *Nibea japonica*. (Mobile phase: water/acetonitrile/trifluoroacetic acid = 55:45:0.1, flow rate: 0.5 mL/min.).

The HPLC gel filtration chromatogram of SNNHs of *Nibea japonica* under the same chromatographic conditions is shown in Figure 4B. The components of less than 1, 1–5, 5–10 and more than 10 kDa accounted for 47.88%, 46.21% 1.04% and 4.76% of the components, respectively [21], which showed that the SNNHs consisted mainly of many small peptides. SNNHs were more water-soluble and antioxidant than the collagen found in the swim bladders of *Nibea japonica*.

2.4. Fractionation of SNNH-1 from SNNHs

Based on the principle of mechanical retention, the enzymatic hydrolysates in the swim bladders of *Nibea japonica* were isolated by ultrafiltration membranes and divided into four parts with different Mw distributions. Subsequently, the DPPH radical scavenging rate of each part was determined. As shown in Figure 5, peptide fractions of less than 1 kDa had the highest DPPH free radical scavenging rate compared to the other ultrafiltration fractions. Similar research has shown that the low-Mw content of protein hydrolysates has higher antioxidant activity [22]. Therefore, fractions with a Mw of less than 1 kDa were chosen for subsequent activity evaluation and named SNNH-1.

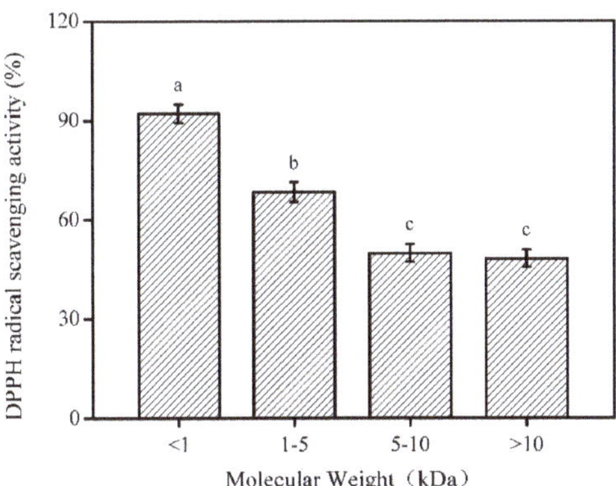

Figure 5. The effect of four fractions of SNNHs on DPPH radical scavenging activity. Results were triplicates of the mean ± SD. Different letters indicate significant differences between groups ($p < 0.05$).

2.5. Amino Acid Composition of SNNH-1

The antioxidant properties of peptides are related to their amino acid composition [23]. The acidic and basic amino acids of the peptide segment help antioxidant peptides obtain better metal ion chelating and free radical scavenging abilities [24]. In addition, peptide chains containing hydrophobic amino acids can better exist at the water–lipid interface, which improves the free radical scavenging ability of the polypeptide [25]. Therefore, the amino acid composition of SNNH-1 was analyzed and the results were expressed in residues per 1000 total residues. Table 3 shows that SNNH-1 contained 7 essential amino acids and 10 non-essential amino acids. The highest amino acid content of SNNH-1 was glycine, alanine, proline, and hydroxyproline, which accounted for 19.23%, 13.34%, 11.87%, and 10.28% of the amino acid content, respectively. The amino acid content of SNNH-1 products was consistent with the collagen peptides from *Nibea japonica* in a previous study (glycine (21.22%), alanine (9.79%), proline (10.78%) and hydroxyproline (9.28%)) [26]. In addition, cysteine was not detected in SNNH-1. It has previously been reported that the activity of the polypeptide formed after protein hydrolysis is related to the content of hydrophobic amino acids contained in the polypeptide [27]. For example, a polypeptide with antioxidant activity has mostly N-terminal hydrophobic amino acids, and the presence of hydrophobic amino acids is positively correlated with its antioxidant activity [28]. The content of the hydrophobic amino acids, proline and alanine, in SNNH-1 was 11.87% and 13.34%, respectively, which is relatively high and suggests that SNNH-1 has strong antioxidant activity.

2.6. Antioxidant Activity of SNNH-1

Free radicals can damage biological membranes because the excessively high oxygen concentration in the lipid bilayer of the membrane is easily attacked by free radicals, which results in lipid peroxidation. Free radicals generated during the process of lipid peroxidation inactivate the transport enzymes on the membrane, result in an imbalance between the internal and external environment of the cell, reduce the fluidity of the membrane, and eventually cause damage to the entire organization and function of the cell [2,29,30].

Radical scavenging activity is a significant interest for the cosmeceutical industry in order to prevent photoaging and ultraviolet damage. To assess the antioxidant activity of SNNH-1, assays for DPPH radical, hydroxyl radical, superoxide anion radical, and 2,2′-azino-bis-3-ethylbenzothiazoline-6-sulfonic acid (ABTS) radical scavenging were used and compared with GSH as an activity control. As shown in Figure 6, SNNH-1 from the swim bladders of *Nibea japonica* has a dose-dependent scavenging effect on DPPH radicals, hydroxyl radicals, superoxide anion radicals and ABTS radicals. SNNH-1 showed high scavenging capabilities of DPPH radicals (Figure 6A), hydroxyl radicals (Figure 6B),

ABTS radicals (Figure 6C) and superoxide anion radicals (Figure 6D). Furthermore, the antioxidant activity of SNNH-1 is close to that of GSH. Therefore, the high free radical scavenging activity of SNNH-1 indicates that it is a potential candidate to be developed as an antioxidant and can be used in anti-aging health products and cosmetics.

Table 3. Composition and contents of amino acids of SNNH-1.

Amino Acid	SNNH-1
Aspartic acid	38
Threonine *	14
Serine	24
Glutamic acid	76
Glycine	340
Alanine	134
Cysteine	0
Valine *	18
Methionine *	14
Isoleucine *	8
Leucine *	23
Tyrosine	7
Phenylalanine *	16
Histidine	6
Lysine *	28
Arginine	59
Proline	105
Hydroxyproline	86
Imino acid	191

* Human-essential amino acids.

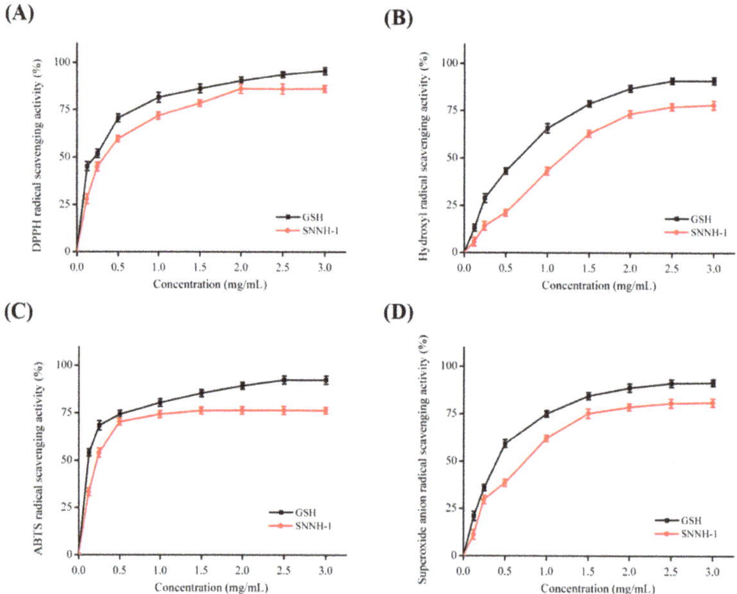

Figure 6. DPPH radical (**A**), hydroxyl radical (**B**), ABTS radical (**C**), and superoxide anion radical (**D**) scavenging activities of SNNH-1. All results were triplicates of the mean ± SD.

2.7. Cytotoxic and Allergenic Potential of SNNH-1

The cytotoxicity and allergenic potential of SNNH-1 were assessed by the MTT method and lactate dehydrogenase (LDH) toxicity test, respectively. As shown in Figure 7A, human umbilical vein endothelial cells (HUVECs) treated with different concentrations of SNNH-1 for 24 h did not show decreased viability. In contrast, SNNH-1 promoted the growth of HUVECs. Therefore, SNNH-1 has no obvious cytotoxic effects in vitro. LDH is a glycolytic enzyme that is widely present in the cytoplasm of cells. Normally, LDH is only in the cytoplasm and measuring LDH levels in cell culture supernatants is a sensitive indicator of cell damage. Increased LDH levels in cell culture supernatants indicates that the cells have been damaged, and that the membrane permeability of the cells has increased. Therefore, the amount of LDH that has leaked out of the membrane from the cytoplasm reflects the degree of cell damage [31]. In addition, the release of LDH can be closely related to allergic reactions and inflammation. In this study, LDH release from SNNH-1-treated cells was lower than that from untreated cells (Figure 7B). These results were consistent with the data presented by Yang et al. [26], who showed that MCPs from *Nibea japonica* skin have the potential to promote cell growth. Therefore, SNNH-1 may be considered as a non-cytotoxic and hypoallergenic material.

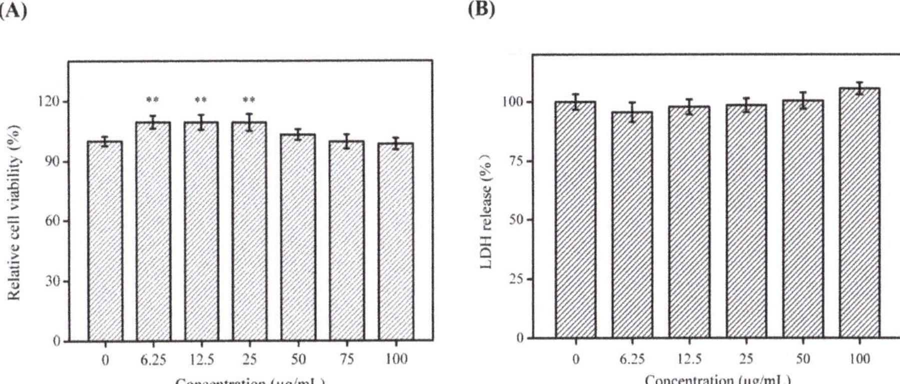

Figure 7. Effects of treatment with different concentrations of SNNH-1 for 24 h on relative cell viability (**A**) and LDH (**B**) in HUVECs. Data are presented as the mean ± SD (n = 6). ** p < 0.01 vs. the Control group.

2.8. Effects of SNNH-1 on the Levels of GSH-Px, SOD, CAT and Malondialdehyde (MDA) in an H_2O_2-Induced HUVECs Injury Model

To evaluate the antioxidant activity of SNNH-1, we investigated the effect of SNNH-1 pretreatment on the levels of GSH-Px, SOD, CAT and MDA after oxidative damage of HUVECs induced by H_2O_2. HUVECs have various functions and are often used to study the relationship between cardiovascular disease and oxygen-free radicals. Therefore, HUVECs were selected as the cell model of oxidative stress in this experiment [6]. H_2O_2 injury is currently the most widely used cell injury model. Previous studies have shown that H_2O_2 can not only attack biofilms and trigger lipid peroxidation reactions, which destroys the integrity of biofilms, but it can also reduce the activity of antioxidant enzymes in cells [32,33]. CAT, SOD and GSH-Px are all important components of enzymes in the cell's antioxidant defense system, and they have very important impacts on the body's oxidation and antioxidant homeostasis. CAT can decompose H_2O_2 into water and oxygen [34]. SOD scavenges superoxide anion free radicals and prevents cell damage. GSH-Px can promote the reaction of hydrogen peroxide and GSH to produce water and oxidized glutathione (GSSG). As shown in Figure 8A–C, H_2O_2 treatment significantly reduced the activities of GSH-Px, SOD and CAT in HUVECs. However, in the SNNH-1 pretreated group, the activities of these three enzymes were increased significantly in a dose-dependent manner. These data show that SNNH-1 inhibits intracellular lipid peroxidation to a certain extent and enhances the cell's antioxidant defense system.

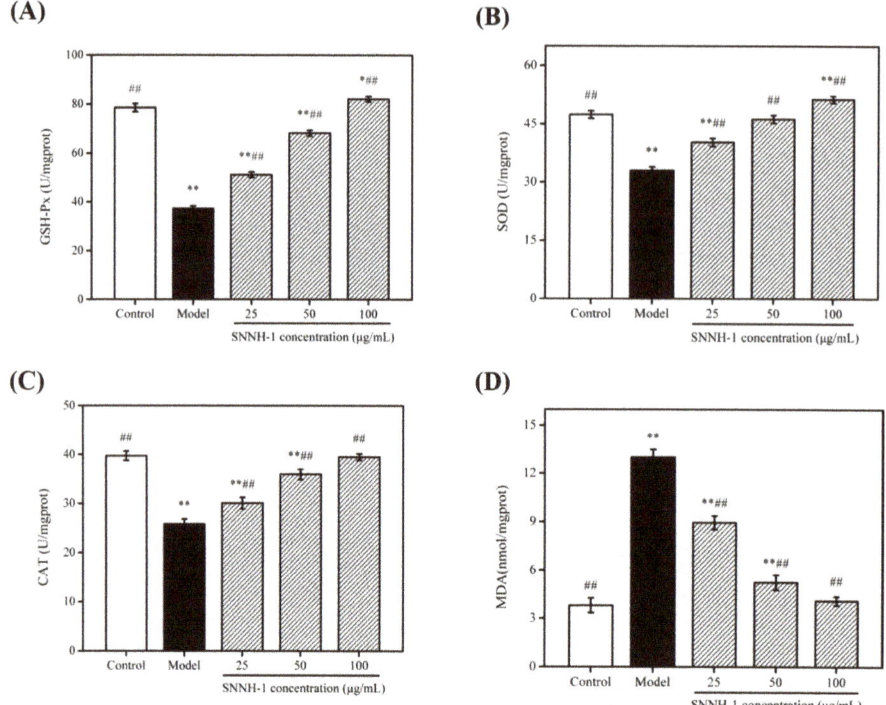

Figure 8. Effect of SNNH-1 at different concentrations (25, 50, and 100 µg/mL) on levels of GSH-Px (**A**), SOD (**B**), CAT (**C**), and MDA (**D**) in an H_2O_2-induced HUVECs injury model. Data are presented as the mean ± SD (*n* = 6). * $p < 0.05$, ** $p < 0.01$ vs. the Control group, ## $p < 0.01$ vs. the Model group (treatment with 600 µM H_2O_2).

MDA is a product of peroxidized lipids and can attack unsaturated fatty acids in the cell membrane, which causes cell damage. Therefore, the degree of lipid peroxidation and cell damage can be judged according to the MDA content in cells [35]. As shown in Figure 8D, the MDA content in the cells was significantly higher after H_2O_2 treatment compared to the control group, which indicates that HUVECs were damaged by H_2O_2 and a large amount of MDA was formed. However, pretreatment of HUVECs with SNNH-1 decreased the amount of MDA with increasing peptide concentrations resulting in lower MDA contents. Furthermore, the MDA level in the high-dose group was similar to the MDA amount in the control group. In addition, Cai et al. [36] reported that FPYLRH (S8) from the swim bladders of Miiuy Croaker (*Miichthys miiuy*) can up-regulate the levels of SOD and GSH-Px, and down-regulate the contents of MDA, suggesting that it plays a protective role in the antioxidant effects on HUVECs against H_2O_2-induced injury.

2.9. Effects of SNNH-1 on ROS Levels in a H_2O_2-Induced HUVECs Injury Model

When the strength of the redox reaction inside the body is beyond the capacity of the body to resist oxidation, then the production of ROS will increase [32,37]. In previous studies, it was shown that H_2O_2 can increase the levels of ROS in cells. In addition, cells in an environment containing free radicals for a long time may cause damage to important biological macromolecules, DNA mutations, damage to organs and tissues, and diseases [35,38]. Vascular endothelial cells are one of the main sources of ROS in the body and these cells are involved in the process of disease. At each stage of disease formation, ROS have a direct impact. Therefore, detecting the content of ROS in cells can directly reflect the antioxidant capacity of the cells, as well as the degree of oxidative damage. Here, the effect of SNNH-1 on ROS levels in HUVECs was studied. As shown in Figure 9A,B, after H_2O_2 treatment, the fluorescence intensity in HUVECs was significantly higher compared to the control group. However, SNNH-1 pretreatment effectively reduced ROS levels in HUVECs. In addition, ROS levels decreased with increasing concentrations of SNNH-1. Furthermore, the level of H_2O_2 produced in HUVECs showed a similar result (Figure 9C). These results indicate that the protective effect of SNNH-1 on H_2O_2-induced HUVECs injury may be due to the inhibition of intracellular ROS production. These results are in accordance with the data presented by Li et al. [22], who found that collagen peptides from sea cucumbers (*Acaudina molpadioides*) could effectively protect cells from H_2O_2-induced damage.

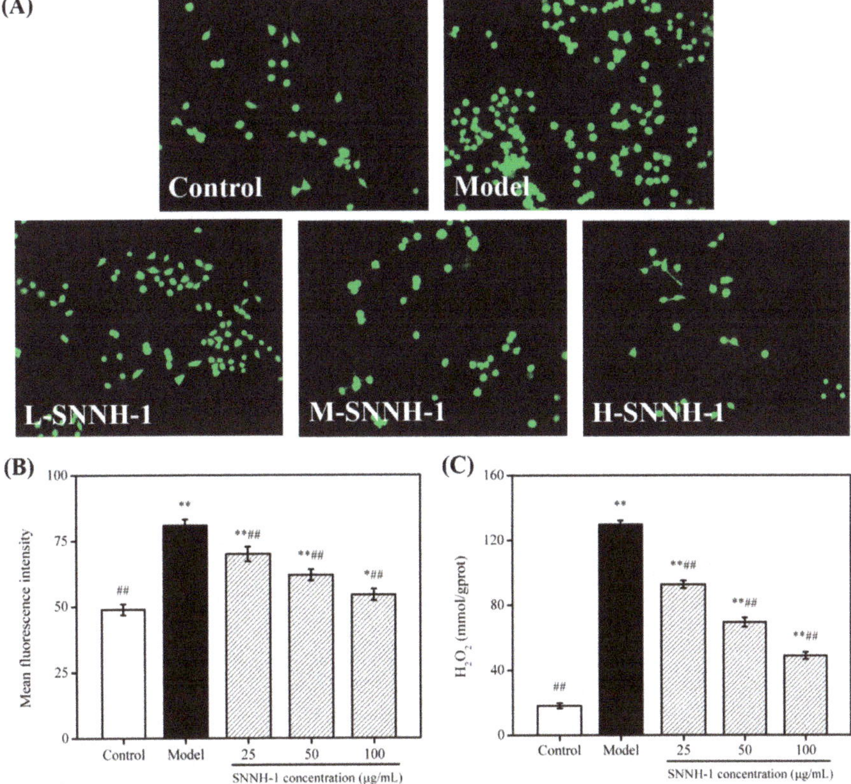

Figure 9. Effect of SNNH-1 on ROS levels in an H_2O_2-induced HUVECs injury model. (**A**) Fluorescence images of different groups observed by a fluorescence microscope; (**B**) mean fluorescence intensity of cells in different groups; (**C**) H_2O_2 levels in HUVECs of different groups. Data are presented as the mean ± SD (n = 3). * $p < 0.05$, ** $p < 0.01$ vs. the Control group, [##] $p < 0.01$ vs. the Model group (treatment with 600 µM H_2O_2).

3. Materials and Methods

3.1. Materials and Chemicals

Swim bladders of *Nibea japonica* were obtained from the Zhejiang Marine Fisheries Research Institution (Zhoushan, China). Five proteases (trypsin (≥250 U/mg), neutrase (≥60,000 U/g), alcalase (≥200 U/mg), pepsin (≥3000 U/mg) and papain (≥500 U/mg)) were purchased from Beijing Asia Pacific Hengxin Co., Ltd. (Beijing, China). DPPH, ABTS, and phenazine methosulfate (PMS) were purchased from Sigma Chemicals (Shanghai, China) Trading Co., Ltd. HUVECs were purchased from the Cell Bank of Type Culture Collection of the Chinese Academy of Sciences (Shanghai, China). All chemicals were of analytical grade.

3.2. Optimization of Preparative Conditions

The swim bladders of *Nibea japonica* were pretreated by the method presented by Chen et al. [20]. Five major factors were selected in the single-factor experiments to set up preliminary ranges of the extraction variables, which include temperature, hydrolysis time, pH, solid:liquid ratio and enzyme concentration.

To further optimize the extraction conditions of the antioxidant peptides of the swim bladder, according to the single factor experiments, the two least influential factors were eliminated. Using the three other factors (extraction temperature, pH, and enzyme concentration), a three-level and three-factor response surface test was designed through the BBD and Design Expert. The activity of the extracted antioxidant peptides was evaluated by DPPH clearance rate. According to preliminary experimental results, the range and level of independent variables (Table 4) were classified.

Table 4. Factors and levels in the response surface design.

Independent Factors	Symbol	Level of Factor		
		−1	0	1
Temperature (°C)	X_1	40	45	50
pH	X_2	6	7	8
Enzyme concentration (U/g)	X_3	700	1000	1300

The BBD in the experiment design contained 15 experimental points (Table 2) and a multiple regression analysis was performed on the response obtained from each experimental design to fit the following quadratic polynomial model:

$$\gamma = \beta_0 + \sum_{i=1}^{k} \beta_i X_i + \sum_{i=1}^{k} \beta_{ii} X_i^2 + \sum \sum_{i<j} \beta_{ij} X_i X_j \tag{1}$$

where γ is the predicted response, β_0 is the intercept, β_i, β_{ii} and β_{ij} are the linear, quadratic, and interaction coefficients, respectively, and both X_i and X_j are independent factors. Each experimental design was performed in triplicate. According to Design Expert 8.0.6, a variance table analysis was generated to determine the influence of the regression coefficients on the linear, quadratic and interaction terms.

3.3. Determination of the Mw Distribution of SNNHs

The Mw distribution of SNNHs was analyzed by gel filtration chromatography using a high-performance liquid chromatography system (Agilent 1260, Palo Alto, CA, USA). Water:acetonitrile:trifluoroacetic acid (55:45:0.1) were adopted as the mobile phase, the flow rate was 0.5 mL/min and the UV wavelength was 220 nm [21]. The column was calibrated with standard materials: peroxidase (40,500 Da), aprotinin (6512 Da), insulin (5807 Da), *Cyclina sinensis* polypeptide

(Arg-Val-Ala-Pro-Glu-Glu-His-Pro-Val-Glu-Gly-Arg-Tyl-Leu-Val, 1751.78 Da) [39], and *Anthopleura anjunae* oligopeptide (Tyr-Val-Pro-Gly-Pro, 531.61 Da) [40]. A standard curve of the log (*Mw*) with retention time was created. The Mw of SNNHs was calculated based on the retention time using the standard curve equation.

3.4. Fractionation of SNNH-1 by Ultrafiltration

A GM-18 Roll film separation system (Bona Biotechnology Co., Ltd., Jinan, China) was used to purify peptides with the highest DPPH free radical scavenging activity with 10, 5 and 1 kDa Mw cut-off membranes. Collected peptides with different Mw were separated and evaluated for their antioxidant activity after lyophilization [41].

3.5. Amino Acid Composition Measurement of SNNH-1

Amino acid analysis was determined based on the method described by Tang et al. [19]. In brief, SNNH-1 was hydrolyzed by dissolving it in 6 M HCl at 110 °C for 24 h without oxygen and then it was vaporized. The hydrolysate was analyzed by a Hitachi L-8800 amino acid analyzer (Hitachi, Tokyo, Japan).

3.6. Antioxidant Activity of SNNH-1

The DPPH, hydroxyl, ABTS and superoxide anion radical scavenging activity of SNNH-1 was performed as in the methods described by Chen et al. [20].

3.7. Cytotoxic and Allergenic Properties of SNNH-1

The cytotoxic and allergen properties of SNNH-1 were determined using HUVECs in accordance with the manufacturers' instructions of the MTT assay kit and the LDH release assay. The MTT and LDH test methods were performed as described by Lin et al. [42].

3.8. Assays for Antioxidant Enzymatic Activity of SNNH-1 in H_2O_2-Induced HUVECs

HUVECs were inoculated into 6-well plates (1×10^5 cells/well) and incubated in a 5% CO_2 incubator at 37 °C for 24 h. SNNH-1 at final concentrations of 0, 25, 50 and 100 µg/mL were added into the protection groups and cultured for another 24 h. Each group was treated with 600 µmol/L H_2O_2 for 4 h. The group treated without SNNH-1 and H_2O_2 was used as the control group. Subsequently, 500 µL of cell lysis buffer was added to each well on ice. Cells were lysed for 30 min, and centrifuged at 12,000 rpm for 10 min at 4 °C. The resulting supernatant was stored at 4 °C. Levels of GSH-Px, SOD, CAT, MDA, and H_2O_2 were determined by using assay kits according to the manufacturers' instructions (Nanjing Jiancheng Bioengineering Institute, Nanjing, China), and protein concentrations were determined using the bicinchoninic acid (BCA) method.

3.9. Determination of the Levels of ROS in H_2O_2-Induced HUVECs

HUVECs were seeded in 6-well plates with a density of 1×10^4 cells/mL for 24 h. SNNH-1 at final concentrations of 25, 50 and 100 µg/mL was added to the protection groups for 24 h and HUVECs were treated with 600 µmol/L H_2O_2 for 4 h in a 5% CO2 incubator at 37 °C. Production of ROS was determined using a ROS Assay Kit (Nanjing Jiancheng Bioengineering Institute, Nanjing, China). Next, a total of 2 mL of 2′,7′-dichlorodihydrofluorescein diacetate (DCFH-DA) fluorescent probe solution (10 µM) was added to the cells, and incubated for 30 min. Then cells were washed three times with serum-free Dulbecco's modified Eagle medium (DMEM) and observed under a fluorescence microscope (Axio Imager A2, Carl Zeiss, Oberkochen, Germany). The fluorescence intensity was analyzed using Image J software.

3.10. Statistical Analysis

Data are presented as the mean ± SD ($n = 3$). Multiple-group comparisons were determined using ANOVA (SPSS 19.0 software, Armonk, New York, NY, USA). $p < 0.05$ was considered statistically significant.

4. Conclusions

In this work, RSM was used to optimize the extraction process of SNNHs from the swim bladders of *Nibea japonica*. It was found that the optimal extraction conditions were a neutrase concentration of 1100 U/g, a temperature of 47.2 °C and a pH of 7.3, which resulted in the maximum DPPH clearance rate of 95.44%. SNNH-1 (Mw < 1 kDa) was obtained from the extracted collagen peptide SNNHs by ultrafiltration. SNNH-1 had good scavenging activities of DPPH, hydroxyl, ABTS, and superoxide anion radicals. In addition, SNNH-1 showed an important protective effect against H_2O_2 injury in HUVECs by promoting cell proliferation, reducing the contents of ROS and MDA, and enhancing the activity of antioxidant enzymes (GSH-Px, SOD, and CAT). These results provide a basis for the future application of SNNH-1 in food processing, pharmaceuticals and cosmetics.

Author Contributions: Z.Y. conceived and designed the experiments. J.Z., X.T. and B.X. carried out the experiment. F.Y. and J.G. performed the statistical analysis of the data. J.Z. wrote the manuscript. All authors have read and agreed to the published version of the manuscript.

Funding: This work was supported by the National Spark Program of China (grant No. 2015GA700044), the Science and Technology Bureau of Zhoushan (grant No. 2012C23023).

Conflicts of Interest: The authors declare no conflict of interest.

References

1. Ward, C.W.; Prosser, B.L.; Lederer, W.J. Mechanical Stretch-Induced Activation of ROS/RNS Signaling in Striated Muscle. *Antioxid. Redox Signal.* **2014**, *20*, 929–936. [CrossRef] [PubMed]
2. Wu, R.B.; Wu, C.; Liu, D.; Yang, X.; Huang, J.; Zhang, J.; Liao, B.; He, H. Antioxidant and anti-freezing peptides from salmon collagen hydrolysate prepared by bacterial extracellular protease. *Food Chem.* **2018**, *248*, 342–356. [CrossRef] [PubMed]
3. Sila, A.; Bougatef, A. Antioxidant peptides from marine by-products: Isolation, identification and application in food systems. *J. Funct. Foods* **2015**, *21*, 10–26. [CrossRef]
4. Jin, Y.; Liu, K.; Peng, J.; Wang, C.; Kang, L.; Chang, N.; Sun, H. Rhizoma Dioscoreae Nipponicae polysaccharides protect HUVECs from H_2O_2-induced injury by regulating PPARγ factor and the NADPH oxidase/ROS–NF-κB signal pathway. *Toxicol. Lett.* **2015**, *232*, 149–158. [CrossRef] [PubMed]
5. Bougatef, A.; Hajji, M.; Balti, R.; Lassoued, I.; Triki-Ellouz, Y.; Nasri, M. Antioxidant and free radical-scavenging activities of smooth hound (*Mustelus mustelus*) muscle protein hydrolysates obtained by gastrointestinal proteases. *Food Chem.* **2009**, *114*, 1198–1205. [CrossRef]
6. Zhao, W.H.; Luo, Q.B.; Pan, X.; Chi, C.F.; Sun, K.L.; Wang, B. Preparation, identification, and activity evaluation of ten antioxidant peptides from protein hydrolysate of swim bladders of miiuy croaker (*Miichthys miiuy*). *J. Funct. Foods* **2018**, *47*, 503–511. [CrossRef]
7. Johansen, J.S.; Harris, A.K.; Rychly, D.J.; Ergul, A. Oxidative stress and the use of antioxidants in diabetes: Linking basic science to clinical pratice. *Cardiovasc. Diabetol.* **2005**, *4*, 5–16. [CrossRef]
8. Huang, Y.; Ruan, G.; Qin, Z.; Li, H.; Zheng, Y. Antioxidant activity measurement and potential antioxidant peptides exploration from hydrolysates of novel continuous microwave-assisted enzymolysis of the Scomberomorus niphonius protein. *Food Chem.* **2017**, *223*, 89–95. [CrossRef]
9. Yu, Z.P.; Yin, Y.G.; Zhao, W.Z.; Wang, F.; Yu, Y.D.; Liu, B.Q.; Liu, J.B.; Chen, F. Characterization of ACE-Inhibitory Peptide Associated with Antioxidant and Anticoagulation Properties. *J. Food Sci.* **2011**, *76*, 1149–1155. [CrossRef]
10. Jara, A.M.R.; Liggieri, C.S.; Bruno, M.A. Preparation of soy protein hydrolysates with antioxidant activity by using peptidases from latex of Maclura pomifera fruits. *Food Chem.* **2018**, *246*, 326–333. [CrossRef]

11. Senphan, T.; Benjakul, S. Antioxidative activities of hydrolysates from seabass skin prepared using protease from hepatopancreas of Pacific white shrimp. *J. Funct. Foods* **2014**, *6*, 147–156. [CrossRef]

12. Xia, Y.; Bamdad, F.; Gaenzle, M.; Chen, L. Fractionation and characterization of antioxidant peptides derived from barley glutelin by enzymatic hydrolysis. *Food Chem.* **2012**, *134*, 1509–1518. [CrossRef] [PubMed]

13. Chi, C.F.; Wang, B.; Hu, F.Y.; Wang, Y.M.; Zhang, B.; Deng, S.G.; Wu, C.W. Purification and identification of three novel antioxidant peptides from protein hydrolysate of bluefin leatherjacket (*Navodon septentrionalis*) skin. *Food Res. Int.* **2015**, *73*, 124–129. [CrossRef]

14. Kumar, N.S.S.; Nazeer, R.A.; Jaiganesh, R. Purification and biochemical characterization of antioxidant peptide from horse mackerel (*Magalaspis cordyla*) viscera protein. *Peptides* **2011**, *32*, 1496–1501. [CrossRef]

15. You, L.; Zhao, M.; Regenstein, J.M.; Ren, J. In vitro antioxidant activity and in vivo anti-fatigue effect of loach (*Misgurnus anguillicaudatus*) peptides prepared by papain digestion. *Food Chem.* **2011**, *124*, 188–194. [CrossRef]

16. Ding, J.F.; Li, Y.Y.; Xu, J.J.; Su, X.R.; Gao, X.; Yue, F.P. Study on effect of jellyfish collagen hydrolysate on anti-fatigue and anti-oxidation. *Food Hydrocoll.* **2011**, *25*, 1350–1353. [CrossRef]

17. Ren, J.; Zhao, M.; Wang, H.; Cui, C.; You, L. Effects of supplementation with grass carp protein versus peptide on swimming endurance in mice. *Nutrition* **2011**, *27*, 789–795. [CrossRef]

18. Yu, F.; Zong, C.; Jin, S.; Zheng, J.; Nan, C.; Ju, H.; Yan, C.; Huang, F.; Yang, Z.; Tang, Y. Optimization of Extraction Conditions and Characterization of Pepsin-Solubilised Collagen from Skin of Giant Croaker (*Nibea japonica*). *Mar. Drugs* **2018**, *16*, 29. [CrossRef]

19. Tang, Y.; Jin, S.; Li, X.; Li, X.; Hu, X.; Yan, C.; Huang, F.; Yang, Z.; Yu, F.; Ding, G. Physicochemical Properties and Biocompatibility Evaluation of Collagen from the Skin of Giant Croaker (*Nibea japonica*). *Mar. Drugs* **2018**, *16*, 222. [CrossRef]

20. Chen, Y.; Jin, H.; Yang, F.; Jin, S.; Liu, C.; Zhang, L.; Huang, J.; Wang, S.; Yan, Z.; Cai, X.; et al. Physicochemical, antioxidant properties of giant croaker (*Nibea japonica*) swim bladders collagen and wound healing evaluation. *Int. J. Biol. Macromol.* **2019**, *138*, 483–491. [CrossRef]

21. Zhang, H.; Yang, P.; Zhou, C.; Li, S.; Hong, P. Marine Collagen Peptides from the Skin of Nile Tilapia (*Oreochromis niloticus*): Characterization and Wound Healing Evaluation. *Mar. Drugs* **2017**, *15*, 102–112.

22. Li, Y.; Li, J.; Lin, S.J.; Yang, Z.S.; Jin, H.X. Preparation of Antioxidant Peptide by Microwave-Assisted Hydrolysis of Collagen and Its Protective Effect Against H_2O_2-Induced Damage of RAW264.7 Cells. *Mar. Drugs* **2019**, *17*, 642. [CrossRef] [PubMed]

23. Sun, W.; Zhao, H.; Zhao, Q.; Zhao, M.; Yang, B.; Wu, N.; Qian, Y. Structural characteristics of peptides extracted from Cantonese sausage during drying and their antioxidant activities. *Innov. Food Sci. Emerg. Technol.* **2009**, *10*, 558–563. [CrossRef]

24. Je, J.; Qian, Z.; Kim, S. Antioxidant peptide isolated from muscle protein of bullfrog, *Rana Catesbeiana Shaw*. *J. Med. Food* **2007**, *10*, 401–407. [CrossRef] [PubMed]

25. Samaranayaka, A.G.P.; Lichan, E.C.Y. Food-derived peptidic antioxidants: A review of their production, assessment, and potential applications. *J. Funct. Foods* **2011**, *3*, 229–254. [CrossRef]

26. Yang, F.; Jin, S.; Tang, Y. Marine Collagen Peptides Promote Cell Proliferation of NIH-3T3 Fibroblasts via NF-kB Signaling Pathway. *Molecules* **2019**, *24*, 4201. [CrossRef]

27. Elias, R.J.; Kellerby, S.S.; Decker, E.A. Antioxidant Activity of Proteins and Peptides. *Crit. Rev. Food Sci. Nutr.* **2008**, *48*, 430–441. [CrossRef]

28. Udenigwe, C.C.; Aluko, R.E. Chemometric Analysis of the Amino Acid Requirements of Antioxidant Food Protein Hydrolysates. *Int. J. Mol. Sci.* **2011**, *12*, 3148–3161. [CrossRef]

29. Luisi, G.; Stefanucci, A.; Zengin, G.; Dimmito, M.P.; Mollica, A. Anti-Oxidant and Tyrosinase Inhibitory In Vitro Activity of Amino Acids and Small Peptides: New Hints for the Multifaceted Treatment of Neurologic and Metabolic Disfunctions. *Antioxidants* **2018**, *8*, 7. [CrossRef]

30. Yang, X.; Zhao, Y.; Qiu, Y.; Chi, C.; Wang, B. Preparation and Characterization of Gelatin and Antioxidant Peptides from Gelatin Hydrolysate of Skipjack Tuna (*Katsuwonus pelamis*) Bone Stimulated by in vitro Gastrointestinal Digestion. *Mar. Drugs* **2019**, *17*, 78. [CrossRef]

31. Xu, Z.; Fang, Y.; Chen, Y.; Yang, W.; Ma, N.; Pei, F.; Kimatu, B.M.; Hu, Q.; Qiu, W. Protective effects of Se-containing protein hydrolysates from Se-enriched rice against Pb^{2+}-induced cytotoxicity in PC12 and RAW264.7 cells. *Food Chem.* **2016**, *202*, 396–403. [CrossRef] [PubMed]

32. Zhu, Z.; Shi, Z.; Xie, C.; Gong, W.; Hu, Z.; Peng, Y. A novel mechanism of Gamma-aminobutyric acid (GABA) protecting human umbilical vein endothelial cells (HUVECs) against H_2O_2-induced oxidative injury. *Comp. Biochem. Physiol. C Toxicol. Pharmacol.* **2019**, *217*, 68–75. [CrossRef] [PubMed]

33. Jin, J.; Ahn, C.; Je, J. Purification and characterization of antioxidant peptides from enzymatically hydrolyzed ark shell (*Scapharca subcrenata*). *Process Biochem.* **2018**, *72*, 170–176. [CrossRef]

34. Chi, C.; Cao, Z.; Wang, B.; Hu, F.; Li, Z.; Zhang, B. Antioxidant and Functional Properties of Collagen Hydrolysates from Spanish Mackerel Skin as Influenced by Average Molecular Weight. *Molecules* **2014**, *19*, 11211–11230. [CrossRef] [PubMed]

35. Wang, L.; Ding, L.; Yu, Z.; Zhang, T.; Ma, S.; Liu, J. Intracellular ROS scavenging and antioxidant enzyme regulating capacities of corn gluten meal-derived antioxidant peptides in HepG2 cells. *Food Res. Int.* **2016**, *90*, 33–41. [CrossRef]

36. Cai, S.Y.; Wang, Y.M.; Zhao, Y.Q.; Chi, C.F.; Wang, B. Cytoprotective Effect of Antioxidant Pentapeptides from the Protein Hydrolysate of Swim Bladders of Miiuy Croaker (*Miichthys miiuy*) against H_2O_2-Mediated Human Umbilical Vein Endothelial Cell (HUVEC) Injury. *Int. J. Mol. Sci.* **2019**, *20*, 5425. [CrossRef]

37. Pelicano, H.; Carney, D.A.; Huang, P. ROS stress in cancer cells and therapeutic implications. *Drug Resist. Updat.* **2004**, *7*, 97–110. [CrossRef]

38. Zheng, L.; Yu, H.; Wei, H.; Xing, Q.; Zou, Y.; Zhou, Y.; Peng, J. Antioxidative peptides of hydrolysate prepared from fish skin gelatin using ginger protease activate antioxidant response element-mediated gene transcription in IPEC-J2 cells. *J. Funct. Foods* **2018**, *51*, 104–112. [CrossRef]

39. Li, W.; Ye, S.; Zhang, Z.; Tang, J.; Jin, H.; Huang, F.; Yang, Z.; Tang, Y.; Chen, Y.; Ding, G.; et al. Purification and Characterization of a Novel Pentadecapeptide from Protein Hydrolysates of *Cyclina sinensis* and Its Immunomodulatory Effects on RAW264.7 Cells. *Mar. Drugs* **2019**, *17*, 30. [CrossRef]

40. Wu, Z.Z.; Ding, G.F.; Huang, F.F.; Yang, Z.S.; Yu, F.M.; Tang, Y.P.; Jia, Y.L.; Zheng, Y.Y.; Chen, R. Anticancer Activity of *Anthopleura anjunae* Oligopeptides in Prostate Cancer DU-145 Cells. *Mar. Drugs* **2018**, *16*, 125. [CrossRef]

41. Liu, Y.; Wan, S.; Liu, J.; Zou, Y.; Liao, S. Antioxidant Activity and Stability Study of Peptides from Enzymatically Hydrolyzed Male Silkmoth. *J. Food Process. Pres.* **2017**, *41*, 1–10. [CrossRef]

42. Lin, X.H.; Chen, Y.Y.; Jin, H.X.; Zhao, Q.L.; Liu, C.J. Collagen Extracted from Bigeye Tuna (*Thunnus obesus*) Skin by Isoelectric Precipitation: Physicochemical Properties, Proliferation, and Migration Activities. *Mar. Drugs* **2019**, *17*, 261. [CrossRef] [PubMed]

Article

Pepsin-Soluble Collagen from the Skin of *Lophius litulo*: A Preliminary Study Evaluating Physicochemical, Antioxidant, and Wound Healing Properties

Wen Zhang, Jiawen Zheng, Xiaoxiao Tian, Yunping Tang, Guofang Ding, Zuisu Yang and Huoxi Jin *

Zhejiang Provincial Engineering Technology Research Center of Marine Biomedical Products, School of Food and Pharmacy, Zhejiang Ocean University, Zhoushan 316022, China; zhangwen1225z@163.com (W.Z.); jwzheng1996@163.com (J.Z.); TIANXIAOXIAO0208@163.com (X.T.); tangyunping1985@zjou.edu.cn (Y.T.); dinggf2007@163.com (G.D.); abc1967@126.com (Z.Y.)
* Correspondence: jinhuoxi@163.com; Tel.: +86-0580-226-0600; Fax: +86-0580-254-781

Received: 26 November 2019; Accepted: 13 December 2019; Published: 16 December 2019

Abstract: The structure of pepsin-solubilized collagen (PSC) obtained from the skin of *Lophius litulon* was analyzed using the sodium dodecylsulphate polyacrylamide gel electrophoresis (SDS-PAGE), Fourier transform infrared spectroscopy (FTIR), and scanning electron microscopy (SEM). SDS-PAGE results showed that PSC from *Lophius litulon* skin was collagen type I and had collagen-specific $\alpha 1$, $\alpha 2$, β, and γ chains. FTIR results indicated that the infrared spectrum of PSC ranged from 400 to 4000 cm^{-1}, with five main amide bands. SEM revealed the microstructure of PSC, which consisted of clear fibrous and porous structures. In vitro antioxidant studies demonstrated that PSC revealed the scavenging ability for 2,2-diphenyl-1-picrylhydrazyl (DPPH), HO·, O$_2^-$·, and ABTS·. Moreover, animal experiments were conducted to evaluate the biocompatibility of PSC. The collagen sponge group showed a good biocompatibility in the skin wound model and may play a positive role in the progression of the healing process. The cumulative results suggest that collagen from the skin of *Lophius litulon* has potential applications in wound healing due to its good biocompatibility.

Keywords: *Lophius litulon* skin; pepsin-solubilized collagen; characterization; antioxidant activity; biocompatibility

1. Introduction

Collagen, a biological macromolecule is one of the most abundant proteins in both invertebrates and vertebrates [1]. Collagen is mainly found in fibrillar connective tissue and widely used in the food, cosmetic, and nutritional health industries. In addition, highly stable collagen fibers formed by cross-linking and self-aggregation can be used as biomaterials to prepare a variety of practical scaffolds [2]. During wound healing, collagen initiates a signaling cascade that produces matrix metalloproteinases by activating integrin, which degrades the collagen matrix and causes keratinocytes to migrate [3]. The degradation products of collagen can be absorbed by cells and have antioxidation and immunomodulation biological effects [3,4]. Due to their excellent biocompatibility, antioxidant abilities, low immunogenicity, and extensive sources, collagen has become a research hotspot and main target of ideal biocompatibility carrier. For example, Itoh et al. studied the biocompatibility, bone conductivity and effectiveness of the new hydroxyapatite/type I collagen (HAp/Col) composite as a carrier of recombinant human bone morphogenesis protein. The results supported the idea that HAp/Col has high bone conduction activity and can induce bone remodeling units [5]. The study of

Lukasiewicz et al. suggested that the collagen coating extracted by acetic acid greatly reduced the visceral adherence to the polypropylene mesh and did not increase complications or cause changes in the binding of the polypropylene mesh [6]. Chen et al. studied the antioxidant activity of ASC and PSC obtained from *Nibea japonica* swim bladders, and pointed out that PSC has potential application value in wound healing [7].

In general, ROS and antioxidants are balanced in healthy individuals. However, this balance can be disorganized in a wound [8]. Increased production of ROS due to injury and oxidative stress may result in poor wound healing [9]. Therefore, it is essential for wound healing to maintain the balance of ROS. Shao et al. studied the lipid peroxidation rate of diabetic rats using Plumbagin, and the results indicated that Plumbagin significantly accelerated wound healing by increasing the levels of antioxidant enzymes such as SOD, CAT, GPx, GR, and GST in diabetic rats [10]. The study of Abood et al. revealed that external use of P. macrocarpa fruit extract can accelerate wound healing and reduce tissue injury by increasing the activity of SOD, CAT, and MDA antioxidant enzymes [11].

Lophius litulon belongs to the Lophiidae and is a kind of deep-sea fish found in the Atlantic, Pacific, and Indian oceans. It is a commercially valuable aquatic resource, and fish fillets, stomachs, intestines, and livers are considered a delicacy in Japan, Korea, and China [12]. However, during processing, *Lophius litulon* skin is often discarded as a by-product, resulting in environmental pollution and waste. Chi et al. isolated three pentapeptides from muscle protein hydrolysates in monkfish (*Lophius litulon*) and evaluated their antioxidant activity [13]. Ma et al. showed that collagen peptides extracted from *Lophius litulon* skin had good antioxidant activity in vitro [14]. However, the physicochemical and biocompatibility properties of collagen extracted from *Lophius litulon* skin have not been reported. Therefore, in this study, we conducted a preliminary study evaluating physicochemical, antioxidant and biocompatibility properties of pepsin-solubilized collagen (PSC) from the skin of *Lophius litulon*.

2. Results and Discussion

2.1. Sodium Dodecylsulphate Polyacrylamide Gel Electrophoresis (SDS-PAGE) Analysis

It can be seen from the SDS-PAGE diagram of PSC (Figure 1) that the extracted components have four peptide chains: γ, β, α1, and α2. There are at least two different α chains (α1 and α2) in collagen type I [15]. It can be seen that the α1 chain is approximately twice as dense as α2, which is in accordance with the characteristics of collagen type I [16]. Furthermore, the PSC from *Lophius litulon* skin is similar to other marine fish source collagens, such as collagens from bigeye tuna skin and *Nibea japonica* skin [16,17].

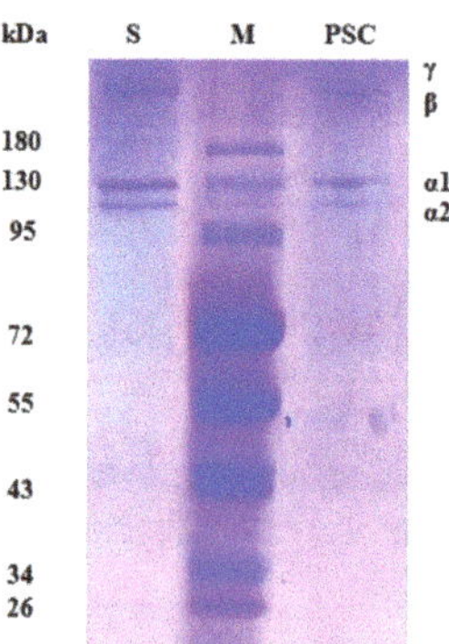

Figure 1. Sodium dodecyl sulfate-polyacrylamide gel electrophoresis (SDS-PAGE) analysis of pepsin-solubilized collagen (PSC) from *Lophius litulon* skin. M: Prestained color protein markers; S: Collagen type I from bovine Achilles tendon was used as standard.

2.2. Amino Acid Composition

We determined the amino acid composition of pepsin-solubilized collagen (PSC) from the skin of *Lophius litulon* and the results were used g/100 g as expression. As shown in Table 1, the collagen from Lophius litulon skin was rich in glycine (Gly), proline (Pro), glutamicacid (Glu), and alanine (Ala). Generally, the content of Gly in collagen was the highest, and the proportion of Gly in the total amino acid components of PSC was 23.85%. The content of hydroxyproline (Hyp) in PSC from the skin of *Lophius litulon* was 1.554 g/100 g, which was close to the content of Hyp in other collagen [18]. Collagen which is rich in hydrophobic amino acids has been shown to demonstrate high antioxidant properties [19,20]. Mcgavin et al. purified the 72 kD protein on receptor(s) for bone sialoprotein (BSP) in Staphylococcus aureus cells, and their amino acid composition analysis showed that the 72 kD protein contained 28.0% hydrophobic amino acids in total [21]. The content of hydrophobic amino acids in the PSC accounted for 22.16% in total, indicating that the collagen has potential antioxidant properties.

Table 1. Amino acid compositions of PSC from *Lophius litulon* skin.

Amino Acids	Content (g/100 g)
Aspartic acid (Asp)	2.529
Threonine (Thr)	1.310
Serine (Ser)	2.522
Glutamic (Glu)	4.708
Priline (Pro)	4.720
Glycine (Gly)	9.684
Alanine (Ala) *	4.009
Valine (Val) *	1.640
Methionine (Met) *	1.044
Isoleucine (Ile)	0.262
Leucine (Leu) *	1.312
Tyrosine (Tyr) *	0.000
Phenylalanine (Phe) *	0.991
Lysine (Lys)	1.462
Histidine (His)	0.931
Arginine (Arg)	3.484
Hydroxyproline (Hyp)	1.554

* represents hydrophobic amino acids.

2.3. Relative Solubility

The solubility of the PSC against pH is shown in Figure 2A. PSC had good solubility in the range of 1 to 4 of pH. When the pH was greater than 4, the solubility of PSC decreased sharply with rising pH. Low solubility of PSC was caught sight of an alkaline pH from 8 to 10. Generally, the solubility of collagen is the lowest near its iso-ionic point [22,23]. Our results indicate that the iso-ionic point of PSC may be approximately pH 6.

Increases in NaCl concentration increase ion strength and enhance hydrophobic interaction between protein chains, thus reducing the solubility of collagen and leading to protein precipitation [24]. Since the average salt concentration of the ocean is 3.5%, the salt concentrations in this study ranged from 0% to 6%. Figure 2B displays changes in collagen solubility with changing NaCl concentration. Small increments of NaCl concentration (≤2 g/100 mL) seemed to have a moderate impact on the collagen solution, leading to a slight decrease in relative solubility (100%–90%). However, solubility was rapidly decreased when NaCl level were higher than 2 g/100 mL. The above results are consistent with PSC-SC from the cartilage of Siberian sturgeon and collagens from the skin of pufferfish [25,26].

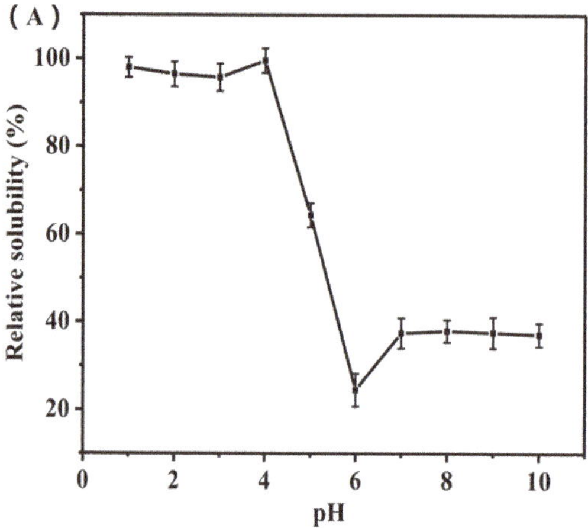

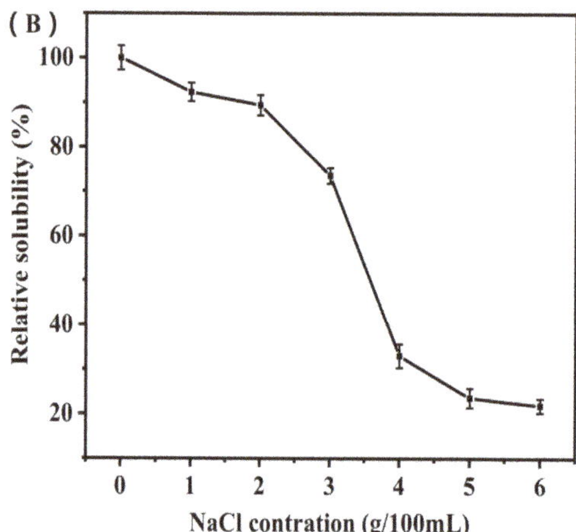

Figure 2. Effect of pH (**A**) and NaCl concentration (**B**) on the solubility of PSC.

2.4. Fourier Transform Infrared (FTIR) Spectroscopy

Figure 3 presents the characteristic peaks conforming with the main absorption bands in the FTIR spectra for PSC. The FTIR spectra for PSC ranged between 400 and 4000 cm^{-1}. These peaks corresponded to five major amide bands, including amide A, amide B, amide I, amide II, and amide III, which were evident in the amino acids' composition and high ratios of proline and hydroxyproline in the collagen molecule. The amide A bands of PSC were measured at a wavelength of 3408.14 cm^{-1}. Amide A bands are commonly associated with a free N–H stretching vibration and shows the presence of hydrogen bands, which usually appear in the range of 3400–3440 cm^{-1} [27]. The amide B bands of PSC were measured at a wavelength of 2959.48 cm^{-1}, consistent with the asymmetric extension of CH$_2$ [25]. The amide I bands of PSC were measured at 1653.66 cm^{-1}, with a characteristic absorption wavelength within 1600–1700 cm^{-1}, which is associated with the stretching vibration of the carbonyl group [28]. The amide II bands of PSC were measured at a wavelengths between 1550–1600 cm^{-1}, which can be owing to a N–H bending vibration coupled with a C–N stretching vibration [29]. Futhermore, the amide III bands (1220–1320 cm^{-1}) were measured at a wavelength of 1239.84 cm^{-1}, which may be affected by N–H deformation and C–N stretching vibration [30]. The FTIR spectra of PSC has significant collagen characteristics and is similar to the skin of *Nibea japonica* [31] and the scale of *Oreochromis niloticas* [32].

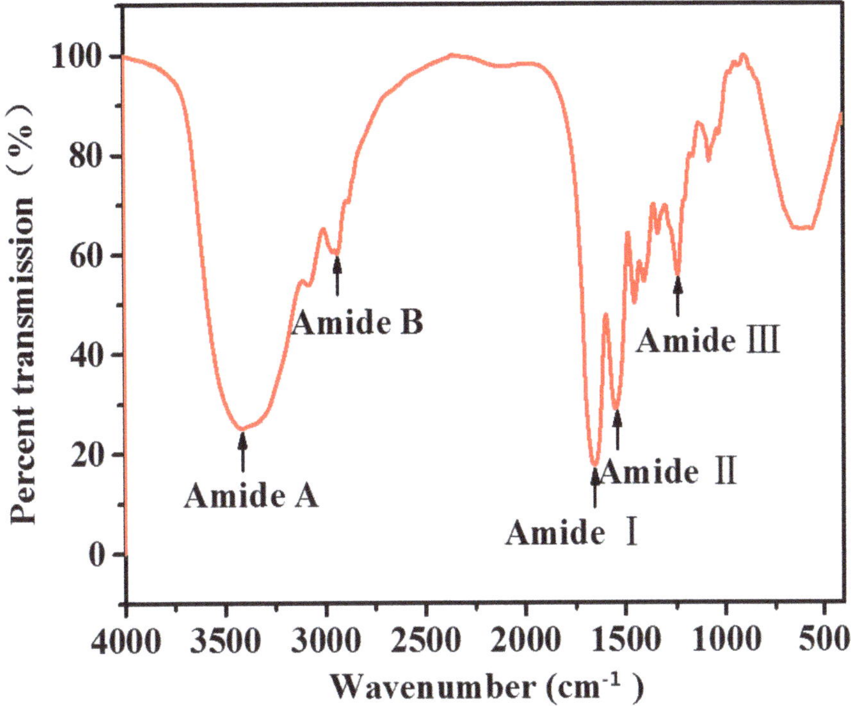

Figure 3. FTIR analysis of PSC from *Lophius litulon* skin.

2.5. Scanning Electron Microscope (SEM) Analysis

In recent years, collagen has proven to be of great value in the field of biomedicine and for various medical applications. Therefore, it is important to understand the surface morphology and micromorphology of collagen. The surface area and ultrastructure of the PSC was observed using SEM microstructure to evaluate its potential applications. SEM images of PSC taken at ×250, ×500, ×1000, and ×2500 are presented in Figure 4. The SEM images of PSC displayed a complex, multilayered, polymeric fibrous meshwork appearance. At ×500 magnification, the fibrous and porous structures were clearly visible. PSC with inter-connected porous structures are similar to other marine collagens, such as collagen from the skin of sole fish and Pacific cod (*Gadus macrocephalus*) [33,34]. The microstructure of PSC indicates that it has great potential for application as a wound excipient or drug carrier in biomedical engineering and medicine.

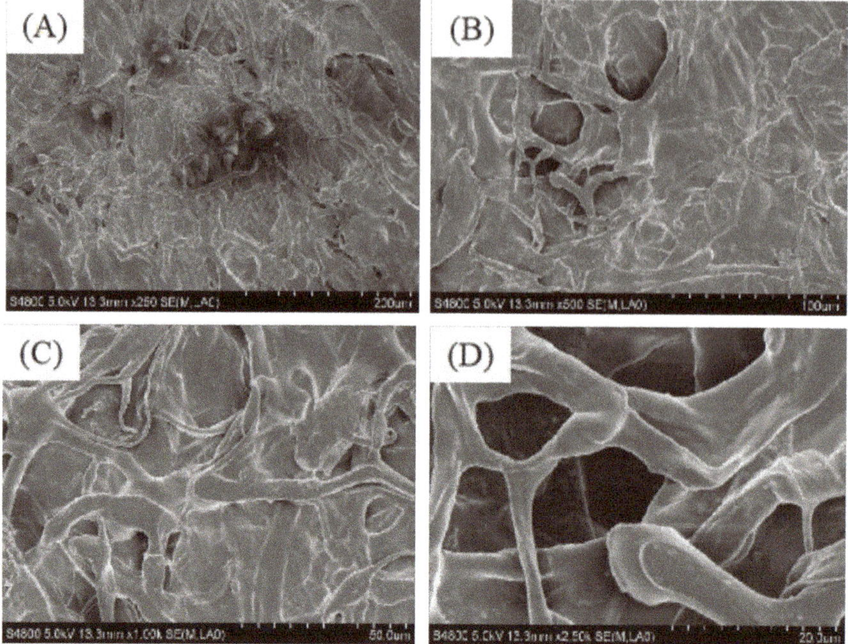

Figure 4. SEM images of PSC from *Lophius litulon* skin. (**A**) ×250, (**B**) ×500, (**C**) ×1000, and (**D**) ×2500.

2.6. Antioxidant Activity

One of the major factors that plays an important part in wound healing is the regulation and oxidation of inflammation. Oxidative stress plays a crucial role in the progression of wound healing [35]. Reactive oxygen species (ROS) are a vital component of oxidative stress. Therefore, the existance of antioxidants is considered to be an essential factor in successful wound healing [36,37]. Besides, a large number of experiments have demonstrated that marine collagen and collagen peptides exhibit antioxidant effects. In this study, we assessed the antioxidant properties of collagen extracted from the skin of *Lophius litulon* using four free radicals: DPPH·, ABTS·, HO·, and O_2^{-}·.

As shown in Figure 5, PSC was able to scavenge DPPH·, ABTS·, HO·, and O_2^{-}·, but its scavenging ability for these four free radicals was lower than those of the positive control (ascorbic acid). Additionally, the scavenging activities of collagen on free radicals was dose dependent. When the concentration of collagen was high, the scavenging activities of free radicals tended to be stable. These phenomena were similar to other previously examined marine collagens. In addition, compared to some other marine collagens, the scavenging activities of free radicals by collagen from *Lophius litulon* skin have certain advantages. For example, the scavenging activities of DPPH·, HO·, and O_2^{-}· by collagen peptides from tilapia skin were lower than those of PSC extracted from *Lophius litulon* skin at the same concentration [38]. The concentrations of ASC and PSC from *Nibea japonica* swim bladders were higher than that of PSC from *Lophius litulon* skin when the clearance rate of ABTS· reached 50%. Moreover, the scavenging activities of both HO· and O_2^{-}· by ASC and PSC from *Nibea japonica* swim bladders did not reach 50% even at concentrations of 10 mg/mL, which were obviously lower than that of PSC from *Lophius litulon* skin [7]. These results indicated that the PSC from *Lophius litulon* skin has a moderate antioxidant capacity.

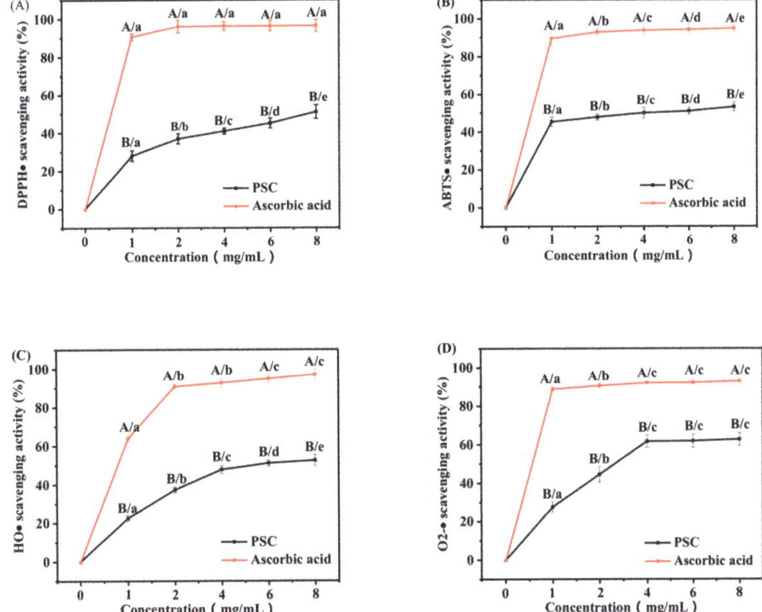

Figure 5. (**A**) 2,2-diphenyl-1-picrylhydrazyl (DPPH)·, (**B**) ABTS·, (**C**) HO·, and (**D**) O_2^{-}· scavenging activities of PSC from *Lophius litulon* skin. (**A–B**) Values with different letters indicated significant differences in different samples at the same concentrations ($p < 0.05$). (a–e) Values with different letters indicated significant differences in the same samples at the different concentrations ($p < 0.05$).

2.7. Estimation of Antioxidant Status in Animal Serum

Malondialdehyde (MDA) is widely used for indicator of lipid peroxidation, and an increase in MDA content suggests lipid peroxidation damage during skin aging [39]. Superoxide dismutase (SOD) is present in every cell of the human body and has been shown to protect cells and tissues against oxidative stress [38]. SOD reduces superoxide anion radicals to generate hydrogen peroxide, which can then be disintegrated into water and oxygen by glutathione peroxidase (GSH-Px) and catalase (CAT), and preventing the formation of hydroxyl radicals [38,40].

As shown in Figure 6A, MDA content gradually decreased as the number of days increased, and the content of the PSC sponge group was always lower than that of the control group. There were significant differences in MDA content between the PSC sponge group and the control group at 3, 7, and 12 days ($p < 0.05$). Figure 6B and 6C show that the content of CAT and SOD in the serum of mice gradually increased over time. Moreover, the content of the PSC sponge group was always higher than that of the control group. In addition, compared with the control group, CAT content in the PSC sponge group was significantly different at 3, 7, and 12 days, and SOD content in the PSC sponge group had a significant difference at days 7 and 12 days ($p < 0.05$). The above results demonstrate that PSC slowed the formation of MDA andenhanced the activity of CAT and SOD.

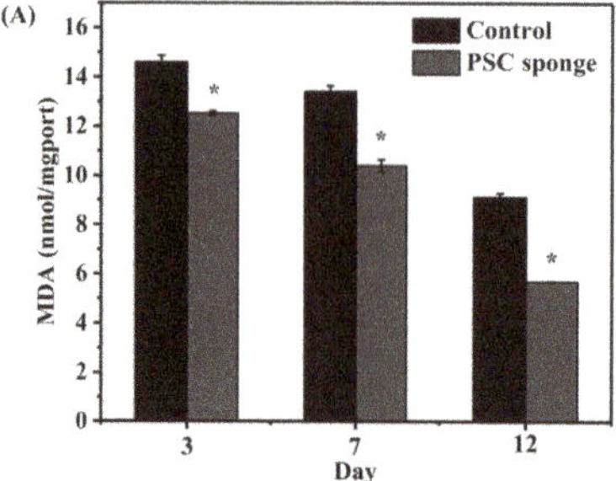
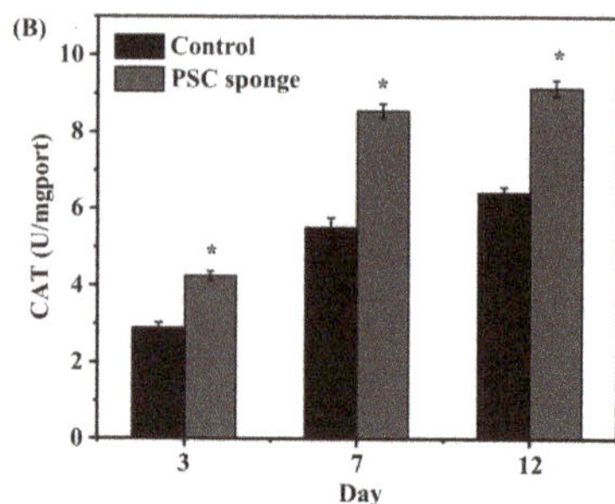
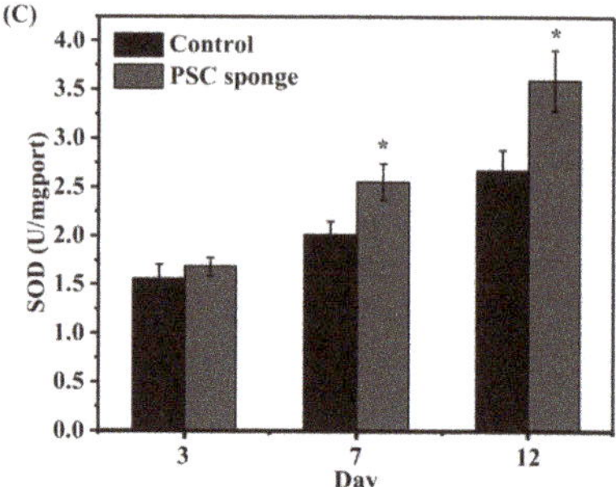

Figure 6. Effects of PSC on serum antioxidant activity in mice. (**A**) MDA, (**B**) CAT, and (**C**) SOD. Data are expressed as the mean ± SD ($n = 5$). * a significant difference when compared to the control ($p < 0.05$).

2.8. Evaluation of Biocompatibility in the Skin Wound Model

Many papers have been reported that the antioxidant defense can reduce ROS production and accelerate wound healing. It was shown that PSC from *Lophius litulon* skin has a certain antioxidant function based on its ability to scavenge free radicals and increase the content of SOD and CAT in the serum of mice. Therefore, the PSC from *Lophius litulon* skin may have potential applications in wound healing. Of course, another prerequisite for application in the wound healing purposes is that the PSC must have good biocompatibility. In this study, a skin wound model was preliminary tested for evaluation of biocompatibility of PSC and its potential for wound healing. Photographs were taken on days 0, 3, 7, and 12 using a digital camera to assess wound status. As presented in Figure 7A, PSC sponge treatment did not significantly cause epithelial necrosis and wound deterioration, indicating that PSC sponge has no significant toxic effect on skin cells and negative stimulating effect on wounds. Furthermore, the wound healing rates were 23.99%, 47.12%, and 56.77% in the control group, respectively, on days 3, 7, and 12, while the wound healing rates were 36.92%, 53.97%, and 88.34% in the PSC sponge group on days 3, 7, and 12 (Figure 7B).

To evaluate the progression of the healing process as a proof that the sponge is biocompatible, histological studies were performed on the wound tissue at 3, 7, and 12 days.

As shown in Figure 8, after 3 days the thickness of epidermal cells in the control group was smaller and the epidermis was partially exfoliated. In addition, a large number of inflammatory cells infiltrated the dermis. Granulation tissue and neovascularization, fibroblast proliferation, and collagen fiber accumulation were virtually absent in the control group. There was a thinner layer of epidermal cells in the PSC sponge group, and some inflammatory cells still existed in the dermis. There was also some granulation tissue formation in the collagen sponge group.

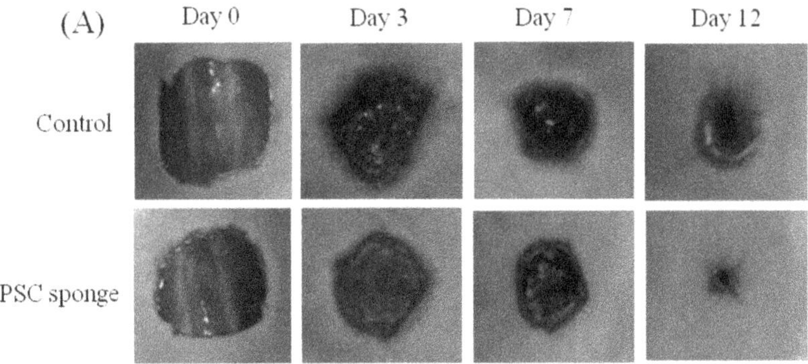

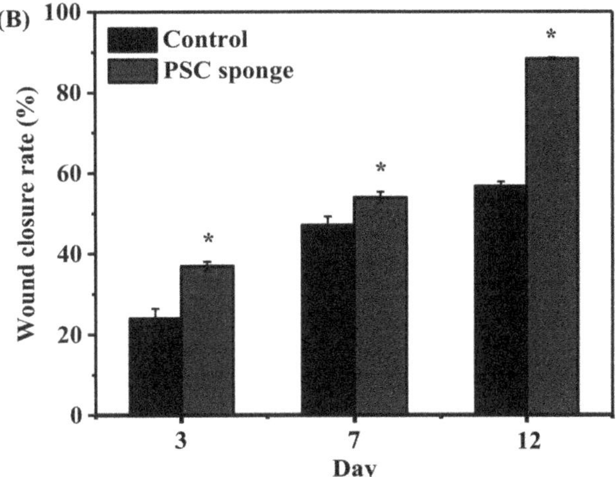

Figure 7. Effect of PSC in wound healing. (**A**) Representative photographs on days 0, 3, 7, and 12. (**B**) Wound contraction (%) on days 3, 7, and 12. Data are expressed as the mean ± SD ($n = 5$). *, a significant difference when compared to the control ($p < 0.05$).

At day 7, 1 to 3 layers of epidermal cells could be observed in the control group, and some areas of epidermal cells had not yet grown. A large number of inflammatory cells were still infiltrated into the dermis. At this time, we observed new granulation tissue and blood vessels, fibroblast proliferation and collagen fiber accumulation. The epidermal cells of PSC sponge treated tissues reached 3–5 layers, and had fewer inflammatory cells. Sebaceous glands and hair follicle tissue increased, fibroblasts proliferated, and collagen fibers accumulated to a greater extent than at day 3.

By day 12, the epidermal cells in the control group had 3-4 layers, and the dermal cells were infiltrated with a large number of inflammatory cells. At this point, the new granulation tissue was thickened, new angiogenesis increased, and sebaceous cells and hair follicle tissue began to form. Meanwhile, in the PSC sponge group, there were 5–8 layers of epidermal cells were observed, and inflammatory cells in the dermis were almost non-existent. The collagen sponge group demonstrated higher epidermal regeneration, granulation tissue regeneration, angiogenesis, fibroblast proliferation, and collagen fiber deposition.

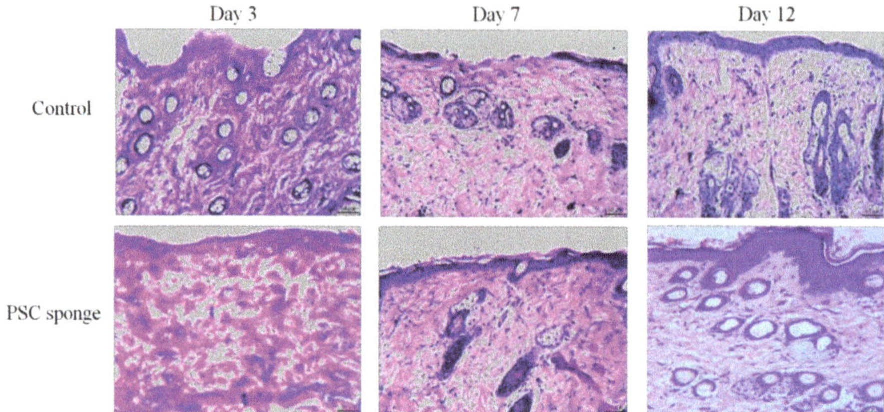

Figure 8. Histopathological observation of wound tissue of skin of the control and PSC sponge groups with H&E staining on days 3, 7, and12 (magnification: ×200).

It was seen from the results that PSC sponge treatment could promote the epidermal cell proliferation and reduce the production of inflammatory cells in the skin wound model. The regeneration of epithelial cells, the reduction in the number of inflammatory cells, the augment in the thickness of granulation tissue, and the deposition of collagen fibers are considered to be signs of progress in the healing process [41]. Therefore, our results indicated that the PSC sponge from the skin of *Lophius litulon* has good biocompatibility and the potential for wound healing.

3. Materials and Methods

3.1. Materials and Chemical Reagents

The pepsin-soluble collagen (PSC) used in this experiment was obtained by laboratory extraction and separation. The prestained color protein marker (cat. no. P0068) was obtained from Beyotime Biotechnology (Shanghai, China). Collagen type I (cat. no. C8060) was provided by Solarbio (Beijing, China). MDA, CAT, SOD and Hematoxylin-eosin (H&E) staining kits were supplied by the Nanjing Jiancheng Bioengineering Institute (Nanjing, China). The enzyme-linked immunosorbent assay (ELISA) kits for detecting IL-6, IL-1β, and tumor necrosis factor (TNF-α) were purchased from Elabscience Biotechnology Co., Ltd. (Wuhan, China).

3.2. SDS-PAGE Analysis

According to the method described by Tang et al. [16], PSC was analyzed using SDS-PAGE. PSC samples were first dissolved in 0.5 M acetic acid and mixed with 5× loading buffer. The mixture was boiled at 100 °C for 5 min and placed in a centrifuge (12,000 rpm, 5 min) to remove the remaining fragments. Electrophoresis was performed using 7.5% gel to estimate the molecular weight of PSC. Collagen type I was used as a positive control.

3.3. Amino Acid Analysis

The amino acid composition of the PSC samples were hydrolyzed using 6 M HCl at 110 °C for 24 h and then vaporized. The hydrolyzates were neutralized in a 25 mL citric acid buffer and analyzed using an amino acid analyzer (L-8800, Hitachi, Tokyo, Japan).

3.4. Relative Solubility

The lyophilized PSC sample (3 mg/mL) was dissolved in 0.5 M acetic acid. The PSC solution (8 mL) was then transferred to a 50 mL centrifuge tube and the pH of PSC solution was adjusted using either 6 M HCl or 6 M NaOH to a final pH of 1.0 to 10.0, and was maintained at 10 mL through deionized water. The mixture was centrifuged at 4 °C at 12,000 rpm for 10 min and the protein content in the supernatant was measured using the BCA kit from the Nanjing Jiancheng Bioengineering Institute.

The influence of NaCl on PSC solubility was determined as follows: the PSC (3 mg/mL) was dissolved in 0.5 M acetic acid, and 5 mL NaCl (in 0.5 M acetic acid) in different solution concentrations (0%, 2%, 4%, 6%, 8%, 10%, and 12%) were added to the solution until the final concentration was either 0%, 1%, 2%, 3%, 4%, 5%, or 6%. The mixed solutions were then stirred for 30 min at 4 °C and then centrifuged (4 °C, 12,000 rpm, 10 min). The protein content in the supernatant was measured as described above. Relative solubility was calculated as the ratio of the protein concentration at the current pH to the protein concentration at the maximum pH.

3.5. FTIR Spectroscopic Analysis

The FTIR spectra of lyophilized PSC samples were received using a FTIR spectrometer (Bruker, Rheinstetten, Germany). The infrared spectra were recorded at a resolution of 1 cm^{-1} and a wavelength range of 4000–400 cm^{-1}. Analysis of the spectral data were analyzed using ORIGIN 8.0 software (Thermo Nicolet, Madison, WI, USA).

3.6. SEM Analysis

The collagen samples were mounted on a blade with two-sided adhesive tape. They were then placed inside a sputter for gold sputtering and the images of sputtered specimens were observed using a scanning microscope (JSM-840, JEOL, Tokyo, Japan) at ×250, ×500, ×1000, and ×2500 magnification.

3.7. Antioxidant Activity

The scavenging activity of PSC for DPPH·, ABTS·, HO·, and O$_2$- radicals was measured according to the procedures described by Chen et al. [7], with ascorbic acid used as a positive control.

3.7.1. DPPH Radical Scavenging Activity

One milliliter of samples containing different concentrations (1–8 mg/mL) were added to different centrifuge tubes (5 mL), and then 250 μL DPPH (0.02%) ethanol and 1.0 mL absolute ethanol were added. The mixture was left in the dark for 30 min and the absorbance (As) was measured at 517 nm. The sample replaced with deionized water was used as the control group (Ac), and the DPPH replaced with ethanol was used as the blank group (Ab). The calculation formula of DPPH· scavenging activity is as follows:

$$\text{DPPH·scavenging activity \%} = \frac{[1 - (\text{As} - \text{Ab})]}{\text{Ac}} \times 100\%$$

3.7.2. ABTS Radical Scavenging Activity

One milliliter of ABTS radical diluent and 1 mL samples with different concentrations (1–8 mg/mL) were added into different centrifuge tubes (5 mL). The mixture was left in the dark for 10 min and the absorbance was measured at 734 nm (As). Samples replaced with deionized water were used as the control group (Ac). The calculation formula of ABTS· scavenging activity is as follows:

$$\text{ABTS· scavenging activity (\%)} = \frac{\text{Ac} - \text{As}}{\text{Ac}} \times 100\%$$

3.7.3. Hydroxyl Radical Scavenging Activity

One milliliter of o-Phenanthroline solution (1.5 mM) was added in different centrifuge tubes (5 mL), and then 1mL samples of different concentrations (1–8 mg/mL) and the 1mL FeSO4 solution (1.5 mM) were added, respectively. Finally, 1.0 mL of 0.03% H_2O_2 solution was added to induced the reaction. The mixture was bathed in water at 37 °C for 90 min and its absorbance was measured at 536 nm (As). Samples replaced with deionized water were used as the control group (Ac), and the 0.03% H2O2 solution replaced with deionized water was used as the blank group (Ab). The calculation formula of HO· scavenging activity is as follows:

$$\text{HO· scavenging activity (\%)} = \frac{[1 - (\text{As} - \text{Ab})]}{\text{Ac}} \times 100\%$$

3.7.4. Superoxide Anion Radical Scavenging Activity

One-milliliter samples of different concentrations (1–8 mg/mL) were added into different centrifuge tubes (5 mL), and then 1 mL Nitro tetrazolium blue chloride (NBT) solution (2.52 mM) and 1 mL NADH (624 µM) was added. At last, 1 mL phenazine methosulfate (PMS) solution (120 µM) was added to induced the reaction. The mixture was bathed in water at 25 °C for 5 min and the absorbance was measured at 560 nm (As). Samples replaced with deionized water were set as control group (Ac). The calculation formula of O_2-· scavenging activity is as follows:

$$\text{O2} - \text{· scavenging activity (\%)} = \frac{\text{As} - \text{Ac}}{\text{Ac}} \times 100\%$$

3.8. Animals Grouping and Wound Creation

Thirty male ICR mice (22–24 g) were provided by the Experiment Animal Center of Zhejiang Province (certificate no. SCXK 2014-0001). All the mice were kept under conventional and uniform conditions at 22 °C. After the mice were given 7 days to adapt to their new environment, they were randomly divided into two groups: the control group and the PSC sponge group. All mice were fed with the SPF grade mouse feed at 4 g/day.

The mice were narcosised with 4% chloral hydrate before undergoing the surgical procedure to create the wound. The hair was then removed from the mice and sterile surgical scissors were used to create a 1 cm diameter wound area. The control group was only treated with 0.9% saline on the wound area. In the PSC sponge group, the wound area was completely covered with a PSC sponge, and the PSC sponge was sterilized using ultraviolet radiation before use. Photographs were taken on days 0, 3, 7, and 12 using a digital camera to assess wound healing status, and Image J software was used to calculate wound area. The formula for wound closure rate (%) was as follows:

$$\text{Wound closure rate (\%)} = \frac{0 \text{ day wound area} - \text{wound area on a particular day}}{0 \text{ day wound area}} \times 100\%$$

On days 3, 7, and 12, five mice in each group were killed. Serum was collected to assess antioxidant levels and the wound area was carefully removed from each mouse. Each wound area was divided into two parts. One part was used for histopathological observation and the other was stored at −80 °C for analysis of inflammatory factor levels.

3.9. Assessment of Serum Antioxidant Levels

Blood was sampled from the eyes of the mice before each group of mice was killed. Serum and plasma were separated in a refrigerated centrifuge (4 °C, 5000 rpm, 5 min). The contents of MAD, CAT and SOD in the serum were measured using mouse MAD, CAT and SOD kits.

3.10. Histomorphological Observation

After collecting the wound tissue, the tissue was fixed in 4% paraformaldehyde solution for 24–48 h, then embedded in paraffin, cut into sections that were 4 microns in size, stained with a H&E staining kit, and sealed with a neutral gel. The histomorphological changes of the wound tissues in each group were observed under an optical microscope (CX31, Olympus) and photographed using a CCD-NC 6051 photographic system.

3.11. Statistical Analysis

The experimental data were analyzed and processed using IBM SPSS 19.0 statistical software (Armonk, NY, USA). All experiments were repeated in triplicate and the figures were expressed as mean ± standard deviation (SD). The data were analyzed using a one-way analysis of variance (ANOVA) test, and $p < 0.05$ values were considered to be statistically significant.

4. Conclusions

We studied the physicochemical properties, antioxidant activity, and biocompatibility of pepsin-solubilized collagen (PSC) obtained from the skin of *Lophius litulon*. The results of FTIR and SEM revealed that PSC had a three-step spiral structure and porous fiber network microstructure. PSC can scavenge the free radicals in a dose-dependent manner and increase the levels of SOD and CAT, suggesting that the PSC from *Lophius litulon* skin showed antioxidant activity. Furthermore, the antioxidant activity of PSC was higher than that of collagen peptides from tilapia skin and collagens from *Nibea japonica* swim bladders [7,38]. Excessive reactive oxygen species (ROS) can lead to a variety of chronic health problems, and increased antioxidant defense can reduce ROS production and accelerate wound healing [42]. At present, considerable attention has been paid to the use of fish-derived collagen and peptides for wound healing. Hu et al. reported that collagen peptides from the skin of Nile tilapia (Oreochromis niloticus) could enhance the process of wound healing [17]. The present finding suggested that the PSC from *Lophius litulon* skin has good biocompatibility and may be used as a biomaterial for wound healing in clinical and cosmetic fields. In the future, we will further explore the effect of PSC on wound healing.

Author Contributions: Z.Y. conceived and designed the experiments. W.Z., J.Z. and X.T. carried out the experiment. W.Z., H.J., Y.T. and G.D. analyzed the data. W.Z. and H.J. wrote this manuscript.

Funding: The research was funded by the Fundamental Research Funds for the Provincial Universities of Zhejiang Province (2019JZ00005) and the National Natural Science Foundation of China (grant No. 41806153).

Conflicts of Interest: The authors declare no conflict of interest.

References

1. Azizur Rahman, M. Collagen of extracellular matrix from marine invertebrates and its medical applications. *Mar. Drugs* **2019**, *17*, 118. [CrossRef]
2. Aravinthan, A.; Park, J.K.; Hossain, M.A.; Sharmila, J.; Kim, H.J.; Kang, C.W.; Kim, N.S.; Kim, J.H. Collagen-based sponge hastens wound healing via decrease of inflammatory cytokines. *3 Biotech* **2018**, *8*, 487. [CrossRef] [PubMed]
3. Kim, K.O.; Lee, Y.; Hwang, J.W.; Kim, H.; Kim, S.M.; Chang, S.W.; Lee, H.S.; Choi, Y.S. Wound healing properties of a 3-D scaffold comprising soluble silkworm gland hydrolysate and human collagen. *Colloids Surf. B* **2014**, *116*, 318–326. [CrossRef] [PubMed]
4. Pal, P.; Srivas, P.K.; Dadhich, P.; Das, B.; Maity, P.P.; Moulik, D. Accelerating full thickness wound healing using collagen sponge of mrigal fish (*Cirrhinus cirrhosus*) scale origin. *Int. J. Biol. Macromol.* **2016**, *93*, 1507–1518. [CrossRef] [PubMed]

5.   Itoh, S.; Kikuchi, M.; Takakuda, K.; Koyama, Y.; Matsumoto, H.N.; Ichinose, S.; Tanaka, J.; Kawauchi, T.; Shinomiya, K. The biocompatibility and osteoconductive activity of a novel hydroxyapatite/collagen composite biomaterial, and its function as a carrier of rhbmp-2. *J. Biomed. Mater. Res.* **2001**, *54*, 445–453. [CrossRef]

6.   Lukasiewicz, A.; Skopinska-Wisniewska, J.; Marszalek, A.; Molski, S.; Drewa, T. Collagen/polypropylene composite mesh biocompatibility in abdominal wall reconstruction. *Plast. Reconstr. Surg.* **2013**, *131*, 731e–740e. [CrossRef]

7.   Chen, Y.Y.; Jin, H.X.; Yang, F.; Jin, S.J.; Liu, C.J.; Zhang, L.K.; Huang, J.; Wang, S.G.; Yan, Z.Y.; Cai, X.W.; et al. Physicochemical, antioxidant properties of giantcroaker (*Nibeajaponica*) swim bladders collagen and wound healing evaluation. *Int. J. Biol. Macromol.* **2019**, *138*, 483–491. [CrossRef]

8.   Donnez, J.; Binda, M.M.; Donnez, O.; Dolmans, M.M. Oxidative stress in the pelvic cavity and its role in the pathogenesis of endometriosis. *Fertil. Steril.* **2016**, *106*, 1011–1017. [CrossRef]

9.   Janda, J.; Nfonsam, V.; Calienes, F.; Sligh, J.E.; Jandova, J. Modulation of ROS levels in fibroblasts by altering mitochondria regulates the process of wound healing. *Arch. Dermatol. Res.* **2016**, *308*, 239–248. [CrossRef]

10.   Shao, Y.; Dang, M.Y.; Lin, Y.K.; Xue, F. Evaluation of wound healing activity of plumbagin in diabetic rats. *Life Sci.* **2019**, *231*, 116422. [CrossRef]

11.   Abood, W.N.; Al-Henhena, N.A.; Abood, A.N.; Al-Obaidi, M.M.J.; Bartan, R.A. Wound-healing potential of the fruit extract of *Phaleria macrocarpa*. *Bosn. J. Basic Med. Sci.* **2015**, *15*, 25–30. [CrossRef] [PubMed]

12.   Zhang, K.; Xu, Z.; Chen, H.X.; Guo, N.; Li, L. Anisakid and raphidascaridid nematodes (Ascaridoidea) infection in the important marine food-fish *Lophius litulon* (Jordan) (*Lophiiformes: Lophiidae*). *Int. J. Food Microbiol.* **2018**, *284*, 105–111. [CrossRef] [PubMed]

13.   Chi, C.F.; Wang, B.; Deng, Y.Y.; Wang, Y.M.; Deng, S.G.; Ma, J.Y. Isolation and characterization of three antioxidant pentapeptides from protein hydrolysate of monkfish (*Lophius litulon*) muscle. *Food Res. Int.* **2014**, *55*, 222–228. [CrossRef]

14.   Ma, H.W.; Yang, H.C.; Fu, W.D.; Liao, M.F.; Zhou, Y.F.; Xiang, X.W.; Chen, M.; Zheng, B. Antioxidant activity of collagen peptides from *Lophius litulon* skin. *Food Sci.* **2014**, *35*, 80–84.

15.   Noitup, P.; Garnjanagoonchorn, W.; Morrissey, M.T. Fish skin type I collagen. *J. Aquat. Food Prod. Technol.* **2005**, *14*, 17–28. [CrossRef]

16.   Lin, X.H.; Chen, Y.Y.; Jin, H.X.; Zhao, Q.L.; Liu, C.J.; Li, R.W.; Yu, F.M.; Chen, Y.; Huang, F.F.; Yang, Z.S.; et al. Collagen Extracted from Bigeye Tuna (*Thunnus obesus*) Skin by Isoelectric Precipitation: Physicochemical Properties, Proliferation, and Migration Activities. *Mar. Drugs* **2019**, *17*, 261. [CrossRef]

17.   Tang, Y.P.; Jin, S.J.; Li, X.; Li, X.Y.; Hu, X.Y.; Chen, Y.; Huang, F.F.; Yang, Z.S.; Yu, F.M.; Ding, G.F. Physicochemical Properties and Biocompatibility Evaluation of Collagen from the Skin of Giant Croaker (*Nibea japonica*). *Mar. Drugs* **2018**, *16*, 222. [CrossRef]

18.   Hu, Z.; Yang, P.; Zhou, C.X.; Li, S.D.; Hong, P.Z. Marine Collagen Peptides from the Skin of Nile Tilapia (*Oreochromis niloticus*): Characterization and Wound Healing Evaluation. *Mar. Drugs* **2017**, *15*, 102. [CrossRef]

19.   Zou, T.B.; He, T.P.; Li, H.B.; Tang, H.W.; Xia, E.Q. The structure-activity relationship of the antioxidant peptides from natural proteins. *Molecules* **2016**, *21*, 72. [CrossRef]

20.   Alemán, A.; Giménez, B.; Montero, P.; Gómez-Guillén, M.C. Antioxidant activity of several marine skin gelatins. *LWT-Food Sci. Technol.* **2011**, *44*, 407–413. [CrossRef]

21.   Mcgavin, M.H.; Krajewskapietrasik, D.; Rydén, C.; Höök, M. Identification of a staphylococcus aureus extracellular matrix-binding protein with broad specificity. *Infect. Immun.* **1993**, *61*, 2479–2485. [PubMed]

22.   Hofmann, M.; Winzer, M.; Weber, C.; Gieseler, H. Low-volume solubility assessment during high-concentration protein formulation development. *J. Pharm. Pharmacol.* **2016**, *70*, 636–647. [CrossRef] [PubMed]

23.   Savre, M. Maillard-Induced Glycation of Whey Protein Using Maltodextrin and Effect on Solubility, Thermal Stability, and Emulsification Properties. Master's Thesis, University of Minnesota, Minneapolis, MN, USA, 2016.

24.   Jeevithan, E.; Bao, B.; Bu, Y.; Zhou, Y.; Zhao, Q.; Wu, W. TypeIIcollagen and gelatin from silvertip shark (*Carcharhinus albimarginatus*) cartilage: Isolation, purification, physicochemical and antioxidant properties. *Mar. Drugs* **2014**, *12*, 3852–3873. [CrossRef] [PubMed]

25. Luo, Q.B.; Chi, C.F.; Yang, F.; Zhao, Y.Q.; Wang, B. Physicochemical properties of acid- and pepsin-soluble collagens from the cartilage of siberian sturgeon. *Environ. Sci. Pollut. Res.* **2018**, *25*, 31427–31438. [CrossRef] [PubMed]

26. Chen, J.D.; Li, M.; Yi, R.Z.; Bai, K.K.; Wang, G.Y.; Tan, R.; Sun, S.S.; Xu, N.H. Electrodialysis Extraction of Pufferfish Skin (*Takifugu flavidus*): A Promising Source of Collagen. *Mar. Drugs* **2019**, *17*, 25. [CrossRef]

27. Abe, Y.; Krimm, S. Normal vibrations of crystalline polyglycine I. *Biopolymers* **1972**, *11*, 1817–1839. [CrossRef]

28. Zou, Y.; Wang, L.; Cai, P.P.; Li, P.P.; Zhang, M.H.; Sun, Z.L.; Sun, C.; Xu, W.M.; Wang, D.Y. Effect of ultrasound assisted extraction on the physicochemical and functional properties of collagen from soft-shelled turtle calipash. *Int. J. Biol. Macromol.* **2017**, *105*, 1602–1610. [CrossRef]

29. Sun, L.; Hou, H.; Li, B.; Zhang, Y. Characterization of acid- and pepsin-soluble collagen extracted from the skin of nile tilapia (*Oreochromis niloticus*). *Int. J. Biol. Macromol.* **2017**, *99*, 8–14. [CrossRef]

30. Sinthusamran, S.; Benjakul, S.; Kishimura, H. Comparative study on molecular characteristics of acid soluble collagens from skin and swim bladder of seabass (*Lates calcarifer*). *Food Chem.* **2013**, *138*, 2435–2441. [CrossRef]

31. Yu, F.M.; Zong, C.H.; Jin, S.J.; Zheng, J.W.; Chen, N.; Huang, J.; Chen, Y.; Huang, F.F.; Yang, Z.S.; Tang, Y.P.; et al. Optimization of extraction conditions and characterization of pepsin-solubilised collagen from skin of giant croaker (*Nibea japonica*). *Mar. Drugs* **2018**, *16*, 29. [CrossRef]

32. El-Rashidy, A.A.; Gad, A.; El-Hay, G.; Abu-Hussein, A.; Habib, S.I.; Badr, N.A.; Hashem, A.A. Chemical and biological evaluation of egyptian nile tilapia (*Oreochromis niloticas*) fish scale collagen. *Int. J. Biol. Macromol.* **2015**, *79*, 618–626. [CrossRef] [PubMed]

33. Arumugam, G.K.S.; Sharma, D.; Balakrishnan, R.M.; Ettiyappan, J.B.P. Extraction, optimization and characterization of collagen from sole fish skin. *Sustain. Chem. Pharm.* **2018**, *9*, 19–26. [CrossRef]

34. Sun, L.; Li, B.; Song, W.; Si, L.; Hou, H. Characterization of pacific cod (*Gadus macrocephalus*) skin collagen and fabrication of collagen sponge as a good biocompatible biomedical material. *Process. Biochem.* **2017**, *63*, 229–235. [CrossRef]

35. Ustuner, O.; Anlas, C.; Bakirel, T.; Ustun-Alkan, F.; Diren Sigirci, B.; Ak, S.; Akpulat, H.A.; Donmez, C.; Koca-Caliskan, U. In Vitro Evaluation of Antioxidant, Anti-Inflammatory, Antimicrobial and Wound Healing Potential of Thymus Sipyleus Boiss. Subsp. Rosulans (Borbas) Jalas. *Mar. Drugs* **2019**, *24*, 3353. [CrossRef] [PubMed]

36. Cao, X.P.; Chen, Y.F.; Zhang, J.L.; You, M.M.; Wang, K.; Hu, F.L. Mechanisms underlying the wound healing potential of propolis based on its, in vitro, antioxidant activity. *Phytomedicine* **2017**, *34*, 76–84. [CrossRef]

37. Vittorazzi, C.; Endringer, D.C.; Andrade, T.U.; Scherer, R.; Fronza, M. Antioxidant, antimicrobial and wound healing properties of *Struthanthus vulgaris*. *Pharm. Biol.* **2016**, *54*, 331–337. [CrossRef]

38. Ren, S.; Li, J.; Guan, H. The antioxidant effects of complexes of tilapia fish skin collagen and different marine oligosaccharides. *J. Ocean Univ. China* **2010**, *9*, 399–407. [CrossRef]

39. Inal, M.E.; Kanbak, G.; Sunal, E. Antioxidant enzyme activities and malondialdehyde levels related to aging. *Clin. Chim.* **2001**, *305*, 75–80. [CrossRef]

40. Yang, Q.M.; Pan, X.H.; Kong, W.B.; Yang, H.; Su, Y.D.; Zhang, L.; Zhang, Y.N.; Yang, Y.L.; Ding, L.; Liu, G.A. Antioxidant activities of malt extract from barley (*Hordeum vulgare* L.) toward various oxidative stress in vitro and in vivo. *Food. Chem.* **2010**, *118*, 84–89.

41. Ashcroft, G.S.; Mills, S.J.; Ashworth, J.J. Ageing and wound healing. *Biogerontology* **2002**, *3*, 337–345. [CrossRef]

42. Majumder, P.; Paridhavi, M. A novel polyherbal formulation hastens diabetic wound healing with potent antioxidant potential: A comprehensive pharmacological investigation. *Nat. Prod. Chem. Res.* **2018**, *6*, 4. [CrossRef]

Article

Biocompatibility of a Marine Collagen-Based Scaffold In Vitro and In Vivo

Dafna Benayahu [1],*, Leslie Pomeraniec [1], Shai Shemesh [2], Snir Heller [2], Yoav Rosenthal [2], Lea Rath-Wolfson [3] and Yehuda Benayahu [4]

[1] Department of Cell and Developmental Biology Sackler Faculty of Medicine, Tel Aviv University, Tel Aviv 69978, Israel; leslieyael.p@gmail.com
[2] Department of Orthopaedic Surgery, Rabin Medical Center, Petach Tikva, and Sackler Faculty of Medicine, Tel Aviv University, Tel Aviv 69978, Israel; shemesh.shai@gmail.com (S.S.); snirheller@gmail.com (S.H.); yoav.rosenthal@gmail.com (Y.R.)
[3] Department of Pathology, Rabin Medical Center, Petach Tikva, and Sackler Faculty of Medicine, Tel Aviv University, Tel Aviv 69978, Israel; leawolfson@gmail.com
[4] School of Zoology, George S. Wise Faculty of Life Science, Tel Aviv University, Tel Aviv 69978, Israel; yehudab@tauex.tau.ac.il
* Correspondence: dafnab@tauex.tau.ac.il; Tel.: +972-3-6406187

Received: 10 June 2020; Accepted: 5 August 2020; Published: 11 August 2020

Abstract: Scaffold material is essential in providing mechanical support to tissue, allowing stem cells to improve their function in the healing and repair of trauma sites and tissue regeneration. The scaffold aids cell organization in the damaged tissue. It serves and allows bio mimicking the mechanical and biological properties of the target tissue and facilitates cell proliferation and differentiation at the regeneration site. In this study, the developed and assayed bio-composite made of unique collagen fibers and alginate hydrogel supports the function of cells around the implanted material. We used an in vivo rat model to study the scaffold effects when transplanted subcutaneously and as an augment for tendon repair. Animals' well-being was measured by their weight and daily activity post scaffold transplantation during their recovery. At the end of the experiment, the bio-composite was histologically examined, and the surrounding tissues around the implant were evaluated for inflammation reaction and scarring tissue. In the histology, the formation of granulation tissue and fibroblasts that were part of the inclusion process of the implanted material were noted. At the transplanted sites, inflammatory cells, such as plasma cells, macrophages, and giant cells, were also observed as expected at this time point post transplantation. This study demonstrated not only the collagen-alginate device biocompatibility, with no cytotoxic effects on the analyzed rats, but also that the 3D structure enables cell migration and new blood vessel formation needed for tissue repair. Overall, the results of the current study proved for the first time that the implantable scaffold for long-term confirms the well-being of these rats and is correspondence to biocompatibility ISO standards and can be further developed for medical devices application.

Keywords: collagen fibers; scaffold; biomedical device; biocompatibility

---

1. Introduction

Collagen is a main extracellular matrix protein that supports the structure mainly of skeletal tissues. Collagen has a main function during the healing process and in cases when the tissue will not heal spontaneously, there is a need to foster it. These bring the collagen to serve both as an essential protein and a major component in biomedical scaffolding for various tissue regeneration approaches. The scaffold facilitates and promotes the autologous stem cells growth and differentiation that will progressively enter the scaffold and replace it by the regenerating tissue. The tissue-engineered scaffolds approach mimics the natural tissue structure and physical properties of the targeted tissue. Such an approach also aims to minimize the use of autologous grafts, which are limited by the availability of the patient's own tissue and may avoid additional surgical procedures. Thus, the challenge is to find a suitable source of collagen, whose extraction and purification procedures will be suitable for medical applications. The overall goal is to find a replacement to synthetic polymer materials used as medical devices, which do not integrate with the body and may trigger an immune response such as chronic inflammation [1]. The use of natural materials as scaffolds is beneficial, and recently it has been shown that certain marine organisms are a promising bio-source to obtain collagen for scaffold formation in a variety of biomedical applications. The great interest in the field is highlighted by a series of studies for biomaterial isolated from different marine species allow assorted applications as biomedical devices [2–11].

Collagens are a heterogeneous family of extracellular matrix proteins, but are highly conserved through evolution. The collagen structure is highly homologous among invertebrates and mammalians, including mice and humans, which allow the collagen to be metabolically compatible through different phylogenetic groups. The collagen family is divided into fibrillar and non-fibrillar categories based on packing and ultrastructure, and this study focus on a fibrillar collagen that was studied in detail and described by us for its molecular and ultrastructure [12]. Marine sources with a high quantity of collagen are bio-compatible with mammalian cells in vitro [13]. The natural materials are biological macromolecules, which are considered to minimize immunological reactions [14–16]. Another advantage of using marine-derived collagen over the mammalian one is that invertebrates can be cultivated [17] under controlled conditions for the purpose of collagen extraction, thus having regulatory and quality control allowing overcoming ethical constraints for use in medical applications.

Most invertebrate collagens appear as extracellular matrices, and their mechanical properties are limited. In this study, the unique collagen fibers were shown to be extracted by a rather simple procedure by pulling out from the coral tissues [6,13]. These unique coral-derived collagen fibers were identified for their mechanical properties [6,18–20], which attribute the bio composite suitability for the formation of scaffolds that can be tailored to meet the mechanical properties of the target tissues. The bio composite developed in this study used collagen fibers which were embedded in alginate hydrogel [6,13], allowing the formation of a three-dimensional scaffold that supports cells growth, as was recently studied for its biocompatibility in vitro, and proved to support cell growth and differentiation [13]. The collagen fibers in the bio-composite provide mechanical and biological cues for cell proliferation and tissue regeneration. An additional molecular cue can be added into the bio composite using nanoparticles (NPs) as a delivery of growth factors, providing biological cues that modulate and promote cell proliferation and differentiation into a desired lineage fate. Based on in vitro studies, the efficiency of cells to endocytose the NPs that were continued to proliferate and differentiate [13,21].

This study further analyses the biocompatibility of the bio composite in vivo in a rat animal model. The experiment demonstrated the healing of a rotator cuff tear, the most common musculoskeletal injury occurring in the shoulder [22]. The bio composite implant served as an augment facilitating tendon repair by relief the load from the healing tendon and eventually allowing both the restoration of this mechanosensitive tissue and the mobility of the operated extremity. In addition, it also served separately as a subcutaneous implant. Following the biocomposite implantation in two sites, the rats were followed up for any sign of the material cytotoxic effects and followed up on their viability, well-being, and functionality. For any new material such as the bio composite presented here, the follow-up is required for biomedical development and resulted in no indication of any toxicity of the material. Thus, the material tested in vivo for biocompatibility to meet the standards established by the Food and Drug Administration Organization for the sub-chronic toxicity (ISO-10993) of transplants, and improved the physiology of the operated rats. The designed biomaterial will allow the future development of bio composite-based products with optimal mechanical properties that will fully integrate with the natural tissue, contributing to its healing processes.

2. Results

The bio-composite production is demonstrated in Figure 1 and a detailed procedure is described in the Materials and Methods section. We analyzed the bio-composite for stability under various storage conditions: (I) dry film was air-dry at room temperature for 14 months, (II) film stored at 4 °C immersed in 70% ethanol for 8 months, (III) film was stored at 37 °C in cell growth media for 6 months. The follow up of dry bio-composite film after 14 months presented no signs of powdering or tearing of the alginate film. The film was immersed in buffer for re-hydration, and the material demonstrated a good stability and the recovery of elasticity, along with the preservation of structure and the organization of the collagen fibers embedded in the alginate. Thus, the dry re-hydrated material was fully functional as a scaffold (Figure 2A–C). Similarly, the composite structure was preserved when the film was maintained at 4 °C in 70% ethanol for 8 months (Figure 3A,D,E). As for the bio composite film immersed in cell growth media at 37 °C for 6 months, visual inspection revealed that the alginate displayed signs of disintegration on its surface, whereas the collagen fibers strengthened the bio-composite and maintained the structure (Figure 2A,F,G). These results confirm the in vitro stability of the bio-composite even when stored for long periods under different conditions, as dry or wet material.

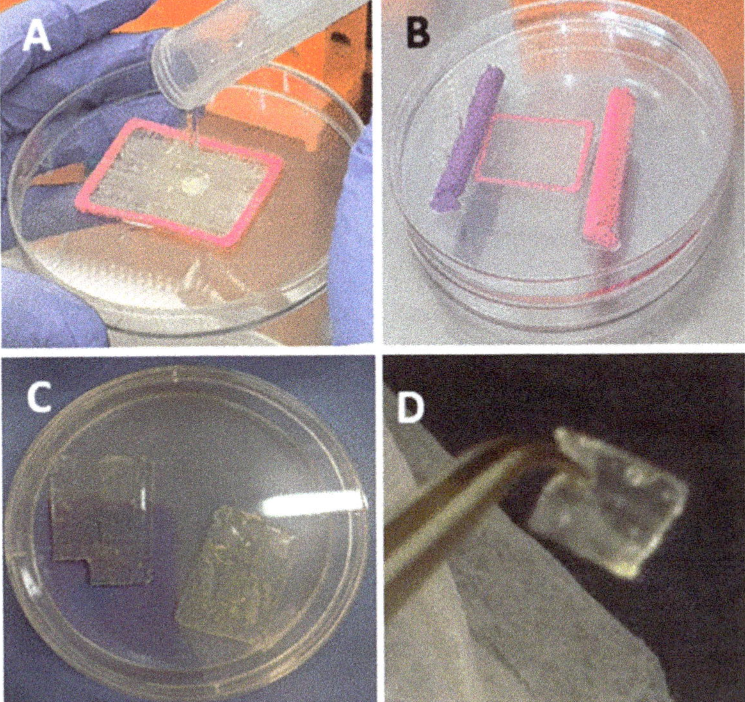

Figure 1. Bio composite preparation. (**A**) Collagen fibers spun around a PLA frame embedded in alginate solution; (**B**) the frame with collagen fibers and alginate inserted in a dialysis bag and immerse in calcium solution to allow alginate gelation; (**C**) the resulting bio-composite device was stored in ethanol solution and washed buffer before use; (**D**) collagen-alginate hydrogel device used for supraspinatus tendon augmentation.

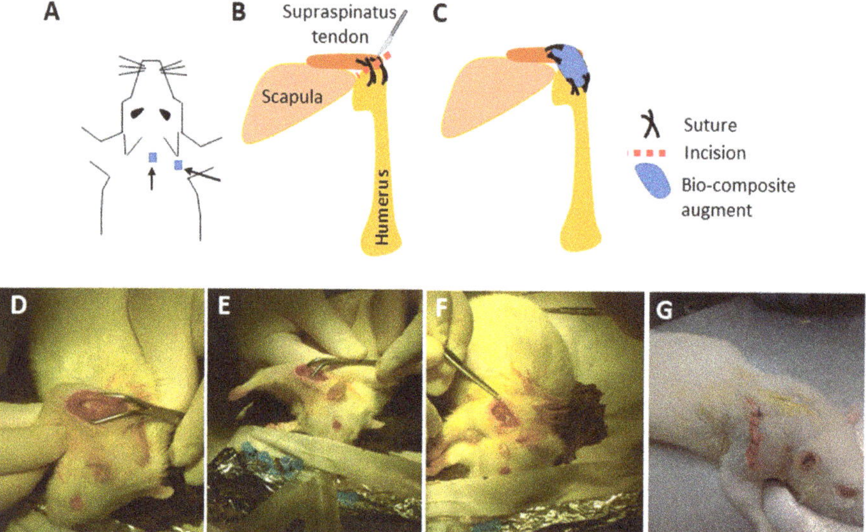

Figure 2. (**A**) Table summarizing the conditions that bio composite films were stored in: storage durations dry or wet material when immersed in 70% ethanol or in growth medium for long periods (at least 6–14 months), at temperatures (4 °C, 25 °C and 37 °C), as indicated in each experiment. (**B–G**) Pictures show the bio composite under various conditions, (**B**) biocomposite dry material, (**C**) rehydrated samples from B in PBS, showing its elasticity. (**D**) and (**E**) magnification of the collagen fibers preserving their structure in the biocomposite. (**F**) Bio-composite film kept at 37 °C in cell growth medium. (**G**) Bio-composite film kept at 4 °C in ethanol.

Figure 3. Surgical procedure. For rotator cuff tearing, repair, and augmentation with collagen-alginate film. (**A**) Implantation sites for subcutaneous and augment implants; (**B**) surgical tearing and repair of the supraspinatus; (**C**) augmentation with bio-composite film; (**D**) exposure of the supraspinatus tendon. (**E**) Suturing the tendon before cutting; (**F**) augmentation of supraspinatus tendon; (**G**) rat after surgical tearing, repair, and augmentation of the rotator cuff tendon.

The bio composite film aimed to serve as scaffold for tissue repair was recently used in a cell culture system and analyzed for growth, migration, and differentiation in 2D and 3D scaffolds in vitro [13]. In the current study, we applied the bio composite film as medical device for tissue repair in vivo at two sites in rat. In one site, the scaffold was used as an augment for aid of repair of a torn rotator cuff supraspinatus tendon, and the other was transplanted in sub-cutaneous site. For tendon repair, a unilateral surgical detachment of the rotator cuff supraspinatus tendon model was utilized. The tendon detachment was then sutured and an augment of bio-composite collagen-alginate was laid on the repair site. The purpose of the transplantation of the augment was in order to contribute to the tendon recovery by relieving the overall load during the healing process. The rat mobility was monitored by video during the rotator cuff recovery period (Figures 3 and 4). It was observed immediately after surgery that the rat lost the proper function of the operated extremity. The rat had impaired motion and avoided leaning on the operated foreleg, while bending and keeping the right front limb close to the chest (Figure 4 and Video S1). The recovery follows up of operated rats included monitoring of animal weight and demonstrated the animal wellbeing along with the weight rising over the weeks (Figure 4). In addition, the Rats' daily follow up in order to confirm their well been there was no signs of depression or suffering. Follow up by physical examination demonstrated healing of the external surgical wound, together with animals' recovery of their mobility (Figure 4B). At first after operation, the rats demonstrated the tendency to step on the left foreleg and hind legs, while bending and keeping the right front limb close to the chest (Figure 4 and Video S1). As time passed, the operated rats displayed comfort stepping on the recovered limb (Video S1) and the healing of the external surgical wound (Figure 4). Altogether, the results demonstrated no adverse responses or any sign of cytotoxicity to the transplant in the operated animals.

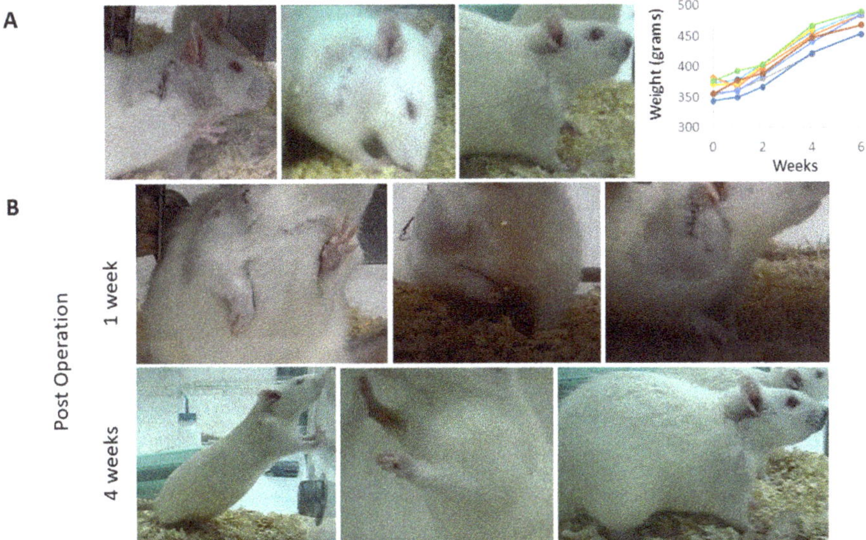

Figure 4. Animal recovery after surgery. (**A**) Recovery follow up of the operated site wound healing and animal weight; (**B**) physical signs for right rotator cuff function impairment 1 week after surgery (middle raw), bending and keeping the right front limb close to the chest, avoid leaning on the operated limb and tendency to step or stand on the left foreleg and hind limbs/short step on the operated foreleg and fast weight shifting to the left foreleg. After 4 weeks, the operated site displayed complete healing, and rats' motility recovery allows them to perform daily activities.

At the end of the experiment, the scaffolds transplanted as augments or at subcutaneous sites and the surrounded tissues were extracted and examined histologically. This analysis was performed in order to assess cells' interaction with the bio-composite and a potential cytotoxic effect or foreign body response. The results visualized that the transplanted bio-composites were integrated with the surrounding tissues, as seen at the macro-level as a square cube surrounded by a fibrotic tissue growing over (Figure 5). The histological analysis revealed that the tissue was formed around and inside the scaffold, (the alginate is visible as am orphic material and marked by A in Figures 6 and 7, and the collagen fibers are marked by arrows). Around the bio composite, the following findings were noted: inflammatory cell recruitment and differentiation to lymphocyte and plasma cells and macrophage activation fused to form giant cells. These findings are expected to be a response to foreign material associated with formation granulation tissue and fibroblast being part of the inclusion process of this material (Figures 6 and 7). The formation of new blood vessels developing through the alginate material was noted (Figure 6). Masson trichrome staining demonstrate that the fibrous tissue found around the bio-composite developed inside the scaffold along (Figure 7).

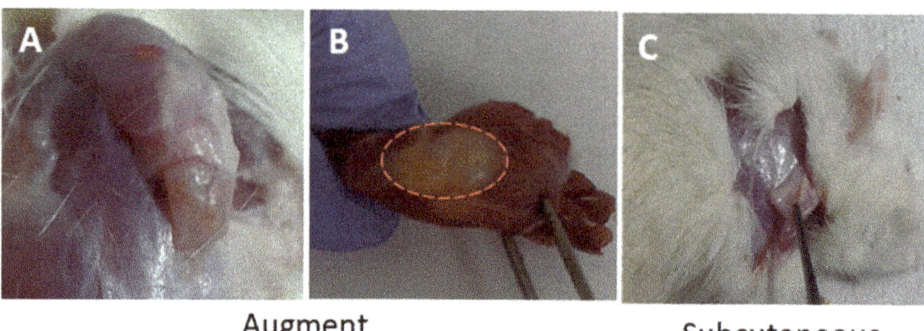

Figure 5. Macro observation of the bio composite film transplants extracted from the augment over rotor cuff (**A–C**). Sub-cutaneous site.

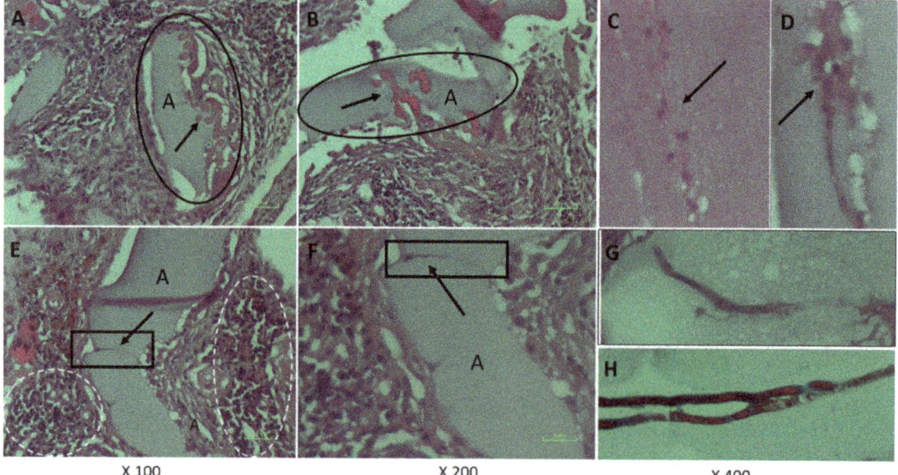

Figure 6. Histopathological examination of the bio-composite transplant retrieved from subcutaneous stained by haematoxylin and eosin shows the alginate as an amorphic material (light pink, in elliptic circle) marked in **A** and the collagen fibers embedded (darker pink, marked by arrow). Around the bio-composite, a fibrous tissue is observed and clusters with lymphocyte and plasma cells infiltration (nuclei stained in blue, marked in a white dashed circle). At (**C,D,G,H**) (arrow or square), a blood vessel that is developing through the alginate material is seen. Magnifications are (**A,E**) × 100; (**B,F**) × 200; (**C,D,G,H**) × 400.

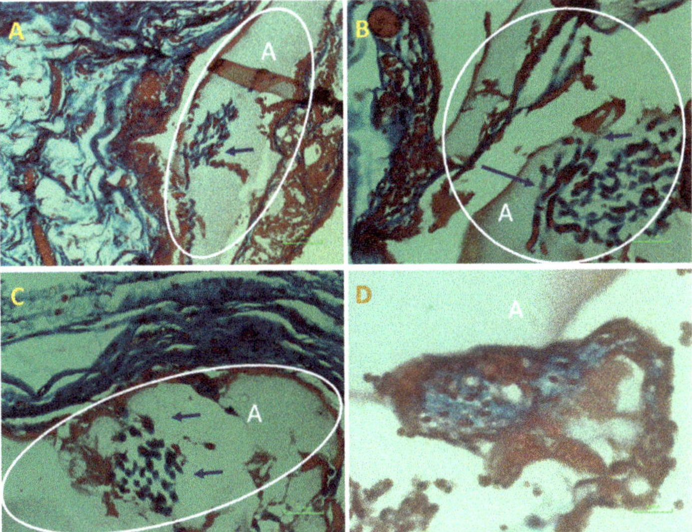

Figure 7. Histopathological examination of the augment bio-composite transplant stained by Masson trichrome seen in (**A–D**). Alginate is an am orphic material (light pink, marked in **A** and the collagen fibers embedded in the alginate (marked by arrow, darker blue). Around the bio-composite, a fibrous tissue (blue) is observed.

3. Discussion

Implant integration with the healing tendon relies on autologous cells function during tissue regeneration, and the augment facilitated the improvement of this recovery step. In the current study, we aimed to evaluate the stability of the bio-composite augment. We analyzed the produced bio composite film both in vitro [13] and in vivo to examine the material stability and its biocompatibility, cytotoxicity, and potential for tissue repair.

First, we evaluated in vitro the stability of the collagen-alginate hydrogel bio composition device under different storage conditions (Figure 2). The bio-composite maintained the natural structure of collagen fibers was stable for at least 14 months as dry material with no sign of film powdering. When the film was hydrated in phosphate buffer to recover its elasticity, no effect on the structure and organization was noted. In addition, the film stored in 70% ethanol at 4 °C displayed the conservation of the bio composite integrity. When the bio-composite film was incubated at 37 °C in growth medium, only superficial hydrogel was disrupted, while the supporting fiber structure in the film and its' shape were kept intact. Thus, the device is stable for long periods (at least 6–14 months) as dry and wet material at a wide range of temperatures (4 °C, 25 °C and 37 °C).

A biocompatibility study recently showed the properties of the bio-composite for cells growth and differentiation in vitro with no sign of cytotoxicity for period of up to 11 weeks [13]. The current study, aimed to follow the scaffold in-vivo where the collagen-alginate scaffold film was transplanted in rats at two locations: (1) subcutaneous and (2) for rotator cuff tendon tearing as an augment. The two experimental sites allowed the biofilm to contact with different tissues. At the subcutaneous site, it was next to the hypodermis of the skin, while the augment was next to the tendon and muscle, a rich vascularized tissue. When the devices were extracted from the operated animals, they were found at the transplanted sites and seen intact. These results indicate that the bio-composite film display stability in vivo up to 14 weeks and proved to be a successful scaffold during the healing process.

The rats were analyzed for their recovery, and the animals displayed the disruption of the left rotator cuff function, as shown in Figure 4. The rats' tendency was to protect the operated limb, bending it close to the chest, and avoiding standing and stepping in this foreleg. During recovery follow up, the operated rats recovered and were back to normal use of the operated limb. After four weeks, all the animals displayed normal mobility with no difference between right (operated) and left rotator cuff and no preference for one of them for activity. This result indicates that the presence of the scaffold augment protects the tendon and tissue regeneration during the healing processes. In these experiments, the rats were with good appetite, increased their weight, were back to normal activity and social non-aggressive behavior, with no signs of stress or suffering along the study, and up to complete curing (Figure 4)—i.e., no sign of cytotoxicity on these rats. Rats were sacrificed and the scaffolds were taken for a histology analysis to evaluate the transplant biocompatibility. The transplants were found to be integrating with the surrounding tissues and encapsulated in the granulation tissue, which is a stage of the healing process and a normal and expected response. A mild immunological response was observed by presence of plasma cells and macrophages, known to secrete pro-fibrogenic factors, which enhance fibro genesis, and is common at this stage. Thus, a fibrous capsule that developed around a transplanted material inhibit macrophages' activity by avoiding their reaching the transplant. Therefore, it is concluded that during the experimental time, the bio-composite device has no negative effect on animals' health, and that this in vivo assay matched the in vitro previous cytotoxicity research [13].

The mechanisms underlying tendon healing encompass the step of macrophages recruitment, followed by fibrous tissue formation and the regulation of ECM remodeling. The presence of fibro-proliferation is proven and angiogenesis lead to tissue remodeling during the wound's healing. Such steps were observed during the current in vivo rotor cuff tendon augmentation and subcutaneous transplant. These findings emphasize the biocompatibility of the bio-composite and its safety, which is a principal quality for clinical use. Therefore, it is concluded that the analyzed material meets the ISO standard for biocompatibility.

4. Material and Methods

4.1. Bio-Composite Preparation and In Vitro Storage and Stability Analysis

4.1.1. Collagen Fiber Isolation

Coral collagen fibers were isolated by mechanical extraction from the soft coral *Sarchophyton ehrenbergi* [6,13,17]. The isolated fibers were manually spun around a thin polylactic acid (PLA) frame to create a dense net of multidirectional fiber bundles. The extracted fibers were washed thoroughly in several solutions—distilled water, 0.1% sodium dodecyl sulphate, 0.5 M ethyl enediamine tetraacetic acid (EDTA), and phosphate buffered saline (PBS)—and immersed in ethanol 70% until bio-composite fabrication.

4.1.2. Alginate/Collagen Bio-Composite Fabrication

Next, 3% (*w*/*v*) alginate solution was produced by dissolving sodium alginate (Protanal LF 10/60, FMC BioPolymer, Philadelphia, PA, USA) in distilled water. The isolated collagen fibers were arranged in various orientations on a frame to provide the mechanical properties to the biocomposite. The collagen fibers were embedded in alginate and then inserted into a dialysis membrane (6000–8000 MWCO, Spectra Por, Spectrum Labs Inc., Rancho Dominguez, CA, USA). The membrane was sealed, flattened, and soaked for 24 h in a calcium-containing solution at physiological concentration (0.02 M $CaCl_2$). Calcium divalent cations mediate cross-linking between the polysaccharide chains of the alginate, which become a hydrogel (Figure 1). The bio-composite was removed from the dialyzed membrane and the frame, and then stored in 70% ethanol at 4 °C. Before implantation, the bio-composite transplants were washed in phosphate buffered saline (PBS).

4.1.3. Bio-Composite In Vitro Storage and Stability

The stability of the bio-composite film was analyzed under various conditions—dry or immersed in 70% ethanol, and in growth medium for long periods (at least 6–14 months) at a wide range of temperatures (4 °C, 25 °C, and 37 °C) as indicated in each experiment.

4.2. Rat In Vivo Study

Bio composites of collagen/alginate were transplanted in Wister male rats 12-weeks old (average weight of 350 gr) with the approval of Tel Aviv University Institutional Animal Care and Use Committee Number 01-18076. The rats underwent a unilateral detachment of the right supraspinatus tendon, and were subjected to repair and augmentation with a bio composite scaffold implant as described in the following surgical procedure of rotator cuff augmentation or sub-cutaneous implantation sites. Rats were anesthetized with 5% isoflurane inhalation (maintained at 2–3% during surgery) and injected with Rimadyl (NSAID) prior to surgery. All the animals were subjected to identical unilateral supraspinatus detachment and repair. The animals were operated on the right shoulder, where a 1.5 cm skin incision was performed over the anterolateral border of the glen humeral joint. The deltoid was exposed and the supraspinatus tendon was identified (Figure 3). The supraspinatus tendon was then cut to its' full thickness, with a blade (No 15) just proximal to the tendon-bone insertion at the greater tubercle (rotator cuff footprint). A horizontal mattress suture was placed; passing horizontally through both sides of the detached tendon and tied over the proximal side using Vicryl 2-0 (Ethicon, Somerville, NJ, USA). Subsequently, an augmentation bio composite film of 5×5 mm^2 was placed over the repaired supraspinatus tendon and sutured down to the supraspinatus with Vicryl 4-0 (Ethicon, Somerville, NJ, USA). Finally, an additional subcutaneous implantation was performed, using a similar bio composite film scaffold that was placed under the skin at the back of the neck. The wound was closed in layers: the subcutaneous layer was closed with Vicryl 3-0 (Ethicon, Somerville, NJ, USA) and the skin was sutured with Vicryl 3-0 (Ethicon, Somerville, NJ, USA). The operated rats were followed up by observation until awakening from anaesthesia and daily monitored after surgery for well-being and physiological recovery. The follow up was carried out by the visual monitoring of the rats' activity (see Video S1 and Figure 4) and weighing. During the experiment, the animals were permitted unrestricted cage activity. The rats were kept in a conventional facility with 12 h light/dark cycles and were fed with standard chow and provided water ad libitum.

4.3. Histology

The rats were sacrificed and the transplanted bio-composite films, which served as tendon augment or for subcutaneous implantation, were extracted and processed for histology. These analysed transplants were removed and fixed in 4% paraformaldehyde, embedded in paraffin and 5 μm thick sections were stained with haematoxylin eosin (HE) or Masson Trichrome (MT) to highlight the fibrous tissue. The slides were observed and photographed on a light microscope (Nikon, Tokyo, Japan).

5. Conclusions

The biocompatibility of the collagen-alginate bio composite film was previously assessed in vitro, showing that the scaffold meets the terms of the ISO 10993-5 standards for cytotoxicity. In the in vitro study, mesenchymal stem cells succeeded in migrating and colonizing the scaffold, grew and proliferated, and formed tissue-like structures along and between the collagen fibers [13]. In the current study, the material was transplanted in vivo as scaffolds in two anatomical sites, exhibiting no toxicity for evaluated by the well-being of the rats (increasing their body weight and activity after the surgery for period of few weeks). The surrounding tissues around the implant were evaluated for inflammation reaction and scarring tissue formation, which indicate the body reaction towards the implant. This study demonstrated the collagen-alginate device's biocompatibility integration with the rats' tissue and well physiology. The scaffold allowed the formation of a 3D structure, enabling cell migration and new blood vessel formation, needed for tissue repair. These results showing the overall well-being of the rats during the in vivo experiment follow up for a period of 14 weeks. The study proved the necessary boundary of biocompatibility of the scaffold as an implantable long-term medical device, corresponding to the ISO standards for implantation and sub-chronic toxicity (ISO 10993-6 and ISO 10993-11). This is essential step that will allow us to further develop the bio composite for various applications and compare the benefits of the new scaffold to other available ones.

Supplementary Materials: The following are available online at http://www.mdpi.com/1660-3397/18/8/420/s1, Video S1: Follow up of animal recovery.

Author Contributions: D.B. and Y.B. initiators of the project for the use of collagen as scaffolding material; Y.B. collected the soft corals for collagen fibers. D.B. and L.P. designed the experiments; L.P. prepared the scaffolds and performed the in vitro experiments; Y.R. assisted in the study design and surgical technique of animal model. S.S., S.H. performed the surgical procedures; D.B. and L.R.-W. analysed and interpreted the histology material: D.B., L.P., S.S. and Y.B. drafted the manuscript. All authors have read and agreed to the published version of the manuscript.

Funding: This research received no external funding.

Conflicts of Interest: The authors declare no conflict of interest.

References

1. Anderson, J.M.; McNally, A.K. Biocompatibility of implants: Lymphocyte/macrophage interactions. *Semin. Immunopathol.* **2011**, *33*, 221–233. [CrossRef] [PubMed]

2. Ehrlich, H.; Heinemann, S.; Heinemann, C.; Simon, P.; Bazhenov, V.V.; Shapkin, N.P.; Born, R.; Tabachnick, R.K.; Hanke, T.; Worch, H. Nanostructural Organization of Naturally Occurring Composites—Part I: Silica-Collagen-Based Biocomposites. *J. Nanomater.* **2008**, *2008*, 623838. [CrossRef]

3. Ehrlich, H. Chitin and collagen as universal and alternative templates in biomineralization. *Int. Geol. Rev.* **2010**, *52*, 661–699. [CrossRef]

4. Ehrlich, H.; Deutzmann, R.; Brunner, E.; Cappellini, E.; Koon, H.; Solazzo, C.; Yang, Y.; Ashford, D.; Thomas-Oates, J.; Lubeck, M.; et al. Mineralization of the metre-long biosilica structures of glass sponges is templated on hydroxylated collagen. *Nat. Chem.* **2010**, *12*, 1084–1088. [CrossRef] [PubMed]

5. Ehrlich, H.; Wysokowski, M.; Żółtowska-Aksamitowska, S.; Petrenko, I.; Jesionowski, T. Collagens of poriferan origin. *Mar. Drugs* **2018**, *16*, 79. [CrossRef] [PubMed]

6. Benayahu, D.; Sharabi, M.; Pomeraniec, L.; Awad, L.; Haj-Ali, R.; Benayahu, Y. Unique collagen fibers for biomedical applications. *Mar. Drugs* **2018**, *16*, 102. [CrossRef] [PubMed]

7. Tziveleka, L.A.; Ioannou, E.; Tsiourvas, D.; Berillis, P.; Foufa, E.; Roussis, V. Collagen from the marine sponges Axinella cannabina and Suberites carnosus: Isolation and morphological, biochemical, and biophysical characterization. *Mar. Drugs* **2017**, *15*, 152. [CrossRef]

8. Rahman, M.A. Collagen of Extracellular Matrix from Marine Invertebrates and Its Medical Applications. *Mar. Drugs* **2019**, *17*, 118. [CrossRef]

9. Domard, A.; Domard, M. Chitosan. In *Polymeric Biomaterials, Revised and Expanded*; CRC Press: Boca Raton, FL, USA, 2001.

10. Masuda, K.; Sah, R.L.; Hejna, M.J.; Thonar, E.J.M.A. A novel two-step method for the formation of tissue-engineered cartilage by mature bovine chondrocytes: The alginate-recovered-chondrocyte (ARC) method. *J. Orthop. Res.* **2003**, *21*, 139–148. [CrossRef]

11. Park, B.K.; Kim, M.M. Applications of chitin and its derivatives in biological medicine. *Int. J. Mol. Sci.* **2010**, 5152–5164. [CrossRef]

12. Orgel, J.P.; Sella, I.; Madhurapantula, R.S.; Antipova, O.; Mandelberg, Y.; Kashman, Y.; Benayahu, D.; Benayahu, Y. Molecular and ultrastructural studies of a fibrillar collagen from octocoral (Cnidaria). *J. Exp. Biol.* **2017**, *220*, 3327–3335. [CrossRef] [PubMed]

13. Pomeraniec, L.; Benayahu, D. Mesenchymal Cell Growth and Differentiation on a New Biocomposite Material: A Promising Model for Regeneration Therapy. *Biomolecules* **2020**, *10*, 458. [CrossRef] [PubMed]

14. Luttikhuizen, D.T.; Harmsen, M.C.; Van Luyn, M.J.A. Cellular and molecular dynamics in the foreign body reaction. *Tissue Eng.* **2006**, *12*, 1955–1970. [CrossRef] [PubMed]

15. Wynn, T.A. Common and unique mechanisms regulate fibrosis in various fibroproliferative diseases. *J. Clin. Investig.* **2007**, *117*, 524–529. [CrossRef] [PubMed]

16. Morais, J.M.; Papadimitrakopoulos, F.; Burgess, D.J. Biomaterials/tissue interactions: Possible solutions to overcome foreign body response. *AAPS J.* **2010**, *12*, 188–196. [CrossRef] [PubMed]

17. Sella, I.; Benayahu, Y. Rearing cuttings of the soft coral *Sarcophyton glaucum* (Octocorallia, Alcyonacea): Towards mass production in a closed seawater system. *Aquac. Res.* **2010**, *41*, 1748–1758. [CrossRef]

18. Sharabi, M.; Mandelberg, Y.; Benayahu, D.; Benayahu, Y.; Azem, A.; Haj-Ali, R. A new class of bio-composite materials of unique collagen fibers. *J. Mech. Behav. Biomed. Mater.* **2014**, *36*, 71–81. [CrossRef]

19. Sharabi, M.; Benayahu, D.; Benayahu, Y.; Isaacs, J.; Haj-Ali, R. Laminated collagen-fiber bio-composites for soft-tissue bio-mimetics. *Compos. Sci. Technol.* **2015**, *117*, 268–276. [CrossRef]

20. Sharabi, M.; Varssano, D.; Eliasy, R.; Benayahu, Y.; Benayahu, D.; Haj-Ali, R. Mechanical flexure behavior of bio-inspired collagen-reinforced thin composites. *Compos. Struct.* **2016**, *153*, 392–400. [CrossRef]

21. Yang, X.; Li, Y.; Liu, X.; Huang, Q.; He, W.; Zhang, R.; Feng, Q.; Benayahu, D. The stimulatory effect of silica nanoparticles on osteogenic differentiation of human mesenchymal stem cells. *Biomed. Mater.* **2016**, *12*, 015001. [CrossRef]

22. Deprés-tremblay, G.; Chevrier, A.; Snow, M.; Hurtig, M.B.; Rodeo, S.; Buschmann, M.D. Rotator cuff repair: A review of surgical techniques, animal models, and new technologies under development. *J. Shoulder Elb. Surg.* **2016**, *25*, 2078–2085. [CrossRef] [PubMed]

 *marine drugs* 

Article

Functionalization of 3D Chitinous Skeletal Scaffolds of Sponge Origin Using Silver Nanoparticles and Their Antibacterial Properties

Tomasz Machałowski [1,2], Maria Czajka [3], Iaroslav Petrenko [2], Heike Meissner [4], Christian Schimpf [5], David Rafaja [5], Jerzy Ziętek [6], Beata Dzięgiel [6], Łukasz Adaszek [6], Alona Voronkina [7], Valentin Kovalchuk [8], Jakub Jaroszewicz [9], Andriy Fursov [2], Mehdi Rahimi-Nasrabadi [10,11], Dawid Stawski [3], Nicole Bechmann [12,13,14,15], Teofil Jesionowski [1,*] and Hermann Ehrlich [2,16,*]

1   Institute of Chemical Technology and Engineering, Faculty of Chemical Technology, Poznan University of Technology, Berdychowo 4, 60965 Poznan, Poland; tomasz.g.machalowski@doctorate.put.poznan.pl
2   Institute of Electronics and Sensor Materials, TU Bergakademie Freiberg, Gustav-Zeuner str. 3, 09599 Freiberg, Germany; iaroslavpetrenko@gmail.com (I.P.); andriyfur@gmail.com (A.F.)
3   Institute of Material Science of Textiles and Polymer Composites, Lodz University of Technology, Zeromskiego 16, 90924 Lodz, Poland; maria.czajka@dokt.p.lodz.pl (M.C.); dawid.stawski@p.lodz.pl (D.S.)
4   Department of Prosthetic Dentistry, Faculty of Medicine and University Hospital Carl Gustav Carus of Technische Universität Dresden, Fetscherstraße 74, 01307 Dresden, Germany; heike.meissner@uniklinikum-dresden.de
5   Institute of Materials Science, TU Bergakademie Freiberg, Gustav-Zeuner str. 5, 09599 Freiberg, Germany; schimpf@iww.tu-freiberg.de (C.S.); rafaja@ww.tu-freiberg.de (D.R.)
6   Department of Epizootiology and Clinic of Infectious Diseases, Faculty of Veterinary Medicine, University of Life Sciences, Akademicka 13, 20612 Lublin, Poland; achantina@op.pl (J.Z.); beatadziegiel@o2.pl (B.D.); lukaszek0@wp.pl (Ł.A.)
7   Department of Pharmacy, National Pirogov Memorial Medical University, Pirogov str. 56, 21018 Vinnitsa, Ukraine; algol2808@gmail.com
8   Department of Microbiology, National Pirogov Memorial Medical University, Pirogov str. 56, 21018 Vinnitsa, Ukraine; valentinkovalchuk2015@gmail.com
9   Materials Design Division, Faculty of Materials Science and Engineering, Warsaw University of Technology, Woloska 141, 02507 Warsaw, Poland; jakubjaroszewicz@wp.pl
10 Chemical Injuries Research Center, Systems Biology and Poisonings Institute, Baqiyatallah University of Medical Sciences, Tehran 1951683759, Iran; rahiminasrabadi@gmail.com
11 Faculty of Pharmacy, Baqiyatallah University of Medical Sciences, Tehran 1951683759, Iran
12 Institute of Clinical Chemistry and Laboratory Medicine, University Hospital Carl Gustav Carus, Technische Universität Dresden, Fetscherstrasse 74, 01307 Dresden, Germany; nicole.bechmann@uniklinikum-dresden.de
13 Department of Medicine III, University Hospital Carl Gustav Carus, Technische Universität Dresden, Fetscherstrasse 74, 01307 Dresden, Germany
14 Department of Experimental Diabetology, German Institute of Human Nutrition Potsdam-Rehbruecke, Arthur-Scheunert-Allee 114, 14558 Nuthetal, Germany
15 German Center for Diabetes Research (DZD), Ingolstaedter Landstrasse 1, 85764 München-Neuherberg, Germany
16 Center for Advanced Technology, Adam Mickiewicz University, Uniwersytetu Poznańskiego 10, 61614 Poznan, Poland
*   Correspondence: Teofil.Jesionowski@put.poznan.pl (T.J.); Hermann.Ehrlich@esm.tu-freiberg.de (H.E.)

Received: 14 May 2020; Accepted: 8 June 2020; Published: 10 June 2020

Abstract: Chitin, as one of nature's most abundant structural polysaccharides, possesses worldwide, high industrial potential and a functionality that is topically pertinent. Nowadays, the metallization of naturally predesigned, 3D chitinous scaffolds originating from marine sponges is drawing focused attention. These invertebrates represent a unique, renewable source of specialized chitin due to their ability to grow under marine farming conditions. In this study, the development of composite material in the form of 3D chitin-based skeletal scaffolds covered with silver nanoparticles (AgNPs) and Ag-bromide is described for the first time. Additionally, the antibacterial properties of the obtained materials and their possible applications as a water filtration system are also investigated.

Keywords: chitin; sponges; 3D scaffolds; AgNPs; antibacterial properties; *Aplysina aerophoba*

1. Introduction

More than 1.2 billion people have no access to clean drinking water [1]. Additionally, 1.8 billion people drinking water from sources susceptible to fecal contamination results in the death of approx. one million children every year [2]. Contaminated potable water, with pathogenic microorganisms (i.e., *Escherichia coli* infection) represents one of the world's most serious health threats [3]. Consequently, the development of new materials, which are capable of mitigating the risk of bacterial contamination in water needs to be an ongoing, crucial area of research [4].

Filtration is a widely used method of treating water and recent, numerous attempts aim to develop effective antibacterial composite-based filtration materials [5,6]. Promising examples with respect to their potential application include synthetic polymers (e.g., polypropylene, polyurethane, polyacrylonitrile) [7–9], natural materials (e.g., chitin, chitosan, cellulose, collagen) [10–12] as well as carbon-based composites [13,14], which have been covered with silver nanoparticles (AgNPs) using diverse techniques. The use of silver compounds to disinfect water and the procedure's resulting death of fungi, molds, bacteria and various spores has been documented since ancient times, as cited by Atiyeh [15]. Today, it is a proven fact that direct contact with silver inactivates cells and microorganisms [16]. The mechanism of action of silver involves the inhibition of microbial respiration through binding of metal particles to the bacterial cell membranes [17]. Consequently, silver impairs the microbial respiratory system. The fundamental factors affecting the superior antimicrobial properties of silver-based composites are the size of Ag particles and their solid phase surface development. As recently reported, the "antimicrobial activity of the smaller Ag nanoparticles may be several orders of magnitude greater than that of the corresponding bulk solid" [18]. Thus, it is not surprising that AgNPs are most commonly used in many antibacterial products to protect health and improve the quality of life [19].

Metallization of chitin [20,21], as one of the most abundant structural polysaccharides in nature [22,23], with production in oceans measuring approx. $10^{12–14}$ tons per year [24], remains a solid trend. Chitin is synthesized by a broad assortment of organisms representing different taxonomic groups, mostly crustaceans [25–29] and insects [30–35]. Nowadays, the functionalization of naturally predesigned chitinous scaffolds with a 3D architecture attracts particular attention [36–41]. In this article, the unique skeletal chitin-based 3D scaffolds (Figure 1) isolated from the cultivated under marine farming *Aplysina aerophoba* marine demosponge were used for the first time as a basic construct for fabrication of an antibacterial water filter. This biomaterial was modified by silver nanoparticle deposition using chemical reduction of silver nitrate and the antibacterial action was investigated.

Figure 1. The optical representation of a decellularized *Aplysina aerophoba* demosponge 3D chitinous skeletal scaffold (**A**). Representation of the cross-section (**A-A**). Polyurethane (PU) scaffolds, traditionally used as water filter material, with high magnification of the fibers (**B**). Microscopic representation of the isolated chitinous skeletal scaffold (**C**) shows high structural similarity to the commonly used PU-based filtration material (**B**). The light brownish color is due to the presence of brominated compounds naturally occurring in the skeletal fibers of the sponge.

2. Results

Chitinous 3D scaffolds represent an intriguing alternative to synthetic analogues [20,21,36]. Due to the high porosity (Figure 2) and structural similarity of the poriferan 3D chitinous scaffolds to synthetically produced porous foams, this biological material is particularly interesting for filtration applications. Based on micro-focused X-ray tomographic (micro-CT) analysis (Figure 2), the porosity of a chitinous scaffold isolated from *A. aerophoba* demosponge was estimated at 98.5% (see Table S1).

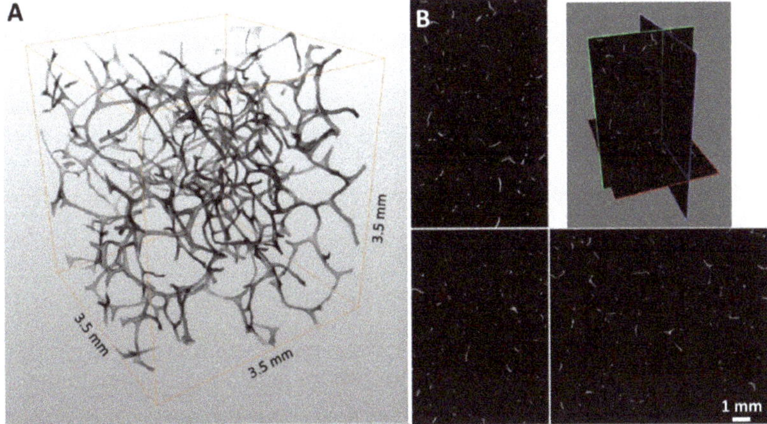

Figure 2. The 3D model (**A**) and cross-sections (**B**) of the 3D chitin–Ag/AgBr composite scaffold obtained by micro-CT.

The alkaline environment of the chemical reduction of $AgNO_3$ promoted the additional release of bromine-derived compounds originally located within the fiber of the chitinous scaffold isolated from *A. aerophoba* demosponge [37]. The clearly visible metallic layer obtained after metallization with Ag is strongly bound to the chitinous fibers even after ultrasonic treatment (Figure 3).

Figure 3. The 3D chitinous scaffolds isolated from the *A. aerophoba* sponge resemble their microarchitecture being covered with nanoparticles of Ag/AgBr (**A,B**). The stereomicroscopy image represents the existence of the tightly bound metallized layer, also taken after 30 min of sonication (**C**).

Ag/AgBr nanoparticles tend to create the spherical-shaped aggregates represented in Figure 4A. The SEM image in Figure 4A shows the chitinous fiber completely covered by Ag/AgBr aggregates, which are exclusively deposited on the surface of the fibers. The highest fraction of aggregates composed of nanoparticles, which constitute approx. 42%, contains particles with a diameter range of 300–400 nm (see Figure S2). At the surface of chitinous fiber, the creation of nanostructured agglomerates with dimensions up to two μm is also observable. EDX-based analysis of the surface of the metallized *A. aerophoba* chitinous scaffold confirms domination of Ag and Br (see Figure 4B and Table S2). Obtained data explains the identification of the Ag-bromide phase within the metallized layers using XRD analysis (see Figure 5).

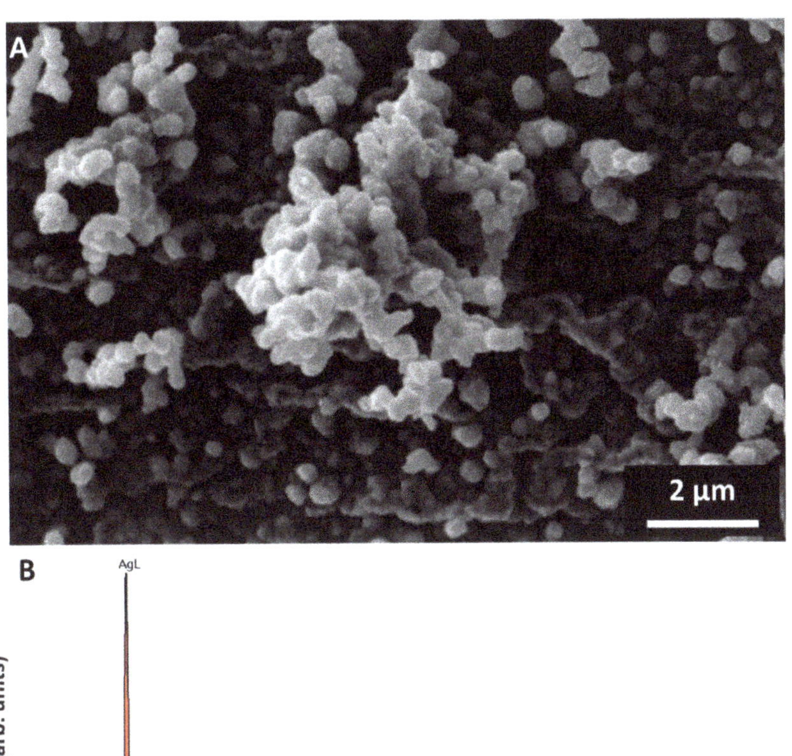

Figure 4. (**A**) SEM image of the surface of the skeletal chitinous scaffold isolated from *A. aerophoba* demosponge covered by the layer of silver/silver bromide nanoparticles. EDX analysis confirms the presence of both Ag and Br within these nanoparticles (**B**). This is in good agreement with the XRD data obtained for the same sample (see Figure 5).

The X-ray diffraction patterns of the skeletal chitin sample before and after reaction with $AgNO_3$ solution are shown in Figure 5. In both cases, the crystal structure of chitin (see [42]) is lost due to pretreatment of the samples before XRD. However, some remainders of the chitin structure are still visible, i.e., the diffraction maximum 021 at $2\theta \approx 12°$ and the 'hump' beginning with the 110 reflection near $2\theta \approx 20°$. The XRD line from chitin are marked in Figure 5. The chitin sample treated with $AgNO_3$ also shows, in addition to the remaining features of the chitin diffraction pattern, peaks of Ag (PDF# 04-004-6434) and AgBr (PDF# 00-006-0438). The presence of metallic Ag confirms the applicability of the synthesis route for the creation of Ag nanoparticles on chitin surface. The Rietveld analysis revealed the Ag lattice parameter of $a = (4.088 \pm 0.001)$ Å, which is practically identical to the tabulated value of $a = 4.089$ Å, and a crystallite size of $D_{iso} = (13 \pm 2)$ nm, confirming the nanocrystalline character of the particles. After Br was identified by the EDX analysis, the remaining peaks in the XRD pattern were successfully assigned to AgBr crystallizing in space group $Fm\bar{3}m$, with a lattice parameter of a = (5.555 ± 0.002) Å. Due to the treatment of the scaffolds with $AgNO_3$, we have also checked for the presence of silver oxides, nitrides and chloride but found no positive match among the database entries. The amount of metallic Ag is approx. 75 vol.%; the amount of AgBr is approx. 25 vol.%, as indicated by the quantitative XRD phase analysis.

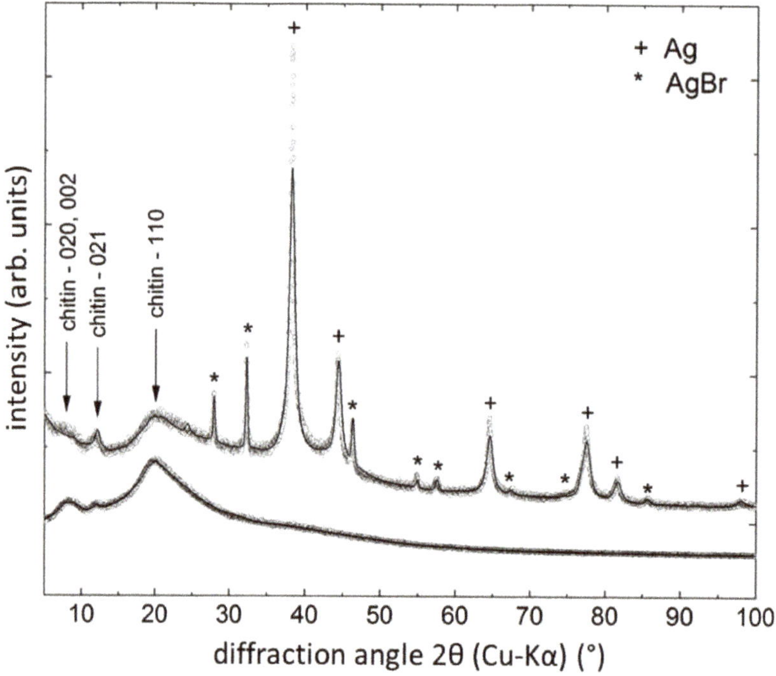

Figure 5. X-ray diffraction patterns (small circles: measured intensities; lines: refinement) of the pure sponge chitin (lower signal intensity) and a chitin sample tightly covered with the Ag and AgBr nanoparticles (upper signal intensity). The sharp diffraction lines in the diffraction pattern belong either to Ag or to AgBr, as indicated.

The assessment of the antibacterial activity of the prepared composite was based on the agar diffusion method. As shown in Figure S3, the inhibition zone of the 3D chitin–Ag/AgBr composite in respect to *E. coli* is greater than that of chitinous skeletal scaffolds before metallization (see Table 1), which indicates the superior antibacterial properties of the created 3D construct. The obtained material contributed to wide zones of inhibition, and the mean value was estimated at 23 and 24 mm for *Escherichia coli and Bacillus subtilis*, respectively (see Figure S3). Interesting results were also obtained for the chitinous scaffold before silver coating with respect to the *E. coli*. The chitinous scaffold used as control was isolated by the method described in Section 4.1. It is mainly composed of Br-containing chitin [37], which is originally responsible for the resistance of this verongiid sponge against predatory microorganisms from marine environments. In this case, the zone of inhibition was 18 mm. However, this effect was not observed in the case of Gram-positive bacterium *B. subtilis* that was comparatively used in this study. Interestingly, the commercially available antibacterial material Suprasorb® A + Ag, which was also used for comparative purposes, did not contribute to any zones of inhibition against both bacterial strains studied.

Table 1. Mean zone of inhibition for both strain (mm) and number of survived *Escherichia coli* strains after 24 h of test tube assay.

Material	Agar Diffusion Method		Test Tube Test	
	E. coli (mm)	*B. subtilis* (mm)	*E. coli* (CFU/100 μL)	*% of Reduction*
Chitin–Ag/AgBr scaffold	23	24	7	99.9
Chitin-based scaffold	18	0	~10^6	0
Suprasorb® A + Ag	0	0	~10^6	0
Control	0	0	~10^6	0

The test tube assay additionally confirmed the antibacterial properties of the chitin–Ag/AgBr scaffold (Figure 6). It was observed that the percentage of the bacteria survival decreased with 24 h and more than 99.9% of the initial bacterial CFU was eliminated (for details, see Table 1 and Figure 6). The rapid increase in *E. coli* degradation was observed between 1 and 3 h (Figure 6A). In contrast to the agar diffusion method, the Br-containing chitinous scaffold before silver coating did not show visible antibacterial potential against *E. coli* even after 24 h.

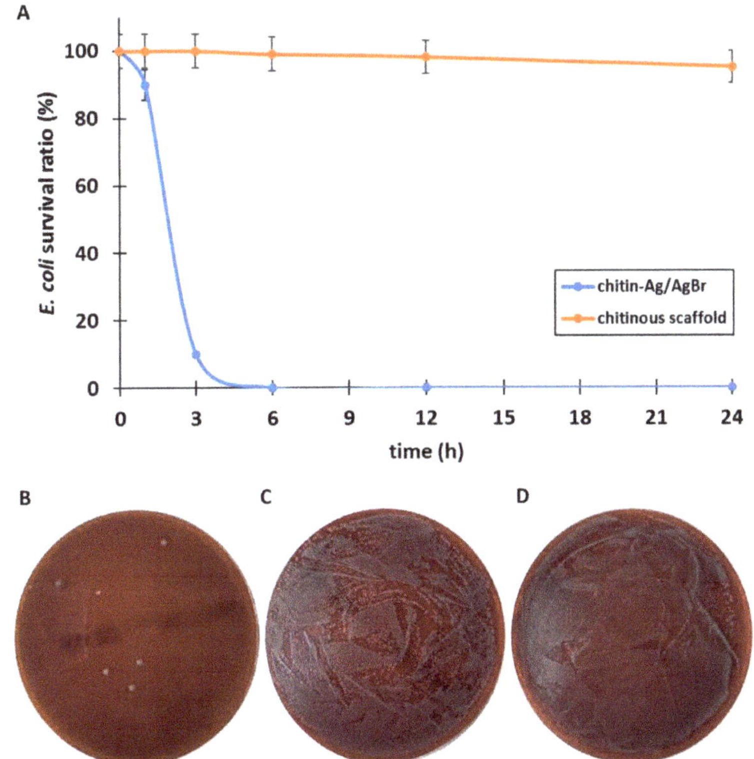

Figure 6. The dynamics of the degradation of live bacteria colonies shown over time, by test tube assay, using a 3D chitin–Ag/AgBr scaffold and a chitinous scaffold before Ag coating, with 5% error bars (**A**). Only seven *E. coli* bacteria colonies survived after 24 h of testing using a chitin–Ag/AgBr scaffold (**B**). Both Br-containing chitinous scaffolds before silver coating (**C**) and commercially available material Suprasorb® A + Ag (**D**) did not show antibacterial activity against *E. coli* even after 24 h.

Results obtained after filtration (Figure S4) clearly indicate that 3D chitin–Ag/AgBr scaffolds possess antibacterial properties against the *E. coli* (ATCC® 25922) strain. In experiments, a superior inhibitory effect was observed after 6 h of filtration with this material. After 24 h, only one colony of this bacterium survived (Figure S4B). Filtration with the Br-containing chitinous scaffold before Ag coating did not cause an observable changes in the number of bacteria. In future, the influence of a diverse chitin–Ag/AgBr filter density has to be examined in relation to enhance antibacterial effect. No increase in absorbance at 419 nm indicated that the silver nanoparticle-based layer was stable on the developed 3D water filter and was not washed away by water flow even after 24 h [43].

3. Discussion

Recently observed rapid development of nanotechnology, especially in the fields of bioinspired materials science and biomimetics, is strictly related to searching for new methods for synthesis of effective hybrid materials and biocomposites with designed properties [44]. Until the discovery of chitin in fibrous skeletons of some verongiid demosponges in 2007 [45], as well as in the verongiid demosponge species *A. aerophoba* in 2010 [46], the commonly accepted opinion was that skeletal fibers of these demosponges are made of a proteinaceous, biological material called spongin [47]. However, this erroneous assumption was experimentally proved based on the solubility of spongin in alkaline solutions, which was not observed with chitin resistant to such chemical conditions [20,36]. It is the resistance of chitin to an alkaline medium up to a concentration of 10% that allows us to look for key ways to use it in specific chemical reactions, including silver metallization, as described in this work. For example, the implementation of such a reaction using spongin-based matrices would not have been possible.

Chitin of poriferan origin also possess characteristic features such as thermostability up to 400 °C [41,48,49], cytocompatibility [50,51] and microporosity [36,52]. Recently, practical applications of demosponge chitinous scaffolds were reported for tissue engineering [39,50,51,53], uranium adsorption [54], bioelectrometallurgy and extreme biomimetics [21] as well as for the photodegradation of organic dyes [49]. Here, we describe, for the first time, antibacterial properties of composite material in the form of 3D chitin-based skeletal scaffolds covered with AgNPs and Ag-bromide. The formation of Ag-bromide on the surface of this specific chitin is due to the presence of bromine in the chitin skeleton of these sponges. We suggest that this compound is formed as a result of alkaline extraction of bromine-containing compounds in the presence of silver ions.

Originally, diverse brominated derivatives (mostly bromotyrosines) [37] are located in the skeletal fibers being intercalated into chitinous layers. They represent an effective form of biochemical defense against harmful pathogens, which constantly fall inside the verongiid sponge, which filters the surrounding water to extract the appropriate feed (i.e., viruses, bacteria, organic micro debris). These unique defense strategies, based on Br-containing secondary metabolite production, allowed this organism to survive more than 500 millions of years [37]. Today, bromotyrosines are recognized as multi-targeted marine drugs with broad fields of applications (i.e., as antitumorigenic and antimetastatic agents [55,56]). In order to isolate pure chitinous scaffolds from verongiid demosponges, diverse methodological approaches based on chemical, electrochemical and enzymatic treatments were reported (see for overview [36,57]). Nonetheless, the alkali treatment of verongiid skeletons is most commonly used [20,36,40] and can be regulated with respect to residual bromine concentration in the chitinous matrices. In our case, secondary metabolite compounds present within scaffolds become a great source of natural bromine. Previously, superb antibacterial properties of Ag [6,16,58] and AgBr [59–61] nanoparticles were already reported, but as separate substances and not within composite materials.

In order to determine antibacterial properties of the designed 3D chitin–Ag/AgBr composite, the agar diffusion method as well as test tube assay was carried out. Moreover, a simple prototype of the filtration set was proposed to assess the determination ability of 3D chitin–Ag/AgBr scaffolds in terms of *E. coli* inactivation. Based on the obtained data, appropriative antibacterial properties against the *E. coli* strain were reported here. Only one bacteria colony from 10^6 CFU/µL survived after 24 h of filtration.

In the 3D composite scaffold developed in this study, under ambient conditions, the content of Ag is approx. 75 vol.% and the amount of AgBr is approx. 25 vol.%. This distinguishes the method described in this study from the metallization of chitin in harsh chemical conditions. For example, previously, poriferan chitin was effectively used as a template for solvothermal and hydrothermal conditions according to an extreme biomimetic approach [41,44,48,49,62]. Element oxide-based composites such as chitin-SiO_2 [62,63], chitin-GeO_2 [41], chitin-ZrO_2 [48,64], chitin-ZnO [65], and chitin-hematite [66] were synthesized under hydrothermal conditions between 65 and 185 °C.

4. Materials and Methods

4.1. Chitin Scaffold Isolation

Air-dried specimens of cultivated *Aplysina aerophoba* (Nardo, 1833) marine demosponges were purchased from BromMarin GmbH (Freiberg, Germany). Chitinous scaffolds were isolated by chemical treatment of specimens in 2 days as follows. At the first stage, selected fragments of *A. aerophoba* skeleton were immersed in pure distilled water at 80 °C for 24 h to remove water-soluble compounds and cells, which were disrupted due to osmotic shock. In the next step, cell-free skeletons, were immersed in 20% acetic acid at room temperature over 4 h in order to remove calcium carbonates. The prefinal stage included treatment with 2.5 M NaOH solution for 6 h at 37 °C, for deproteinization and partial depigmentation. Finally, Br-containing skeletal 3D chitinous scaffolds were neutralized by distilled water and stored at 4 °C.

4.2. Fabrication of Silver-Coated 3D Chitinous Scaffolds

In order to obtain AgNps on the surface of skeletal chitinous scaffolds isolated from the cultivated *A. aerophoba* demosponge, chemical reduction of $AgNO_3$ was carried out. For this purpose, a selected fragment of the skeletal scaffold (10 × 30 mm) was immersed in 30 mL of 1 M $AgNO_3$ solution for 1 h at room temperature. Then, 15 mL of 0.8 M NaOH was added to the above mixture and formation of a precipitate was observed. In the next step, a concentrated ammonia solution was added dropwise until total precipitate dissolution was achieved. After 15 min, 1 mL of ethanol and 25 mL of a mixture of 0.08 M glucose and 0.04 M citric acid were added. The obtained construct was washed several times by deionized water and treated by ultrasound (60 kHz, 300 W) for 30 min at room temperature in order to remove non-attached nanoparticles of silver.

4.3. Characterization of Obtained Materials

4.3.1. Digital Microscopy

Corresponding samples were observed and analyzed by an advanced imaging and measurement system consisting of a Keyence VHX-6000 digital optical microscope (Osaka, Japan) and the swing-head zoom lenses VH-Z20R (magnification up to 200×) and VH-Z100UR (magnification up to 1000×).

4.3.2. Micro-CT Analysis

Scaffolds were scanned using a micro-focused X-ray tomographic system (MICRO XCT-400, Xradia–Zeiss, Pleasanton, CA, USA) at 40 kV and 200 µA. For each sample, 1500 projection images were recorded with an exposure time of 12 sec and a magnification objective of 20×. The volume was reconstructed with the instrument software and was then exported to the CTAn (Bruker Billerica, MA, USA) program for further 3D image analysis. Voxel size was the same for all samples (2 × 2 × 2 µm).

4.3.3. Infrared Spectroscopy

Attenuated total reflectance Fourier transform infrared spectroscopy (ATR-FTIR), was used for the qualitative characterization and identification of the isolated materials. The samples were analyzed by a Nicolet 210c spectrometer (Thermo Fisher Scientific, Waltham, MA, USA).

4.3.4. UV–VIS Spectroscopy

The absorption at 419 nm was recorded with a spectrophotometer (SPECORD S10, Carl Zeiss, Germany). Measurements were recorded for determination of the presence of silver nanoparticles in the solution after filtration [43]. A quartz cuvette was used with a path length of 1 cm (Quartz SUPRASIL®, Hellma Analytics, Müllheim, Germany) and operated at a resolution of 5 nm.

4.3.5. Scanning Electron Microscopy (SEM) and Energy-Dispersive X-ray Spectroscopy (EDX)

Samples were prepared for analysis by freeze-drying, followed by covering with Au using a Cressington Sputter Coater S150B (BOC Edwards, Wilmington, MA, USA). Scanning electron microscopy was performed using a Hitachi S-4700-II (Hitachi, Ltd., Tokyo, Japan) field emission microscope. The elements were analyzed by energy-dispersive X-ray spectroscopy in the EDX analysis system from EDAX and a XL30ESEM Philips scanning electron microscope (Philips, Amsterdam, The Netherlands). Observations were carried out under high vacuum, 20 kV voltage, and 6500× and 8000× magnification. The particle diameter was determined from 100 measurements using representative micrographs by the software (ImageJ, National Institutes of Health, Bethesda, MD, USA).

4.3.6. X-ray Diffraction

X-ray diffraction measurements were performed with the purpose of phase identification and quantification. A FPM RD7 diffractometer equipped with a sealed X-ray tube with Cu anode operated at 40 kV/30 mA was used. The diffractometer worked in the symmetrical diffraction geometry and the powderized sample was fixed with ethanol on a Si[510]-cut zero-background holder. The diffracted beam passed a graphite monochromator to eliminate unwanted radiation components before being registered by a proportional counter. The measurements covered the angular range from 5° to 150° 2θ, with a step size of 0.02° and a dwell time of 12 s per point. Data analysis was performed by database search (ICSD PDF-4+) with the Panalytical HighScore+ program. Rietveld refinement of the diffraction patterns was performed by the MAUD software package [67].

4.4. Antibacterial Activity Studies

4.4.1. Determination of the Zone of Inhibition

The chitin–Ag/AgBr (1) (Figure S2) composite scaffolds were tested for their antibacterial activity using an agar diffusion method. For this purpose, 0.01 g of sample was prepared. For comparison, chitinous scaffold before silver coating (2) and commercially available antibacterial wound dressing Suprasorb® A + Ag (3) were used. *E. coli* (ATCC® 25922) and *B. subtilis* B9 (Collection of Department of Epizootiology and Clinic of Infectious Diseases, University of Life Sciences in Lublin, Poland) were taken as model Gram-negative and Gram-positive bacteria (1.5 on the McFarland scale). Bacteria strain was applied with agar Mueller–Hinton medium. Afterwards, plates were incubated at 37 °C for 24 h. The diameter of inhibitory zone surrounding material pieces was measured in mm for each specimen. Tests performed on three agar plates (for each material) determined a mean value.

4.4.2. Test Tube Antibacterial Assay

For the test tube assay, the *E. coli* ATCC® 25,922 suspension (0.8 on the McFarland scale, approx. 10^6 CFU/µL) in 0.9% sodium chloride was prepared in sterilized test tubes. An amount of 0.01 g of sample (No. 1–3 see Section 4.4.1.) was put into the separated test tubes and immersed in 1.5 mL of suspension described above at 37 °C for 1, 3, 6, 12 and 24 h. The scaffolds were taken out from the test tubes. Then, 100 µL of suspension was plated on Columbia LabAgar plates + 5% sheep blood and incubated at 37 °C for 24 h for determination of number of survive bacterial colonies from the control sample. These measurements were repeated three times and a mean value was evaluated.

4.4.3. Determination of Antibacterial Properties—Filtration Test

The antibacterial activity of the obtained chitin–Ag/AgBr (1) was also determined using the filtration method. The *E. coli* ATCC® 25,922 suspension (0.8 on the McFarland scale, approx. 10^6 CFU/µL) was kept in a storage 1000 mL container filled with 0.9% sodium chloride and filtered at a flow rate of 330 mL/min. The suspension was pumped into the sterilized filtration cartridge, which consists of a 0.5 g of filter inside 50 cm^3 falcon tubes (Figure S1). Changes in the survival rate of the bacteria colony were determined for 1, 3, 6, 12 and 24 h analogous to the test described above.

5. Conclusions

The fundamental difference between chitin-based 3D scaffolds originating from verongiid sponges and other, previously reported chitinous matrices is the possibility to regulate the content of Br during the preparation process by changing the time of alkaline treatment of original skeletal 3D constructs. Therefore, the experimental approach described in this study enables unique nanostructured Ag/AgBr composite in the form of a nanolayer, which remains strongly bound to the surface of the organic matrix and is responsible for selective antibacterial activities. The results open the path to using chitin-based skeletal matrices in the form of acellular scaffolds. Due to the natural ability of the *A. aerophoba* demosponges to regenerate their skeletons and to grow at low depths under marine farming conditions at large scales, their potential for applications in bioinspired materials science and technologies increases dramatically. Future research, dedicated to the optimization of such naturally derived, already prefabricated materials using other metal ions as well as alternative reductants (e.g., lignosulfonates) also for special biomedical applications is strongly indicated.

Supplementary Materials: The following are available online at http://www.mdpi.com/1660-3397/18/6/304/s1, Figure S1: The protype of the filtration set used in this study. Figure S2: Percentage of Ag/AgBr aggregates in individual fractions as a function of diameter ranges. Figure S3: Agar diffusion method results. The antimicrobial activity of chitin–Ag/AgBr scaffolds (1), chitinous scaffold before metallization (2) against *E. coli* (A). Lack of antimicrobial activity represent by Suprasorb® A + Ag (3) against *E. coli* (A). The antimicrobial activity of chitin–Ag/AgBr scaffold (1) against *B. subtilis* (B). Figure S4: Dynamics of the degradation of live bacteria colonies as a function of time of filtration with 5% error bars (A). Filtration clearly indicates that chitin–Ag/AgBr scaffolds possess antibacterial properties against the *E. coli* (ATCC® 25922) strain (blue line). Br-containing chitinous scaffolds before AgNP coating did not reflect any antibacterial effect. The number of survived bacteria colonies was uncountable, even after 24 h (orange line). Only one colony of *E. coli* survived after 24 h of filtration using chitin–Ag/AgBr scaffold (B). Chitinous scaffold before silver coating did not show antibacterial potential against *E. coli* even after 24 h (C). Table S1: Results of 3D quantitative analysis (micro-CT). Table S2. Results of local chemical analyses (SEM-EDX) of selected areas of the *A. aerophoba* chitinous scaffold after metallization.

Author Contributions: Conceptualization, V.K., M.R.-N. and H.E.; data curation, M.R.-N.; investigation, T.M., M.C., H.M., C.S., J.Z., B.D., A.V., J.J. and A.F.; methodology, T.M., M.C., H.M., C.S., D.R., Ł.A., V.K. and D.S.; resources, J.Z., B.D., A.V. and D.S.; software, I.P. and J.J.; supervision, T.J.; validation, I.P., D.R. and T.J.; visualization, Ł.A.; writing—original draft, T.M., N.B. and H.E. All authors have read and agreed to the published version of the manuscript.

Funding: This work was partially supported by the DFG Project HE 394/3, SMWK Project no. 02010311 (Germany), as well as by the Ministry of Science and Higher Education (Poland). M.R.N. is supported by Alexander von Humboldt Fellowship. T.M. is supported by DAAD (Personal ref. no. 91734605).

Acknowledgments: Special thank for Anna Wilczyńska for preparing the agar plate photos.

Conflicts of Interest: The authors declare no conflict of interest.

References

1. Batra, S.; Adhikari, P.; Ghai, A.; Sharma, A.; Sarma, R.; Suneetha, V. Study and design of portable antimicrobial water filter. *Asian J. Pharm. Clin. Res.* **2017**, *10*, 268. [CrossRef]
2. Khan, S.T.; Malik, A. Engineered nanomaterials for water decontamination and purification: From lab to products. *J. Hazard. Mater.* **2019**, *5*, 295–308. [CrossRef] [PubMed]
3. Fewtrell, L. *Silver: Water Disinfection and Toxicity*; Centre for Research into Environment and Health, World Health Organization: Geneva, Switzerland, 2014; pp. 1–50.

4. Wang, L.; Hu, C.; Shao, L. The antimicrobial activity of nanoparticles: Present situation and prospects for the future. *Int. J. Nanomed.* **2017**, *12*, 1227–1249. [CrossRef] [PubMed]

5. Pronk, W.; Ding, A.; Morgenroth, E.; Derlon, N.; Desmond, P.; Burkhardt, M.; Wu, B.; Fane, A.G. Gravity-driven membrane filtration for water and wastewater treatment: A review. *Water Res.* **2019**, *149*, 553–565. [CrossRef] [PubMed]

6. Karumuri, A.K.; Oswal, D.P.; Hostetler, H.A.; Mukhopadhyay, S.M. Silver nanoparticles attached to porous carbon substrates: Robust materials for chemical-free water disinfection. *Mater. Lett.* **2013**, *109*, 83–87. [CrossRef]

7. Phong, N.T.P.; Thanh, N.V.K.; Phuong, P.H. Fabrication of antibacterial water filter by coating silver nanoparticles on flexible polyurethane foams. *J. Phys. Conf. Ser.* **2009**, *187*, 012079. [CrossRef]

8. Picca, R.A.; Paladini, F.; Sportelli, M.C.; Pollini, M.; Giannossa, L.C.; Di Franco, C.; Panico, A.; Mangone, A.; Valentini, A.; Cioffi, N. Combined approach for the development of efficient and safe nanoantimicrobials: The case of nanosilver-modified polyurethane foams. *ASC Biomater. Sci.* **2017**, *3*, 1417–1425. [CrossRef]

9. Shi, Y.; Li, Y.; Zhang, J.; Yu, Z.; Yang, D. Electrospun polyacrylonitrile nanofibers loaded with silver nanoparticles by silver mirror reaction. *Mater. Sci. Eng. C* **2015**, *51*, 346–355. [CrossRef] [PubMed]

10. Heinemann, S.; Heinemann, C.; Ehrlich, H.; Meyer, M.; Baltzer, H.; Worch, H.; Hanke, T. A novel biomimetic hybrid material made of silicified collagen: Perspectives for bone replacement. *Adv. Eng. Mater.* **2007**, *9*, 1061–1068. [CrossRef]

11. Cooper, A.; Oldinski, R.; Ma, H.; Bryers, J.D.; Zhang, M. Chitosan-based nanofibrous membranes for antibacterial filter applications. *Carbohydr. Polym.* **2013**, *30*, 254–259. [CrossRef] [PubMed]

12. Jain, S.; Bhanjana, G.; Heydarifard, S.; Dilbaghi, N.; Nazhad, M.M.; Kumar, V.; Kim, K.; Kumar, S. Enhanced antibacterial profile of nanoparticle impregnated cellulose foam filter paper for drinking water filtration. *Carbohydr. Polym.* **2018**, *202*, 219–226. [CrossRef] [PubMed]

13. Karwowska, E. Antibacterial potential of nanocomposite-based materials—A short review. *Nanotechnol. Rev.* **2017**, *6*, 243–254. [CrossRef]

14. Zeng, X.; Mccarthy, D.T.; Deletic, A.; Zhang, X. Silver/reduced graphene oxide hydrogel as novel bactericidal filter for point-of-use water disinfection. *Adv. Funct. Mater.* **2015**, *25*, 4344–4351. [CrossRef]

15. Atiyeh, B.S.; Costagliola, M.; Hayek, S.N.; Dibo, S.A. Effect of silver on burn wound infection control and healing: Review of the literature. *Burns* **2007**, *33*, 139–148. [CrossRef] [PubMed]

16. Cho, K.-H.; Park, J.-E.; Osaka, T.; Park, S.-G. The study of antimicrobial activity and preservative effects of nanosilver ingredient. *Electrochim. Acta* **2005**, *51*, 956–960. [CrossRef]

17. Imani, R.; Talaiepour, M.; Dutta, J.; Ghobadinezhad, M.R.; Hemmasi, A.H.; Nazhad, M.M. Production of antibacterial filter paper from wood cellulose. *BioResources* **2011**, *6*, 891–900.

18. Nguyen, V.Q.; Ishihara, M.; Nakamura, S.; Hattori, H.; Ono, T.; Miyahira, Y.; Matsui, T. Interaction of silver nanoparticles and chitin powder with different sizes and surface structures: The correlation with antimicrobial activities. *J. Nanomater.* **2013**, *2013*, 467534. [CrossRef]

19. Klippstein, R.; Fernandez-Montesinos, P.M.; Castillo, R.; Zaderenko, A.P.; Pozo, D. Silver nanoparticles interactions with the immune system: Implications for health and disease. In *Silver Nanoparticles*; Pozo, D., Ed.; InTech: Rijeka, Croatia, 2010; pp. 310–325.

20. Wysokowski, M.; Petrenko, I.; Stelling, A.L.; Stawski, D.; Jesionowski, T.; Ehrlich, H. Poriferan chitin as a versatile template for extreme biomimetics. *Polymers* **2015**, *7*, 235–265. [CrossRef]

21. Petrenko, I.; Bazhenov, V.V.; Galli, R.; Wysokowski, M.; Fromont, J.; Schupp, P.J.; Stelling, A.L.; Niederschlag, E.; Stöker, H.; Kutsova, V.Z.; et al. Chitin of poriferan origin and the bioelectrometallurgy of copper/copper oxide. *Int. J. Biol. Macromol.* **2017**, *104*, 1626–1632. [CrossRef] [PubMed]

22. Crini, G. Historical review on chitin and chitosan biopolymers. *Environ. Chem. Lett.* **2019**, *17*, 1623–1643. [CrossRef]

23. Ehrlich, H. Chitin and collagen as universal and alternative templates in biomineralization. *Int. Geol. Rev.* **2010**, *52*, 661–699. [CrossRef]

24. Yadav, M.; Goswami, P.; Paritosh, K.; Kumar, M.; Pareek, N.; Vivekanand, V. Seafood waste: A source for preparation of commercially employable chitin/chitosan materials. *Bioresour. Bioprocess.* **2019**, *6*, 8. [CrossRef]

25. Duan, B.; Huang, Y.; Lu, A.; Zhang, L. Recent advances in chitin based materials constructed via physical methods. *Prog. Polym. Sci.* **2018**, *82*, 1–33. [CrossRef]

26. Zhang, J.; Feng, M.; Lu, X.; Shi, C.; Li, X.; Xin, J.; Yue, G.; Zhang, S. Base-free preparation of low molecular weight chitin from crab shell. *Carbohydr. Polym.* **2018**, *190*, 148–155. [CrossRef] [PubMed]

27. Hamed, I.; Özogul, F.; Regenstein, J.M. Industrial applications of crustacean by-products (chitin, chitosan, and chitooligosaccharides): A review. *Trends Food Sci. Technol.* **2016**, *48*, 40–50. [CrossRef]

28. Seear, P.J.; Tarling, G.A.; Burns, G.; Goodall-Copestake, W.P.; Gaten, E.; Özkaya, Ö.; Rosato, E. Differential gene expression during the moult cycle of Antarctic krill (*Euphausia superba*). *BMC Genom.* **2010**, *11*, 582. [CrossRef] [PubMed]

29. Zeng, J.B.; He, Y.S.; Li, S.L.; Wang, Y.Z. Chitin whiskers: An overview. *Biomacromolecules* **2012**, *13*, 1–11. [CrossRef] [PubMed]

30. Chandran, R.; Williams, L.; Hung, A.; Nowlin, K.; Lajeunesse, D. SEM characterization of anatomical variation in chitin organization in insect and arthropod cuticles. *Micron* **2016**, *82*, 74–85. [CrossRef] [PubMed]

31. Merzendorfer, H.; Zimoch, L. Chitin metabolism in insects: Structure, function and regulation of chitin synthases and chitinases. *J. Exp. Biol.* **2003**, *206*, 4393–4412. [CrossRef] [PubMed]

32. Vincent, J.F.V.; Wegst, U.G.K. Design and mechanical properties of insect cuticle. *Arthropod Struct. Dev.* **2004**, *33*, 187–199. [CrossRef] [PubMed]

33. Rudall, K.M. The chitin/protein complexes of insect cuticles. *Adv. Insect Phys.* **1963**, *1*, 257–313.

34. Liu, S.; Sun, J.; Yu, L.; Zhang, C.; Bi, J.; Zhu, F.; Qu, M.; Jiang, C.; Yang, Q. Extraction and characterization of chitin from the beetle *Holotrichia parallela* motschulsky. *Molecules* **2012**, *17*, 4604–4611. [CrossRef] [PubMed]

35. Davies, G.J.G.; Knight, D.P.; Vollrath, F. Chitin in the silk gland ducts of the spider *Nephila edulis* and the silkworm *Bombyx mori*. *PLoS ONE* **2013**, *8*, e73225. [CrossRef] [PubMed]

36. Klinger, C.; Żółtowska-Aksamitowska, S.; Wysokowski, M.; Tsurkan, M.V.; Galli, R.; Petrenko, I.; Machałowski, T.; Ereskovsky, A.; Martinović, R.; Muzychka, L.; et al. Express method for isolation of ready-to-use 3D chitin scaffolds from *Aplysina archeri* (Aplysineidae: Verongiida) Demosponge. *Mar. Drugs* **2019**, *17*, 131. [CrossRef] [PubMed]

37. Kovalchuk, V.; Voronkina, A.; Binnewerg, B.; Schubert, M.; Muzychka, L.; Wysokowski, M.; Tsurkan, M.V.; Bechmann, N.; Petrenko, I.; Fursov, A.; et al. Naturally drug-loaded chitin: Isolation and applications. *Mar. Drugs* **2019**, *17*, 574. [CrossRef] [PubMed]

38. Binnewerg, B.; Schubert, M.; Voronkina, A.; Muzychka, L.; Wysokowski, M.; Petrenko, I.; Djurović, M.; Kovalchuk, V.; Tsurkan, M.; Martinovic, R.; et al. Marine biomaterials: Biomimetic and pharmacological potential of cultivated *Aplysina aerophoba* marine demosponge. *Mater. Sci. Eng. C* **2020**, *109*, 110566. [CrossRef] [PubMed]

39. Schubert, M.; Binnewerg, B.; Voronkina, A.; Muzychka, L.; Wysokowski, M.; Petrenko, I.; Kovalchuk, V.; Tsurkan, M.; Martinovic, R.; Bechmann, N.; et al. Naturally prefabricated marine biomaterials: Isolation and applications of flat chitinous 3D scaffolds from *Ianthella labyrinthus* (Demospongiae: Verongiida). *Int. J. Mol. Sci.* **2019**, *20*, 5105. [CrossRef] [PubMed]

40. Wysokowski, M.; Machałowski, T.; Petrenko, I.; Schimpf, C.; Rafaja, D.; Galli, R.; Zietek, J.; Pantovic, S.; Voronkina, A.; Ivanenko, V.K.V.N.; et al. 3D chitin scaffolds of marine Demosponge origin for biomimetic mollusk hemolymph-associated biomineralization *ex-vivo*. *Mar. Drugs* **2020**, *18*, 123. [CrossRef] [PubMed]

41. Wysokowski, M.; Motylenko, M.; Beyer, J.; Makarova, A.; Stöcker, H.; Walter, J.; Galli, R.; Kaiser, S.; Vyalikh, D.; Bazhenov, V.V.; et al. Extreme biomimetic approach for developing novel chitin-GeO$_2$ nanocomposites with photoluminescent properties. *Nano Res.* **2015**, *8*, 2288–2301. [CrossRef]

42. Machałowski, T.; Wysokowski, M.; Tsurkan, M.V.; Galli, R.; Żółtowska-Aksamitowska, S.; Petrenko, I.; Czaczyk, K.; Kraft, M.; Bertau, M.; Bechmann, N.; et al. Spider chitin: An ultrafast microwave-assisted method for chitin isolation from *Caribena versicolor* spider molt cuticle. *Molecules* **2019**, *24*, 3736. [CrossRef]

43. Jain, P.; Pradeep, T. Potential of silver nanoparticle-coated polyurethane foam as an antibacterial water filter. *Biotechnol. Bioeng.* **2005**, *90*, 59–63. [CrossRef] [PubMed]

44. Wysokowski, M.; Behm, T.; Born, R.; Bazhenov, V.V.; Meißner, H.; Richter, G.; Szwarc-Rzepka, K.; Makarova, A.; Vyalikh, D.; Schupp, P.; et al. Preparation of chitin-silica composites by in vitro silicification of two-dimensional *Ianthella basta* demosponge chitinous scaffolds under modified Stöber conditions. *Mater. Sci. Eng. C* **2013**, *33*, 3935–3941. [CrossRef] [PubMed]

45. Ehrlich, H.; Maldonado, M.; Spindler, K.D.; Eckert, C.; Hanke, T.; Born, R.; Goebel, C.; Simon, P.; Heinemann, S.; Worch, H. First evidence of chitin as a component of the skeletal fibers of marine sponges. Part I. Verongidae (demospongia: Porifera). *J. Exp. Zool.* **2007**, *308B*, 347–357. [CrossRef] [PubMed]

46. Ehrlich, H.; Ilan, M.; Maldonado, M.; Muricy, G.; Bavestrello, G.; Kljajic, Z.; Carballo, J.L.; Schiaparelli, S.; Ereskovsky, A.; Schupp, P.; et al. Three-dimensional chitin-based scaffolds from Verongida sponges (Demospongiae: Porifera). Part I. Isolation and identification of chitin. *Int. J. Biol. Macromol.* **2010**, *47*, 141–145. [CrossRef] [PubMed]

47. Jesionowski, T.; Norman, M.; Zółtowska-Aksamitowska, S.; Petrenko, I.; Joseph, Y.; Ehrlich, H. Marine spongin: Naturally prefabricated 3D scaffold-based biomaterial. *Mar. Drugs* **2018**, *16*, 88. [CrossRef]

48. Wysokowski, M.; Motylenko, M.; Bazhenov, V.V.; Stawski, D.; Petrenko, I.; Ehrlich, A.; Behm, T.; Kljajic, Z.; Stelling, A.L.; Jesionowski, T.; et al. Poriferan chitin as a template for hydrothermal zirconia deposition. *Front. Mater. Sci.* **2013**, *7*, 248–260. [CrossRef]

49. Wysokowski, M.; Szalaty, T.J.; Jesionowski, T.; Motylenko, M.; Rafaja, D.; Koltsov, I.; Stöcker, H.; Bazhenov, V.V.; Ehrlich, H.; Stelling, A.L.; et al. Extreme biomimetic approach for synthesis of nanocrystalline chitin-(Ti,Zr)O$_2$ multiphase composites. *Mater. Chem. Phys.* **2017**, *188*, 115–124. [CrossRef]

50. Mutsenko, V.; Gryshkov, O.; Rogulska, O.; Lode, A.; Petrenko, A.Y.; Gelinsky, M.; Glasmache, B.; Ehrlich, H. Chitinous scaffolds from marine sponges for tissue engineering. In *Marine-Derived Biomaterials for Tissue Engineering Applications Chitinous*; Springer Nature: Singapore, 2019; pp. 285–307.

51. Mutsenko, V.V.; Gryshkov, O.; Lauterboeck, L.; Rogulska, O.; Tarusin, D.N.; Bazhenov, V.V.; Schütz, K.; Brüggemeier, S.; Gossla, E.; Akkineni, A.R.; et al. Novel chitin scaffolds derived from marine sponge *Ianthella basta* for tissue engineering approaches based on human mesenchymal stromal cells: Biocompatibility and cryopreservation. *Int. J. Biol. Macromol.* **2017**, *104*, 1955–1965. [CrossRef] [PubMed]

52. Petrenko, I.; Khrunyk, Y.; Voronkina, A.; Kovalchuk, V.; Fursov, A.; Tsurkan, D.; Ivanenko, V. Poriferan chitin: 3D scaffolds from nano- to macroscale. A review. *Lett. Appl. NanoBioScience* **2020**, *9*, 1004–1014.

53. Rogulska, O.Y.; Mutsenko, V.V.; Revenko, E.B.; Petrenko, Y.A.; Ehrlich, H.; Petrenko, A.Y. Culture and differentiation of human adipose tissue mesenchymal stromal cells within carriers based on sea sponge chitin skeletons. *Stem Cell Day* **2013**, *23*, 267–270.

54. Schleuter, D.; Günther, A.; Paasch, S.; Ehrlich, H.; Kljajić, Z.; Hanke, T.; Bernhard, G.; Brunner, E. Chitin-based renewable materials from marine sponges for uranium adsorption. *Carbohydr. Polym.* **2013**, *92*, 712–718. [CrossRef] [PubMed]

55. Bechmann, N.; Ehrlich, H.; Eisenhofer, G.; Ehrlich, A.; Meschke, S.; Ziegler, C.G.; Bornstein, S.R. Anti-tumorigenic and anti-metastatic activity of the sponge-derived marine drugs aeroplysinin-1 and isofistularin-3 against pheochromocytoma in vitro. *Mar. Drugs* **2018**, *16*, 172. [CrossRef] [PubMed]

56. Drechsel, A.; Helm, J.; Ehrlich, H.; Pantovic, S.; Bornstein, S.R.; Bechmann, N. Anti-tumor activity vs. normal cell toxicity: Therapeutic potential of the bromotyrosines aerothionin and homoaerothionin *in vitro*. *Mar. Drugs* **2020**, *18*, e236. [CrossRef] [PubMed]

57. Nowacki, K.; Stępniak, I.; Machalowski, T.; Wysokowski, M.; Petrenko, I.; Schimpf, C.; Rafaja, D.; Ziętek, J.; Pantović, S.; Voronkina, A.; et al. Electrochemical method for isolation of chitinous 3D scaffolds from cultivated *Aplysina aerophoba* marine demosponge and its biomimetic application. *Appl. Phys. A* **2020**, *126*, 368. [CrossRef]

58. Mulongo, G.; Mbabazi, J.; Nnamuyomba, P.; Hak-Chol, S. Water bactericidal properties of nanosilver-polyurethane composites. *Nanosci. Nanotechnol.* **2011**, *1*, 40–42. [CrossRef]

59. Suchomel, P.; Kvitek, L.; Panacek, A.; Prucek, R.; Hrbac, J. Comparative study of antimicrobial activity of AgBr and Ag Nanoparticles (NPs). *PLoS ONE* **2015**, *10*, e0119202. [CrossRef] [PubMed]

60. Liu, Z.; Guo, W.; Guoa, C.; Liu, S. Fabrication of AgBr nanomaterials as excellent antibacterial agent. *RSC Adv.* **2015**, *5*, 72872–72880. [CrossRef]

61. Padervand, M.; Elahifard, M.R.; Meidanshahi, R.V.; Ghasemi, S.; Haghighi, S.; Gholami, M.R. Investigation of the antibacterial and photocatalytic properties of the zeolitic nanosized AgBr/TiO composites. *Mater. Sci. Semicond. Process.* **2012**, *15*, 73–79. [CrossRef]

62. Wysokowski, M.; Piasecki, A.; Bazhenov, V.V.; Paukszta, D.; Born, R.; Schupp, P.; Petrenko, I.; Jesionowski, T. Poriferan chitin as the scaffold for nanosilica deposition under hydrothermal synthesis conditions. *J. Chitin Chitosan Sci.* **2013**, *1*, 26–33. [CrossRef]

63. Bazhenov, V.V.; Wysokowski, M.; Petrenko, I.; Stawski, D.; Sapozhnikov, P.; Born, R.; Stelling, A.L.; Kaiser, S.; Jesionowski, T. Preparation of monolithic silica-chitin composite under extreme biomimetic conditions. *Int. J. Biol. Macromol.* **2015**, *76*, 33–38. [CrossRef] [PubMed]

64. Ehrlich, H.; Simon, P.; Motylenko, M.; Wysokowski, M.; Bazhenov, V.V.; Galli, R.; Stelling, A.L.; Stawski, D.; Ilan, M.; Stöcker, H.; et al. Extreme Biomimetics: Formation of zirconium dioxide nanophase using chitinous scaffolds under hydrothermal conditions. *J. Mater. Chem. B* **2013**, *1*, 5092–5099. [CrossRef] [PubMed]

65. Wysokowski, M.; Motylenko, M.; Stöcker, H.; Bazhenov, V.V.; Langer, E.; Dobrowolska, A.; Czaczyk, K.; Galli, R.; Stelling, A.L.; Behm, T.; et al. An extreme biomimetic approach: Hydrothermal synthesis of β-chitin/ZnO nanostructured composites. *J. Mater. Chem. B* **2013**, *1*, 6469–6476. [CrossRef] [PubMed]

66. Wysokowski, M.; Motylenko, M.; Walter, J.; Lota, G.; Wojciechowski, J.; Stöcker, H.; Galli, R.; Stelling, A.L.; Himcinschi, C.; Niederschlag, E.; et al. Synthesis of nanostructured chitin–hematite composites under extreme biomimetic conditions. *RSC Adv.* **2014**, *4*, 61743–61752. [CrossRef]

67. Lutterotti, L.; Matthies, S.; Wenk, H. MAUD: A friendly Java program for Material Analysis Using Diffraction. *Comm. Power Diffr. Newsl.* **1999**, *21*, 1–20.

 *marine drugs*

Article

Physicochemical and Functional Properties of Type I Collagens in Red Stingray (*Dasyatis akajei*) Skin

Junde Chen [1],*, Jianying Li [1,2], Zhongbao Li [2,3,*], Ruizao Yi [1], Shenjia Shi [1,2], Kunyuan Wu [1], Yushuang Li [1] and Sijia Wu [1]

[1] Technical Innovation Center for Utilization of Marine Biological Resources, Third Institute of Oceanography, Ministry of Natural Resources, Xiamen 361005, China. jyLi9188@163.com (J.L.); yiruizao@163.com (R.Y.); papertiger92@163.com (S.S.); yuzuruhanyu@163.com (K.W.); wwwwaway@163.com (Y.L.); gagasaid@163.com (S.W.)

[2] Fisheries College, Jimei University, Xiamen 361021, China

[3] Fujian Provincial Key Laboratory, Marine Fishery Resources and Eco-environment, Jimei University, Xiamen 361021, China

* Correspondence: jdchen@tio.org.cn (J.C.); lizhongbao@jmu.edu.cn (Z.L.); Tel.: +86-592-215527 (J.C.)

Received: 27 August 2019; Accepted: 27 September 2019; Published: 28 September 2019

Abstract: Collagen is widely used in the pharmaceutical, tissue engineering, nutraceutical, and cosmetic industries. In this study, acid-soluble collagen (ASC) and pepsin-soluble collagen (PSC) were extracted from the skin of red stingray, and its physicochemical and functional properties were investigated. The yields of ASC and PSC were 33.95 ± 0.7% and 37.18 ± 0.71% (on a dry weight basis), respectively. ASC and PSC were identified as type I collagen by Sodium Dodecyl Sulfate Polyacrylamide Gel Electrophoresis (SDS-PAGE) analysis, possessing a complete triple helix structure as determined by UV absorption, Fourier transform infrared, circular dichroism, and X-ray diffraction spectroscopy. Contact angle experiments indicated that PSC was more hydrophobic than ASC. Thermal stability tests revealed that the melting temperature of PSC from red stingray skin was higher than that of PSC from duck skin, and the difference in the melting temperature between these two PSCs was 9.24 °C. Additionally, both ASC and PSC were functionally superior to some other proteins from terrestrial sources, such as scallop gonad protein, whey protein, and goose liver protein. These results suggest that PSC from red stingray skin could be used instead of terrestrial animal collagen in drugs, foods, cosmetics, and biological functional materials, and as scaffolds for bone regeneration.

Keywords: collagen; fish skin; structure; functional properties; thermal properties; rheology properties

---

1. Introduction

Collagen, the most abundant protein in vertebrates, is a major structural protein in the connective tissue of animal skin and bone, and constitutes about 30% of the total proteins [1]. Twenty-eight types of collagens have been identified, each possessing a unique amino acid sequence, structure, and function [2]. Among the collagen types, type I is the most promising in terms of its marketable prospects [3]. It possesses a stable triple helix structure, which is formed by three α subunits and a right-handed helix. The hydrogen bond between glycine and an amide group in the adjacent α chains is the key to a stable collagen triple helix. In type I collagens, a very short terminal region (telopeptide) is also present, but it does not participate in the triple helix structure [4]. The intra- and intermolecular covalent cross-linking in this region is mainly formed by the oxidative deamination of the ε-NH_2 groups of lysine and hydroxyllysine residues [5]. The structural characteristics of collagen have allowed researchers to isolate it from raw materials with acetic acid, resulting in acid-soluble collagen (ASC) [6]. Pepsin is also used to digest peptide chains in the telopeptide region of collagen molecules, and the obtained collagen is referred to as pepsin-soluble collagen (PSC) [7]. These two types of collagen have distinct physicochemical properties.

Type I collagen is highly biocompatible and biodegradable and is widely used in the pharmaceutical, tissue engineering, health care product, and cosmetic industries [8]. However, the industrial application of collagen depends heavily on its thermal stability [9]. The rheological properties, water and oil affinity, emulsifying capacity, and foaming capacity of collagen are also important factors that determine the industrial use of type I collagen, and these functional properties are closely related to its structure [10]. Traditionally, collagen is isolated from the skin or bone of pigs, cows, and chickens [11]. Recently, outbreaks of bovine spongiform encephalopathy, foot-and-mouth disease, and avian influenza have made consumers wary of collagen and collagen derivatives from terrestrial animals. Owing to religious reasons, pig collagen is also banned from sale in certain areas [12]. These issues have led to a drastic decline in the market demand of collagen and its derivatives from terrestrial animals. Therefore, a safe alternative collagen source is necessary.

The red stingray (*Dasyatis akajei*) is a fish resource that is abundant in the Northwest Pacific Ocean, mainly along the coastal regions of Japan, South Korea, and China [13]. The skin of the fish is rich in collagen and is an excellent source of raw materials that can be used for the development of collagen products. As such, the use of red stingray skin as an alternative source of collagen may be an effective way to obtain high-value-added products. However, only a few studies have systematically investigated collagen in red stingray skin. Thus, this study aims to isolate and characterize the ASC and PSC found in red stingray skin.

2. Results and Discussion

2.1. Yield

ASC and PSC isolated from red stingray skin reached yields of 33.95 ± 0.7% and 37.18 ± 0.71% (on a dry weight basis), respectively. The yields of red stingray skin collagen were higher than samples from Savigny skin collagen (ASC, 3.89%; PSC, 6.74%; on a dry weight basis) [14], loach skin collagen (ASC, 22.42%; PSC, 27.32%; on a dry weight basis) [15], balloon skin collagen (ASC, 4%; PSC, 19.5%; on a dry weight basis) [16], and duck collagen (28.37 ± 0.58, on a dry weight basis) [17]. Additionally, the yield of PSC from red stingray skin was higher than ASC, indicating that the breakdown of covalent cross-links by pepsin digestion of the terminal peptides of collagen results in the dissolution of more collagen triple helix structures by acetic acid, thus providing a higher collagen yield [18].

2.2. Structural Characterization

2.2.1. Sodium Dodecyl Sulfate Polyacrylamide Gel Electrophoresis (SDS-PAGE)

The SDS-PAGE protein patterns of red stingray collagens were compared using the profile of commercial rat rail type I collagen, which was used as a reference for the collagen type I protein pattern. Both ASC and PSC contained two different types of α chains (α1 and α2). The gray level ratios of α1/α2 in ASC and PSC were 2.36 and 2.38, respectively, with ratios of α1/α2 close to 2, suggesting that both ASC and PSC belonged to type I collagen [19]. Additionally, these values were similar to type I collagen from the skin of tilapia (α1/α2 = 2.3) [20] and the skin of bighead carp (α1/α2 = 2.13) [19], but higher than type I collagen from rat tail tendon (α1/α2 = 1.84). These results indicated ratios of α1/α2 collagen from different sources might be correlated with the sequence of collagen subunit, temperature of normal habit, and age. Electrophoretic positions of ASC chains (α1-MW, 129 kDa; α2-MW, 119 kDa) were similar to the positions observed for PSC (α1-MW, 129 kDa; α2-MW, 119 kDa), but lower than the positions observed for rat collagen (α1-MW, 137 kDa; α2-MW, 123 kDa). These results indicate that the evolution of collagen genes were potentially the leading cause for the structural differences between terrestrial and aquatic animals [21]. Compared to rat tail tendon collagen, the ASC and PSC from red stingray skin contained more β chains (dimer) and γ chains (trimer), indicating a greater extent of intra- and intermolecular cross-linkage in red stingray ASC and PSC compared to collagen from rat tail tendon (Figure 1) [22].

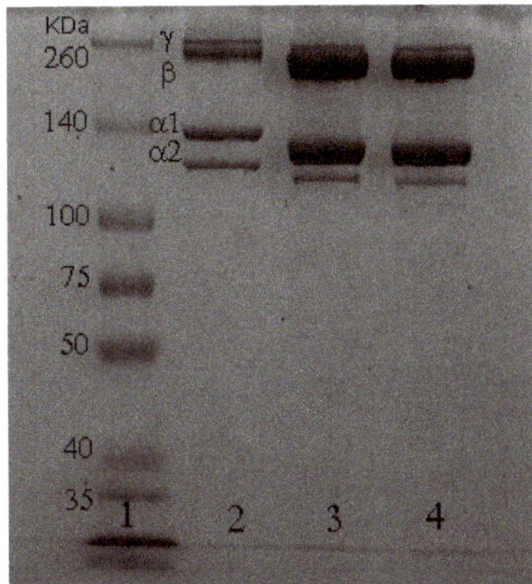

Figure 1. SDS-PAGE patterns of acid-soluble collagen (ASC) and pepsin-soluble collagen (PSC) from the red stingray skin. Lane 1: Standard protein marker (spectra multicolor broad range protein ladder); Lane 2: rat tail type I collagen; Lane 3: ASC collagen; Lane 4: PSC collagen.

2.2.2. Secondary Structure

UV, Fourier transform infrared (FTIR), circular dichroism (CD), and X-ray diffraction (XRD) spectroscopy were used to identify the secondary structure of red stingray collagens. As shown in Figure 2A, the maximum absorption peaks of both types of collagens appeared at 228 nm. This strong absorption peak was due to electrons absorbing light at 228 nm, as the outer electrons of C=O, COOH, and CO-NH$_2$ involved in the peptide bonds of ASC and PSC transit from lower to higher energy states (i.e., n→π*) [9]. ASC and PSC exhibited no absorption peak at 280–300 nm, indicating the absence of tryptophan and tyrosine in the two collagens, as these amino acids exhibit strong absorption in this wavelength range [16].

The characteristic peaks in the FTIR spectra of ASC and PSC correspond to the five amide modes: amide-A, -B, -I, -II, and -III (Figure 2B). The amide-A bands of ASC and PSC appeared at 3320.26 and 3319.96 cm^{-1}, respectively, which is due to the N-H stretching vibration of the two collagens and the present hydrogen bonds [23]. According to Doyle et al., the stretching vibration of free N-H appears in the range 3400–3440 cm^{-1}, but when the peptide segment of a collagen molecule containing an N-H group participates in hydrogen bonding, the stretching vibration of N-H moves to ~3300 cm^{-1} [23]. The amide-B bands of ASC and PSC appeared at 2936.89 and 2935.55 cm^{-1}, respectively, due to the asymmetric stretching vibration of -CH$_2$ in ASC and PSC. Muyonga et al. (2004) demonstrated the interactions between collagen amide-I, -II, and -III with collagen C=O stretching, COO$^-$ coupling, -NH bending, and C-H stretching [24]. The amide-I, -II, and -III bands of ASC appeared at 1660.19, 1550.68, and 1237.87 cm^{-1}, respectively. Compared to ASC, amide-I, -II, and -III bands in PSC appeared at 1659.38, 1551.02, and 1238.79 cm^{-1}, respectively, suggesting a higher degree of hydrogen bonding in PSC. The absorbance of the amide-III band of ASC was 0.98, and the band at 1450–1454 cm^{-1} of PSC was 0.96. It was also found in other marine species The absorption ratio between the amide-III and 1450 cm^{-1} peaks was near to 1 (1.01 for salmon and 0.97 for codfish) [25]. With values close to 1, these absorbance values indicate an intact triple helix structure in both collagens [26].

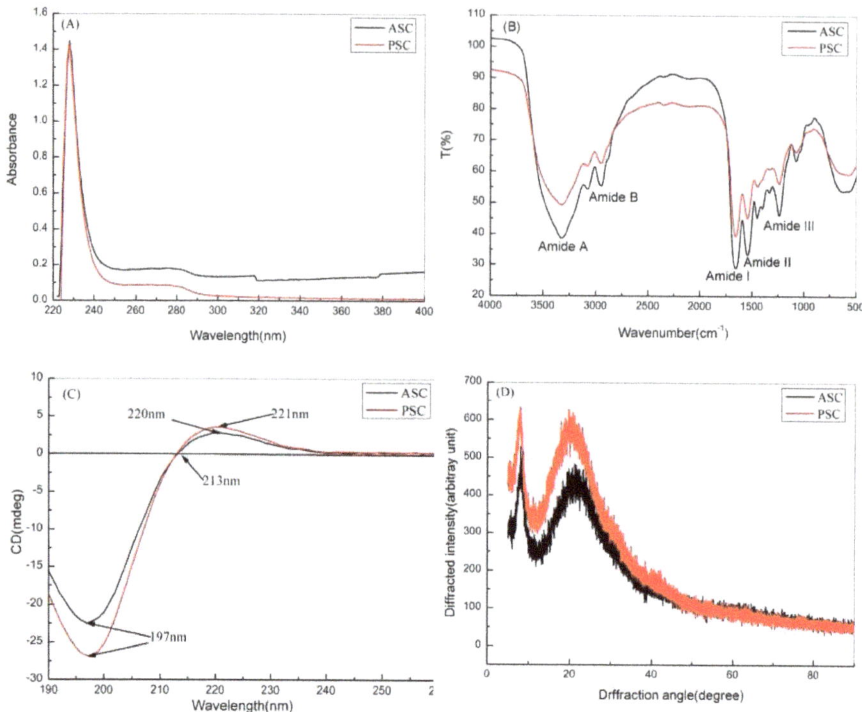

Figure 2. Secondary structure characterization of red stingray collagens. (**A**) UV spectra; (**B**) Fourier transform infrared (FTIR) spectra; (**C**) circular dichroism (CD) spectra; (**D**) X-ray spectra.

Both species exhibited maximum positive cotton effects at 220 and 221 nm, and maximum negative cotton effects at 197 nm. Zero rotation was observed at 213 nm (Figure 2C). These are characteristics of native collagen, indicating that the ASC prepared in this study contained undissociated collagen helices [27]. According to Feng et al. (1996), the ratio of positive peaks to negative peaks (Rpn) in CD spectra can be used to identify triple helical conformation; the Rpn of collagen with a triple helical structure ranges from 0.12 to 0.15 [28]. The ratio between the absolute values of the positive and negative CD peaks of both ASC and PSC was 0.13, further confirming the intact triple helix structure of the collagens.

Bragg's Law [$d = \lambda/(2\sin\theta)$], where $\lambda = 0.154$ is the X-ray wavelength and θ the Bragg angle, was used to analyze the XRD results. Using the first diffraction peaks of ASC and PSC (peak A1) at 8.05° and 7.94°, respectively, the distances between the molecular chains of ASC and PSC were 1.10 and 1.11 by Bragg's Law. The distance between PSC molecular chains was greater than that between ASC molecular chains, indicating that pepsin selectively cleaves the terminal peptide sequences of collagen fibers, weakens the interaction between collagen molecules, and increases the distance between collagen molecular chains [29]. Using the second diffraction peaks of ASC and PSC (peak A2) at 21.12° and 19.68°, respectively, the intermolecular distances in ASC and PSC were found to be 0.42 and 0.45 by Bragg's Law. The intermolecular distance in PSC was greater than in ASC, suggesting that PSC could be potentially more suitable for drug delivery than ASC (Figure 2D) [20].

2.3. *Physicochemical Properties*

2.3.1. Zeta Potential

Both collagens carried positive charges at pH 3–6 and negative charges at pH 7–10. At zero zeta potential, the isoelectric points (pIs) of ASC and PSC were 6.71 and 6.41, respectively. At pH levels above or below the pI of collagen, the repulsion between collagen peptide chains increased, thereby increasing its net solubility (Figure 3). The pI of collagen is related to its amino acid sequence and amino acid residue distribution [20]. The pIs of collagens from different sources were essentially different. The pI values of collagens from red stingray skin were similar to those from brownbanded bamboo shark skin (pI = 6.21, ASC; 6.56, PSC) [30], but higher than those from striped catfish skin (pI = 4.72, ASC; 5.43, PSC) [31] and channel catfish skin (pI = 5.34, ASC; 5.52, PSC) [32].

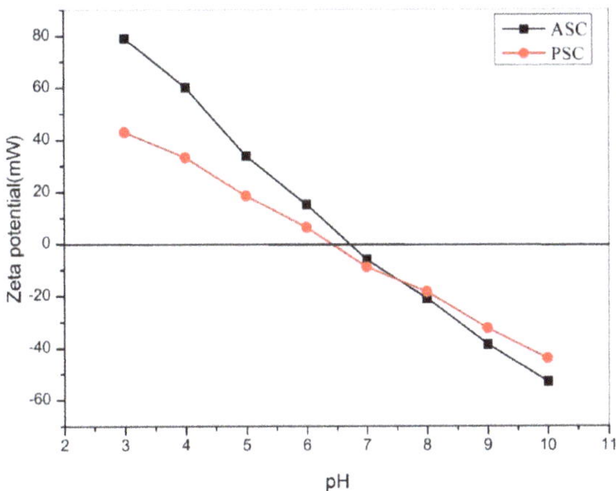

Figure 3. Zeta potential of ASC and PSC at different pH levels.

2.3.2. Water Contact Angle (WCA)

The static WCA of ASC was 100.3° ± 2.31°, and that of PSC was 94.96° ± 0.59°. With values greater than 90°, the contact angles indicate hydrophobicity in both ASC and PSC [33]. The larger static WCA of ASC suggests the presence of a greater number of hydrophobic groups on the surface of ASC than on PSC. According to Zhang et al. [34], the contact angle hysteresis in dynamic contact angle experiments was defined as the difference between advancing and receding angles, which could be used to further characterize the wettability and minimum adhesion at material surface; a larger hysteresis indicates greater hydrophilicity of the material. The contact angle of ASC was 30.67° ± 1.89°, which was smaller than PSC at 42.00° ± 1.14° (Figure 4). The dynamic contact angle experiment reaffirmed the greater hydrophobicity of ASC.

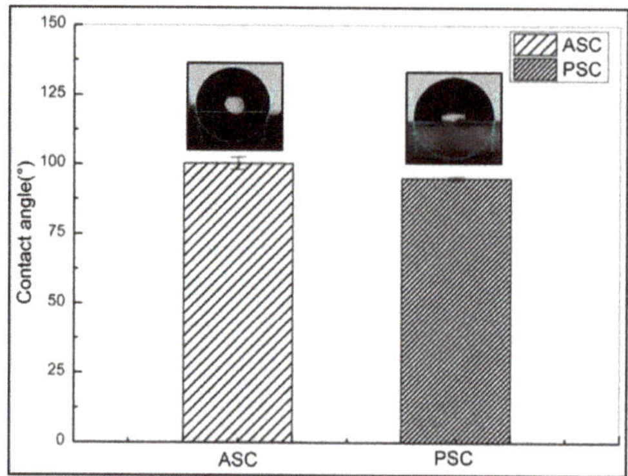

Figure 4. Contact angle data of ASC and PSC from red stingray skin. Values represent means ± standard deviations (SD) of duplicate assays (*n* = 3).

2.3.3. Thermal Properties

The thermal stability of ASC and PSC were characterized by thermal denaturation temperature (T_d) and melting temperature (T_m). T_d is the temperature at which the collagen triple helix structure dissociates into random coils in a solution [29]. As reported in Chen et al. [9], the stability of the collagen triple helix is related to the hydrogen bonding in the molecular chains. As the temperature rises, the hydrogen bonds in the collagen molecules are destroyed and the triple helix structure dissociates and turns into random coils. As a result, the viscosity of collagen decreases. As the temperatures increases, the temperature at which the relative viscosity of collagen is 50% is referred to as the thermal denaturation temperature of collagen [35]. The thermal denaturation temperatures of ASC and PSC from red stingray skin were 23.82 °C and 24.46 °C, respectively (Figure 5A). These values were similar to collagens from the swim bladders of miiuy croaker (T_d = 24.7 °C, ASC; 26.7 °C, PSC) [35], the skins of tiger puffer (T_d = 28.0 °C, ASC; 25.5 °C, PSC) [36], and of Axenillidae (T_d = 24.3 °C) [4], but lower than the skins of giant croaker (T_d ≈ 34.5 °C, ASC; 34.5 °C, PSC) [37] and of river puffer (T_d = 29.5 °C, ASC; 27.5 °C, PSC) [36]. Furthermore, these values were higher than those for the skins of pacific cod (T_d = 14.5 °C, ASC; 16.8 °C, PSC) [8] and Spanish mackerel (T_d = 15.12 °C, ASC; 14.66 °C, PSC) [38]. T_m is defined as the temperature at which the physical form of collagen changes from solid to liquid [39]. At temperatures above T_m, collagen loses its primary structure. DSC performed on the ASC and PSC from red stingray skin provided melting temperatures of 85.25 °C and 95.46 °C, respectively (Figure 5B). The differences between the T_m and T_d of ASC and PSC were 61.43 °C and 71 °C, respectively. This could be related to the degree of hydration in the collagens and the number and nature of covalent cross-links [40]. Collagens from terrestrial animals are generally more stable than those from aquatic animals. This stability is also affected by the body temperature of a given organism and the temperature of their habitat [24]. Interestingly, the melting temperature of PSC from red stingray skin was close to that of duck feet (92.48 °C), but higher than both duck skin (86.22 °C) and duck tendon (88.46 °C) [41]. Considering the importance of thermal stability in the application of collagen, we propose that PSC derived from red stingray skin is an excellent possible alternative to terrestrial sources of collagen.

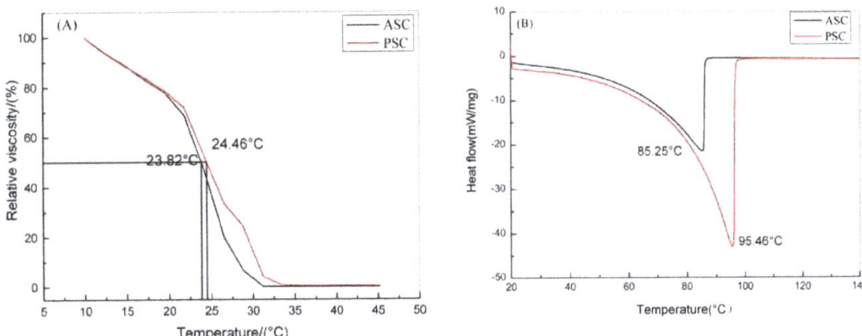

Figure 5. Fractional viscosity (**A**) and differential scanning calorimetry (**B**) of ASC and PSC.

2.3.4. Rheological Characterization of Collagen Solutions

As the concentration of ASC and PSC increased from 5 to 25 mg/mL, complex viscosity (η^*) increased, indicating that concentration affects the stability of the network structure in ASC and PSC solutions (Figure 6). Additionally, the η^* of ASC and PSC decreased as frequency increased, suggesting a shear-thinning flow behavior of the collagen solutions (Figure 6A,C). The loss tangents (tanδ) of ASC and PSC were greater than 1, indicating that the ASC and PSC solutions exhibit viscous behavior (Figure 6B,D). In the frequency range 0.1–10 Hz, tanδ decreased gradually as the ASC and PSC concentration increased. This suggests that as the collagen solution becomes more concentrated, the density of collagen fibers increased and the collagen molecules become more entangled. The motion of collagen molecular chains could not respond rapidly enough to changes in external force; thus, the tanδ of collagen solution gradually decreased [42].

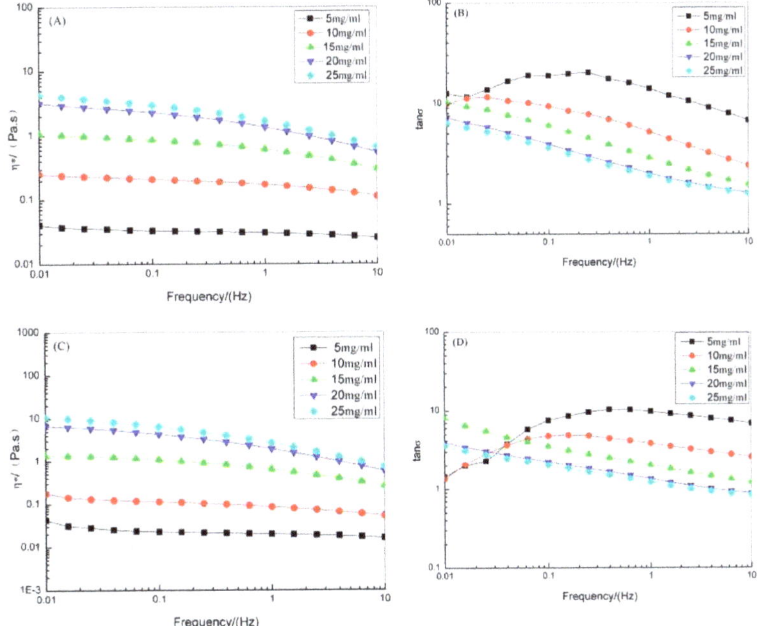

Figure 6. Rheological properties for red stingray collagens solutions with different concentrations. (**A**) complex viscosity (η^*) of ASC; (**B**) loss tangents (tanδ) of ASC; (**C**) η^* of PSC; (**D**) tanδ of PSC.

The η* of the PSC solution decreased gradually when the temperature increased to 20–45 °C and the solution exhibited shear-thinning flow behavior (Figure 7A,C). The η* of the ASC solution exhibited almost no change as the temperature increased to 20–25 °C. As the temperature rose to 30 °C, the η* of the ASC solution decreased drastically. At temperatures above 30 °C, the variation in η* of the ASC solution was no longer regular. The η* values of ASC and PSC exhibited varied rheological behavior as the temperature increased, which is because ASC and PSC have different inter- and intramolecular hydrogen bonds [43]. The tanδ of the ASC solution was greater than 1 at 20–40 °C and the solution exhibited viscous behavior (Figure 7B,D). At temperatures of 20–25 °C, the tanδ of the PSC solution was greater than 1 and the solution exhibited viscous behavior. At temperatures of 30–40 °C and oscillation frequencies of 0.01–0.6 Hz, the tanδ of the PSC solution was less than 1 and the solution exhibited elastic behavior. At temperatures above 30 °C, the change in tanδ of the ASC and PSC solutions lost regularity. This suggests that below its denaturation temperature, collagen could maintain its natural triple helix structure, but above its denaturation temperature, the non-covalent bonds keeping the triple helix structure stable were destroyed. Collagen molecules dissociate, water molecules are released, and the system de-swells. The network structure of the system is thereby destroyed and the molecular structure becomes random and irregular [44].

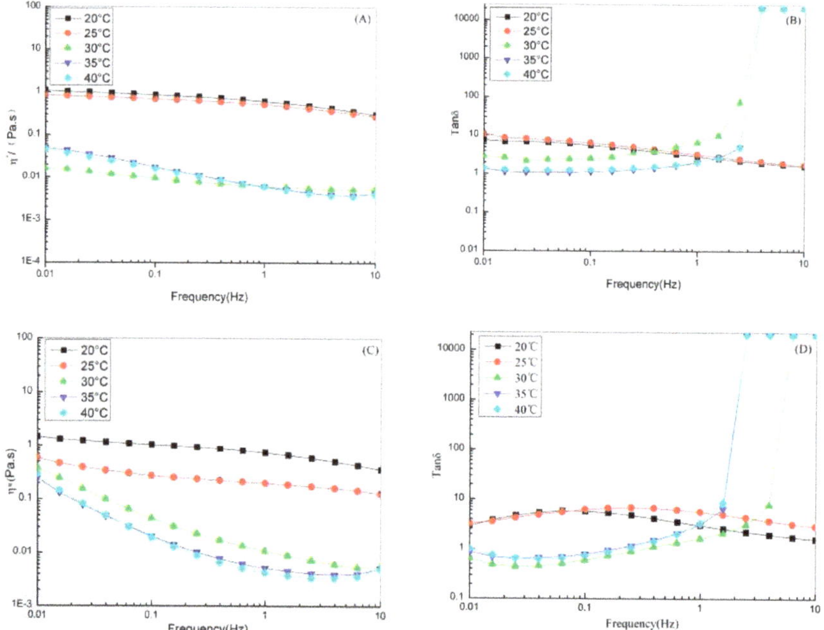

Figure 7. Rheological properties for red stingray collagens solutions at different temperature. (**A**) η* of ASC; (**B**) tanδ of ASC; (**C**) η* of PSC; (**D**) tanδ of PSC.

2.4. Functional Properties

2.4.1. Water Absorption Capacity (WAC) and Oil Absorption Capacity (OAC)

The WACs of ASC and PSC were 20.76 ± 0.55 and 28.48 ± 0.69 mL/g, respectively, and were significantly different (p <0.05). The collagens from red stingray skin had a higher WAC than scallop gonad protein (6.5 mL/g) [45], glutelin (<11 mL/g) [46], and green tea water-insoluble protein (<3 mL/g) [47]. These differences may arise due to the solubility, particle size, micromorphology of proteins, and the physicochemical environment in which the proteins are found [48]. The OAC of red stingray ASC (41.41 ± 0.47 mL/g) was higher than the OAC of PSC (32.40 ± 0.36 mL/g), which was significantly different (p <0.05). As demonstrated by Jitngarmkusol et al., the OAC of a protein is related to its non-polar amino acid residues. The hydrophobic interaction between the non-polar amino acids of protein molecules and the hydrocarbon chains of oil determines the OAC of the protein [49]. The OAC values indicated that red stingray ASC contains more non-polar amino acid residues than PSC. This result is consistent with the results of the contact angle experiments, where ASC had more hydrophobic groups than PSC. Collagens from red stingray skin had a higher OAC than scallop gonad protein (5.2 mL/g) [45], glutelin (<2.3 mL/g) [46], and green tea water-insoluble protein (<5 mL/g) [47]. As reported by Maruyama et al., proteins with higher OAC give better shape retention in food, such as meat or candy products [50]; thus, the collagens from red stingray skin could be used in the meat or candy industries. Additionally, both ASC and PSC exhibited higher OAC than WAC, indicating a larger number of hydrophobic groups in red stingray collagens than hydrophilic groups. This is in agreement with the results of the contact angle experiments.

2.4.2. Foaming and Emulsifying Properties

The foaming and emulsifying properties of collagen samples were characterized by foaming capacity (FC), foam stability (FS), emulsifying activity index (EAI), and emulsifying stability index (ESI). The FC of ASC and PSC varied between 47.62 ± 3.50 and 146.67 ± 2.89 and between 71.43 ± 3.20 and 151.67 ± 5.77%, respectively (Figure 8A). The FCs of red stingray skin collagens were higher than those of scallop gonad protein isolates (25–90%) [45], casein (3.95 ± 0.07–10.15 ± 0.21%) [51], and HBC 19 rice bran protein concentrate (5.2 ± 0.28–10.03 ± 0.39%) [51]. The FS of ASC and PSC varied between 12.50 ± 4.17 and 72.00 ± 3.69% and between 5.32 ± 3.55 and 61.85 ± 1.63%, respectively (Figure 8B). The FS values of collagens were higher than those of casein (0.17 ± 0.002–0.54 ± 0.61%), basmati 386 rice bran protein concentrate (0.65 ± 0.02–2.50 ± 0.03%), and HBC 19 rice bran protein concentrate (3.67 ± 0.09–4.30 ± 0.16%) [51]. The FC and FS of collagens decreased at pH 6 near the pI. The low FC and FS could be due to the poor solubility and weak electrostatic repulsion among the collagen molecules, which was insufficiently strong to prevent collagen aggregation molecules. As collagen molecules aggregated, the interaction between protein and water necessary for foaming was weakened and the FC and FS of collagens lowered [46].

The EAI values of ASC and PSC varied between 61.06 ± 0.83 and 117.11 ± 0.25 m^2/g and between 80.19 ± 0.04 and 105.53 ± 0.41 m^2/g, respectively (Figure 8C). Meanwhile, the ESI of ASC and PSC varied between 1.96 ± 0.05 and 80.36 ± 0.27 min and between 12.68 ± 0.14 and 111.91 ± 0.57 min, respectively (Figure 8D). The EAI and ESI of ASC and PSC decreased at pH 6–7. At this pH, which is close to the pI, the collagens' solubility decreased, as did the electrostatic charge on the collagen molecules. The decreased repulsive intensity increased the possibility of oil droplet aggregation. As oil droplets aggregated, the interaction between oil and water necessary for foaming was weakened and the EAI and ESI of collagens lowered [52]. The EAIs of red stingray skin ASC and PSC were higher than jackfruit seed protein isolates (9.09 ± 0.04–9.80 ± 0.04 m^2/g) [53], whey protein (39.69–65.63 m^2/g) [54], and goose liver protein (2.3 ± 0.2–3.2 ± 0.1 m^2/g) [55]. These results suggest that there is potential for red stingray skin to be used as an alternative to terrestrial protein sources. Considering the importance of foaming and emulsifying properties in the food industry, the results of this study suggest that the ASC and PSC from red stingray skin could be applied in baking, beverages, and minor food ingredients.

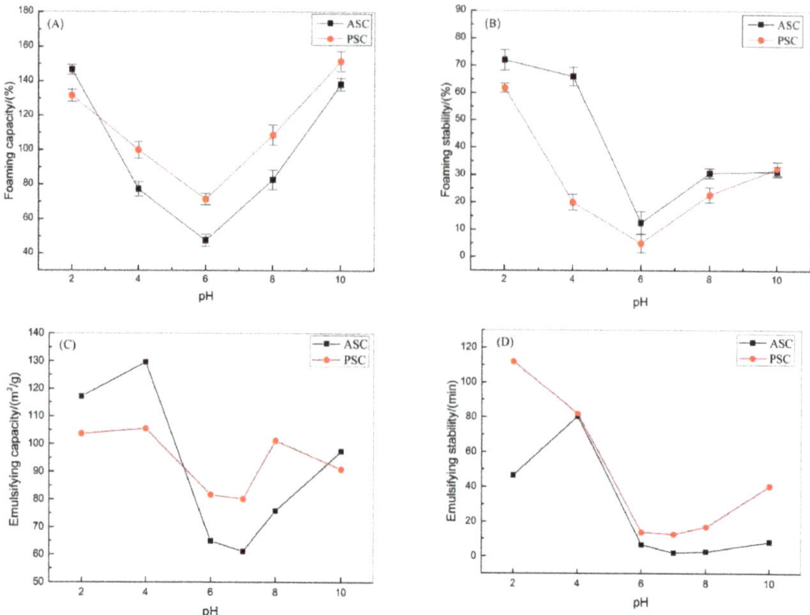

Figure 8. Foaming and emulsifying properties of collagens isolates from red stingray at different pH values. (**A**) foaming capacity (FC); (**B**) foam stability (FS); (**C**) emulsifying activity index (EAI); (**D**) emulsifying stability index (ESI). Values represent means ± standard deviations (SD) of duplicate assays ($n = 3$).

2.5. Cell Proliferation

Cell proliferation experiments were performed to determine the biocompatibility of NIH3T3 fibroblasts. A higher optical density (OD) value indicated a greater cell proliferation rate [9]. As shown in Figure 9, a significant increase in the OD value in the negative control and collagen samples were found from days 1 to 5 ($p < 0.05$). Additionally, there was a significant difference in OD value between the negative control and collagen samples on day 1 (negative control$_{1d}$, 0.213 ± 0.011; ASC sample$_{1d}$, 0.289 ± 0.020; PSC sample$_{1d}$, 0.293 ± 0.016; $p < 0.05$), suggesting the potential use of ASC and PSC in drugs, foods, cosmetics, and as biomedical materials [56]. However, no significant difference was found between negative control and collagen samples at days 2, 3, 4, and 5. This may be caused by the fact that the denaturation temperatures of ASC and PSC were 23.82 °C and 24.46 °C, while the fibroblasts and collagens were cultured at 37 °C. These phenomena might degrade the collagens' intact triple helix structure and then lead to loss of collagen activity.

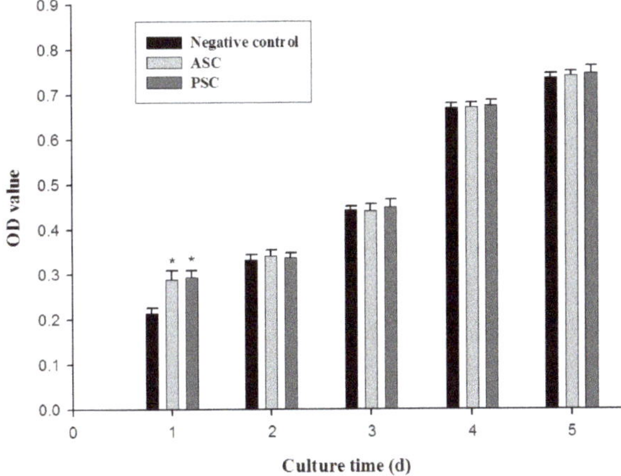

Figure 9. Cell proliferation assay detecting the effects of collagens isolates from red stingray on cell proliferation values represent means ± standard deviations (SD) of duplicate assays (*n* = 3).

3. Materials and Methods

3.1. Materials

Red stingray (*Dasyatis akajei*) skin was purchased from an aquatic product processing plant located in Zhangzhou, China. All red stingray experiments were executed according to the protocol approved by Institutional Animal Care and Use Committee of Third Institute of Oceanography, Ministry of Natural Resources. (date of animal research approval, 04 December 2018). Meat and fat layers were removed from the skin by hand. After washing with water, the skin was stored in polyethylene bags and kept at −20 °C for future analyses. *N,N,N′,N′*-tetramethylethylenediamime (TEMED), sodium dodecyl sulphate (SDS), and Coomassie blue R-250 were obtained from Bio-Rad Laboratories (Hercules, CA, USA). High-molecular-weight markers, pepsin, trifluoroacetic acid (TFA), and formic acid were obtained from Sigma Chemical Co. (St. Louis, MO, USA). All other chemicals and reagents used in this study were analytical grade.

3.2. Extraction of Collagens

3.2.1. Extraction of ASC

ASC from stingray skin was extracted following the methods previously described by Zhao et al. [35] with modifications. Thawed skin was treated with 10 volumes of 0.1 mol/L $NaHCO_3$ for 6 h to remove non-collagenous proteins and pigment. The skin was then stirred and extracted with 0.5 mol/L acetic acid at a 1:10 (*w/v*) ratio for 24 h. Extracted ASC was centrifuged at 20,000× g for 15 min. The supernatant of the ASC extract underwent salt-induced precipitation with 2–5% (*w/v*) NaCl. The precipitate was collected after centrifugation at 10,000× g for 20 min and re-dispersed at a 1:9 (*w/v*) ratio in 0.5 mol/L acetic acid. The solution was dialyzed against a 20-fold volume of 0.1 mol/L acetic acid for 24 h, followed by 24 h of dialysis against distilled water. All processes were carried out below 5 °C. ASC was lyophilized and then stored at −20 °C until use.

3.2.2. Extraction of PSC

PSC from the stingray skin was extracted following the methods previously described by Zhao et al. [35] with modifications. The crushed fish skin was extracted at 4 °C for 24 h in 0.5 M acetic acid containing 1% pepsin (*w/w*) at a 1:10 (*w/v*) ratio. After extraction, the PSC was centrifuged, salt-precipitated, dialyzed, and freeze-dried using the same methods as used for ASC.

3.2.3. Yield

The extraction yields of ASC and PSC were calculated using Equation (1)

$$\text{Yield}/(\%) = \frac{m_1}{m_2} \times 100 \tag{1}$$

where m_1 is the weight of lyophilized collagen and m_2 that of initial dry fish skin.

3.3. Structural Characterization

3.3.1. SDS-PAGE Analysis

SDS-PAGE was performed following the methods previously described by Laemmli (1970) [57] with modifications using a mini-protein vertical slab electrophoresis system (Bio-Rad Laboratories, Hercules, CA, USA). Samples were dispersed separately in 5% SDS solution. The mixtures were incubated at 85 °C for 1 h and centrifuged at 8500× g for 5 min to remove insoluble substances. The dispersed samples were mixed with sample buffer (0.5 M Tris-HCl, pH 6.8) containing 5% SDS (*w/v*) and 20% glycerol (*v/v*) at a 1:1 (*v/v*) ratio. Samples were loaded onto polyacrylamide gels (8% separating matrix, 3% stacking matrix) and electrophoretically separated under a constant current of 20 mA/gel. The gel was then treated with 25% methanol (*v/v*) and 5% acetic acid (*v/v*) for 30 min and stained for 30 min with 0.1% Coomassie Brilliant Blue R-250 dye (*w/v*) (Bio-Rad Laboratories, Hercules, CA, USA) in 30% methanol (*v/v*) and 10% acetic acid (*v/v*), followed by 30 min of de-staining in 30% (*v/v*) methanol and 10% (*v/v*) acetic acid. The staining ratio of α1 to α2 were analyzed using Quantity One 4.6.0 (Bio-Rad Laboratories, Hercules, CA, USA). High-molecular-weight markers were used to estimate the molecular weight of the samples.

3.3.2. Spectroscopy Analysis

UV spectroscopy was performed following the methods previously described by Veeruraj et al. [58] with modifications. Samples were dispersed in 0.5 M acetic acid to make 1 mg/mL collagen solutions. The solutions were centrifuged at 18,000 rpm at 4 °C for 5 min to collect the supernatant. The supernatant was placed in a quartz cell to measure the UV-absorption spectra of the ASC and PSC values on a UH5300 UV-vis spectrophotometer (Hitachi Corporation, Osaka, Japan). Using 0.5 M acetic acid solution as the blank, samples were scanned at 400 nm/min with 1-nm data intervals across wavelengths of 220–400 nm.

FTIR spectroscopy was performed following the methods previously described by Chen et al. [20] with modifications. Samples were mixed thoroughly with KBr powder in a moisture-free environment. Samples were subjected to FTIR analysis using a horizontal ATR trough plate crystal cell (PIKE technology Inc., Madison, WI, USA) equipped with a Bruker Model VERTEX 70 FTIR spectrometer (Bruker Co., Ettlingen, Germany). The spectrum was obtained using 32 scans per sample over a range of 500–4000 cm^{-1} at a resolution of 4 cm^{-1}. Analyses of the spectral data were carried out using the OPUS v6.5 data-collection software (Bruker Co., Ettlingen, Germany).

CD spectroscopy was performed following the methods previously described by Zhang et al. [59]. Samples were dispersed separately in 0.5 M acetic acid to make 0.1 mg/mL collagen solutions. The solutions were centrifuged at 18,000 rpm at 4 °C for 10 min. The supernatant was collected and placed in a quartz cell to measure the CD spectra of ASC and PSC on a Chirascan CD spectropolarimeter (Applied Photophysics Ltd., Leatherhead, England). The CD spectra were obtained at 15 °C with a time-per-point of 1 s and a wavelength interval of 190–260 nm.

XRD spectroscopy was performed following the methods previously described by Chen et al. [20] with modifications. The crystal structures of lyophilized samples were analyzed using a DX-1000 XRD instrument (Dandong Fangyuan Instrument Co., Ltd., Dandong, China). The X-ray source was Cu $K\alpha$ with a tube voltage of 40 kV and tube current of 25 mA, scanning the 2θ interval at 10° and 50° with an angular speed of 0.06°/s.

3.4. Physicochemical Properties

3.4.1. Zeta Potential

The zeta potential of the samples was analyzed following the methods previously described by Chen et al. [20] with modifications. Samples were dispersed in 0.5 mol/L acetate solution to a final concentration of 0.05/100 g, and then stirred at 4 °C for 6 h. The zeta potential of ASC samples was measured by a ZetaPALS zeta potential analyzer (Brookhaven Instruments Co., Holtsville, NY, USA). The pH of the samples (20 mL) was adjusted using a BIZTU titration instrument (Brookhaven Instruments Co., Holtsville, NY, USA) across a pH range of 1–11 with 1 mol/L KOH and 1 mol/L nitric acid. pI was determined from a pH exhibiting zero potential.

3.4.2. WCA

Analyses of the static water contact angle were conducted following the methods previously described by Deng et al. [60] with modifications. Samples were fixed on a glass slide and then a drop of ultrapure water was applied to the sample surface. After the droplet was allowed to equilibrate for 3 s, the contact angle of the sample was determined with a DSA 20 contact angle measuring instrument (Krüss Co., Ltd., Hamburg, Germany). Five independent readings were recorded at different locations on the sample to calculate the average contact angle.

Analyses of the dynamic water contact angle were conducted following the methods previously described by Deng et al. [60] with modifications. Samples were fixed on glass slides and 2 µL of ultrapure water were dispensed on the sample surface, forming the initial droplet. The dynamic contact angle of the sample was determined with a DSA 20 contact angle measuring instrument (Krüss Co., Ltd., Hamburg, Germany). The needle of the DSA 20 was placed in the droplet and 6 µL of deionized water were added to the droplet. Changes in the contact angle between the droplet on the sample surface and the sample were video-recorded. Three measurements were recorded on each sample at different positions and the average was calculated.

3.4.3. Thermal Properties

The T_d values of the samples were measured following the methods previously described by Zhao et al. [35] with modifications. Samples were dispersed in 0.5 M acetic acid to make 20 mg/mL collagen solutions. The viscosity changes of the collagens at different temperatures (10–45 °C) were measured by an MCR 302 rheometer at a heating rate of 1 °C/min. The temperature at which the relative viscosity was 50% was recorded as the T_d of the samples. T_d was calculated using Equation (2).

$$TD = \frac{A}{B} \times 100 \tag{2}$$

where A is the sample viscosity (Pa.s) and B is the sample viscosity (Pa.s) at 10 °C.

T_m was assessed following the methods previously described by Huang et al. [61]. Samples were dispersed separately in 0.05 M acetic acid at a 1:40 (*w/v*) ratio and stored at 4 °C for 48 h. Roughly 5–10 mg of swollen collagen samples were added to an empty aluminum pan for differential-scanning-calorimetry (DSC) measurements using the aluminum pan as the blank (the enthalpy of the fusion of indium is 28.451 J/g and the melting point is 156.4 °C). The settings of the differential scanning calorimeter were as follows: temperature range, 20–140 °C; heating rate, 1 °C/min; nitrogen flow in the sample chamber, 50 mL/min. The peak on the DSC spectrum is the T_s of collagen.

3.4.4. Rheological Properties

The rheological behavior was analyzed following the methods previously described by Wang et al. [62] with modifications. Oscillatory rheological experiments were conducted using an MCR 302 rheometer (Anton Paar, Gratz, Austria). During all experiments, samples were placed in the rheometer, which was equipped with a cone plate geometry (40 nm diameter, angle 2°) at a gap set of 40 μm. Dynamic frequency sweeps for collagen solutions with different concentrations (5, 10, 15, 20, and 25 mg/mL) were performed from 0.01 to 10 Hz at 25 °C. The effect of temperature on the viscoelastic behavior of the collagens was investigated by inducing a steady-state flow in 15 mg/mL sample solutions with frequencies of 0.01–10 Hz at different temperatures while performing the scan. Each sample was allowed to equilibrate for 10 min at initial temperatures of 0 °C, 25 °C, 30 °C, 35 °C, and 40 °C before measurements were recorded. The functional relationship between η*, tanδ, and oscillation frequency (0.01–10 Hz) was determined.

3.5. Functional Properties

3.5.1. WAC and OAC

WAC and OAC were evaluated following the methods previously described by Wani et al. [63] with modifications. Samples were dispersed in ultrapure water/corn oil and mixed thoroughly in a vortex mixer for 2 min. The solution was kept at 25 °C for 30 min and centrifuged at 5,000 rpm/min for 30 min. The supernatant was collected and measured for its volume. WAC and OAC were calculated using Equations (3) and (4).

$$WAC/(\mathrm{mL/g}) = (V_1 - V_2)/m \tag{3}$$

where V_1 is the volume of water added during the experiment (mL), V_2 the volume of water left after centrifugation (mL), and m the sample mass (g);

$$OAC/(\mathrm{mL/g}) = (V_1 - V_2)/m \tag{4}$$

where V_1 is the volume of oil added during the experiment (mL), V_2 the volume of oil left after centrifugation (mL), and m the sample mass (g).

3.5.2. FC and FS

FC and FS were assessed following the methods previously described by Celik et al. [64] with modifications. The sample was dispersed in 0.5 M acetic acid to make a 0.5% (*w/v*) collagen solution. The pH values of 20 mL aliquots of the solution were adjusted to 2, 4, 6, 8, or 10 with 1 M HCl or 1 M NaOH. Each solution was homogenized with a homogenizer at 25 °C at 16,000 rpm/min for 120 s. Then, the sample was quickly transferred to a 250-mL measuring cup. The total volume of the sample was recorded after 30 s. FC and FS were calculated using the following equations:

$$FC/(\%) = (V_1 - V_2) \times 100/V_2 \tag{5}$$

$$FS/(\%) = (V_1 - V_3) \times 100/(V_1 - V_2) \tag{6}$$

where V_1 (mL) is the volume of the sample after homogenization, and V_2 (mL).is the volume before homogenization. V_3 (mL) is the foam volume of the sample after standing at room temperature for 2 h.

3.5.3. Emulsifying Properties

The EAI and ESI of the samples were determined following the methods previously described by Celik et al. [64]. The sample was dispersed in 0.5 M acetic acid to make a 0.5% (*w/v*) collagen solution. The pH values of 6-mL aliquots of the prepared solution were adjusted to 2, 4, 6, 8, or 10 with 1 M HCl or 1 M NaOH. Corn oil (2 mL) was added to each aliquot and homogenized in a homogenizer for 60 s at 16,000 rpm/min. At 0 and 10 min after homogenization, the emulsion was transferred to a 2-mL centrifugal tube and diluted 100 times with 0.1% (*w/v*) sodium dodecyl sulfate (50-μL sample in 5 mL of 0.1% sodium dodecyl sulfate) with thorough mixing. Absorptions were measured at 500 nm using 0.1% sodium dodecyl sulfate as the blank. EAI and ESI were calculated using the following equations:

$$EAI/(m^2/g) = (2 \times 2.303 \times A_0)/(0.25 \times protein\ weight(g)) \tag{7}$$

$$ESI(min) = (A_{10} \times \Delta t)/(A_0 - A_{10}) \tag{8}$$

where A_0 and A_{10} are the absorptions of the emulsions measured at 0 and 10 min, respectively, and $\Delta t = 10$ min.

3.6. Cell Proliferation

Cell proliferation experiments were performed following the methods previously described by Chen et al. [9]. Collagen samples were dispersed in 0.5 M acetic acid to make 0.5 mg/mL collagen solutions. The solutions were added into cell culture plates and then naturally air-dried. The cell culture plates were sterilized via UV treatment and seeded with NIH3T3 fibroblasts. Fibroblasts were incubated with DMEM cell culture medium at 37 °C. Fibroblasts seeded in cell culture plates without collagen were used as negative control, and the cells were grown for 1, 2, 3, 4, and 5 days. At each aforementioned time point, 3-(4,5-dimethylthiazol-2-y)-2,5-diphenyltetrazolium bromide (MTT) solution was added to the cell culture medium and incubated for 4 h at 37 °C. The culture medium was aspirated and then DNSO was added in the plates. After that, the absorbance (OD value) was measured at 490 nm.

3.7. Statistical Analyses

Statistical analyses were conducted using SPSS software (SPSS 17.0, Inc., Chicago, 1L, USA). One-way ANOVA analyses followed by Dunnett's test for multiple comparisons of treatment means with a control were used. Statistical significance was defined as $p < 0.05$.

4. Conclusions

In this study, ASC and PSC were isolated from the skin of red stingray and the yield of PSC was found to be higher than ASC. SDS-PAGE analyses confirmed that both collagens belonged to type I collagen. UV, FTIR, CD, and XRD spectroscopy revealed the intact triple helix structure of ASC and PSC, which native collagens possess. The pI of ASC and PSC were 6.71 and 6.41, respectively, as per the zeta-potential analyzer results. Rheological tests showed that both ASC and PSC solutions exhibited shear-thinning flow behavior. In the thermal stability tests, PSC from red stingray skin was found to have a higher melting temperature than duck skin PSC. Contact angle experiments demonstrated the superior hydrophobicity of PSC to ASC. Functional property analyses indicated better functional properties of ASC and PSC than other proteins from terrestrial sources in terms of WAC, OAC, foaming properties, and emulsifying properties. Collectively, these results support the potential use of collagens from red stingray skin as an alternative for terrestrial collagen sources, as well as their potential use in drugs, foods, and cosmetics, and as biological functional materials.

Author Contributions: J.C. completed the study design, most of the experiments, and manuscript preparation; Z.L. and R.Y. took part in data analysis; J.L., S.S., K.W., Y.L., and S.W. took part in experiments and data analysis.

Funding: This work was supported by grants from the Scientific Research Foundation of the Third Institute of Oceanography, SOA (2019010), and the National Natural Science Foundation of China (41676129, 41106149).

Conflicts of Interest: The authors declare no conflict of interest.

References

1. Foegeding, E.A.; Laneir, T.C.; Hultin, H.O. *Characteristics of Edible Muscle Tissues*; Fennema, O.R., Ed.; Food chemistry, Marcel Dekker Inc.: New York, NY, USA, 1996; pp. 902–906.

2.   Gara, S.K.; Grumati, P.; Urciuolo, A.; Bonaldo, P.; Kobbe, B.; Koch, M.; Paulsson, M.; Wagener, R. Three novel collagen VI chains with high homology to the α3 chain. *J. Biol. Chem.* **2008**, *283*, 10658–10670. [CrossRef] [PubMed]

3.   Chen, J.D.; Li, L.; Yi, R.Z.; Gao, R.; He, J.L. Release kinetics of Tilapia scale I collagen peptides during tryptic hydrolysis. *Food Hydrocoll.* **2018**, *77*, 931–936. [CrossRef]

4.   Tziveleka, L.A.; Ioannou, E.; Tsiourvas, D.; Berillis, P.; Foufa, E.; Roussis, V. Collagen from the marine sponges Axinella cannabina and Suberites carnosus: isolation and morphological, biochemical, and biophysical characterization. *Mar. Drugs* **2017**, *15*, 152. [CrossRef] [PubMed]

5.   Asghar, A.; Henrickson, R.L. *Chemical, Biochemical, Functional and Nutritional Characteristics of Collagen in Food Systems*; Chichester, C.O., Mrata, E.M., Schweigert, B.S., Eds.; Advances in food research, Academic Press: London, UK, 1982; pp. 232–372.

6.   Nalinanon, S.; Benjakul, S.; Kishimura, H. Collagens from the skin of arabesque greenling (*Pleurogrammus azonus*) solubilized with the aid of acetic acid and pepsin from albacore tuna (*Thunnus alalunga*) stomach. *J. Sci. Food Agr.* **2010**, *90*, 1492–1500. [CrossRef] [PubMed]

7.   Moreno, H.M.; Montero, M.P.; Gómez-Guillén, M.C.; Fernández-Martín, F.; Mørkøre, T.; Borderías, J. Collagen characteristics of farmed *Atlantic salmon* with firm and soft fillet texture. *Food Chem.* **2012**, *134*, 678–685. [CrossRef]

8.   Sun, L.L.; Li, B.F.; Song, W.K.; Si, L.L.; Hou, H. Characterization of Pacific cod (*Gadus macrocephalus*) skin collagen and fabrication of collagen sponge as a good biocompatible biomedical material. *Process Biochem.* **2017**, *63*, 229–235. [CrossRef]

9.   Chen, J.D.; Li, M.; Yi, R.Z.; Bai, K.K.; Wang, G.Y.; Tan, R.; Sun, S.S.; Xu, N.H. Electrodialysis extraction of pufferfish skin (*Takifugu flavidus*): a promising source of collagen. *Mar. drugs* **2019**, *17*, 25. [CrossRef]

10.   Shevkani, K.; Singh, N.; Kaur, A.; Rana, J.C. Structure and functional characterization of kidney bean and field pea protein isolates: A comparative study. *Food Hydrocoll.* **2015**, *43*, 679–689. [CrossRef]

11.   Tamilmozhi, S.; Anguchamy, V.; Arumugam, M. Isolation and characterization of acid and pepsin-solubilized collagen from the skin of sailfish (*Istiophorus platypterus*). *Food Res. Int.* **2013**, *54*, 1499–1505. [CrossRef]

12.   Yu, F.M.; Zong, C.H.; Jin, S.J.; Zheng, J.W.; Chen, N.; Huang, J.; Chen, Y.; Huang, F.F.; Yang, Z.S.; Tang, Y.P.; et al. Optimization of extraction conditions and characterization of pepsin-solubilised collagen from skin of giant croaker (*Nibea japonica*). *Mar. Drugs* **2018**, *16*, 29. [CrossRef]

13.   Li, N.; Song, N.; Cheng, G.P.; Gao, T.X. Genetic diversity and population structure of the red stingray, *Dasyatis akajei* inferred by AFLP marker. *Biochem. Syst Ecol.* **2013**, *51*, 130–137. [CrossRef]

14.   Liu, A.; Zhang, Z.H.; Hou, H.; Zhao, X.; Li, B.F.; Zhao, T.F.; Liu, L.Y. Characterization of acid-and pepsin-soluble collagens from the cuticle of *Perinereis Nuntia* (Savigny). *Food Biophys.* **2018**, *13*, 274–283. [CrossRef]

15.   Wang, J.; Pei, X.L.; Liu, H.Y.; Zhou, D. Extraction and characterization of acid-soluble and pepsin-soluble collagen from skin of loach (*Misgurnus anguillicaudatus*). *Int. J. Biol. Macromol.* **2018**, *106*, 544–550. [CrossRef] [PubMed]

16.   Huang, Y.R.; Shiau, C.Y.; Chen, H.H.; Huang, B.C. Isolation and characterization of acid and pepsin-solubilized collagens from the skin of balloon fish (*Diodon holocanthus*). *Food Hydrocoll.* **2011**, *25*, 1507–1513. [CrossRef]

17.   Huda, N.; Seow, N.H.; Normawati, M.N.; Aisyah, N.M.N. Preliminary study on physicochemical properties of duck feet collagen. *Int.J. Poult. Sci.* **2013**, *12*, 615–621. [CrossRef]

18.   Iswariya, S.; Velswamy, P.; Uma, T.S. Isolation and characterization of biocompatible collagen from the skin of puffer fish (*Lagocephalus inermis*). *J. Polym. Environ.* **2018**, *26*, 2086–2095. [CrossRef]

19.   Liu, D.S.; Liang, L.; Regenstein, J.M.; Zhou, P. Extraction and characterization of pepsin-solubilised collagen from fins, scales, skins, bones and swim bladders of bighead carp (*Hypophthalmichthys nobilis*). *Food Chem.* **2012**, *133*, 1441–1448. [CrossRef]

20.   Chen, J.D.; Li, L.; Yi, R.Z.; Xu, N.H.; Gao, R.; Hong, B.H. Extraction and characterization of acid-soluble collagen from scales and skin of tilapia (*Oreochromis niloticus*). *LWT-Food Sci. Technol.* **2016**, *66*, 453–459. [CrossRef]

21.   Exposito, J.Y.; Larroux, C.; Cluzel, C.; Valcourt, U.; Lethias, C.; Degnan, B.M. Demosponge and sea anemone fibrillar collagen diversity reveals the early emergence of A/C clades and the maintenance of the modular structure of type V/XI collagens from sponge to human. *J. Biol. Chem.* **2008**, *283*, 28226–28235. [CrossRef]

22. Jeevithan, E.; Wu, W.H.; Wang, N.P.; Lan, H.; Bao, B. Isolation purification and characterization of pepsin soluble collagen isolated from silvertip shark (*Carcharhinus albimarginatus*) skeletal and head bone. *Process Biochem.* **2014**, *49*, 1767–1777. [CrossRef]

23. Doyl, B.B.; Bendit, E.G.; Blout, E.R. Infrared spectroscopy of collagen and collagen-like polypeptides. *Biopolymers* **1975**, *14*, 937–957. [CrossRef] [PubMed]

24. Muyonga, J.H.; Cole, C.G.B.; Duodu, K.G. Characterisation of acid soluble collagen from skins of young and adult Nile perch (*Lates niloticus*). *Food Chem.* **2004**, *85*, 81–89. [CrossRef]

25. Alves, A.L.; Marques, A.L.P.; Martins, E.; Silva, T.H.; Reis, R.L. Cosmetic potential of marine fish skin collagen. *Cosmetics* **2017**, *4*, 39. [CrossRef]

26. Plepis, A.M.D.G.; Goissis, G.; Das-Gupta, D.K. Dielectric and pyroelectric characterization of anionic and native collagen. *Polym. Eng. Sci.* **1996**, *36*, 2932–2938. [CrossRef]

27. Usha, R.; Ramasami, T. Structure and conformation of intramolecularly cross-linked collagen. *Colloid. Surface. B.* **2005**, *41*, 21–24. [CrossRef] [PubMed]

28. Feng, Y.B.; Melacini, G.; Taulane, J.P.; Goodan, M. Acetyl-terminated and template-assembled collagen-based polypeptides composed of Gly-Pro-Hyp sequence. 2. Synthesis and conformational analysis by circular dichroism ultraviolet absorbance, and optical rotation. *J. Am. Chem. Soc.* **1996**, *118*, 10351–10358. [CrossRef]

29. Zhang, Y.; Liu, W.T.; Li, G.Y.; Shi, B.; Miao, Y.Q.; Wu, X.H. Isolation and partial characterization of pepsin-soluble collagen from the skin of grass carp (*Ctenopharyngodon idella*). *Food Chem.* **2007**, *103*, 906–912. [CrossRef]

30. Kittiphattanabawon, P.; Benjakul, S.; Visessanguan, W.; Kishimura, H.; Shahidi, F. Isolation and characterisation of collagen from the skin of brownbanded bamboo shark (*Chiloscyllium punctatum*). *Food Chem.* **2010**, *119*, 1519–1526. [CrossRef]

31. Singh, P.; Benjakul, S.; Maqsood, S.; Kishimura, H. Isolation and characterization of collagen extracted from the skin of striped catfish (*Pangasianodon hypophthalmus*). *Food Chem.* **2011**, *124*, 97–105. [CrossRef]

32. Tan, Y.Q.; Chang, S.K.C. Isolation and characterization of collagen extracted from channel catfish (*Ictalurus punctatus*) skin. *Food Chem.* **2018**, *242*, 147–155. [CrossRef]

33. Cui, W.G.; Cheng, L.Y.; Li, H.Y.; Zhou, Y.; Zhang, Y.G.; Chang, J. Preparation of hydrophilic poly (L-Lactide) electrospun fibrous scaffolds modified with chitosan for enhanced cell biocompatibility. *Polymer* **2012**, *53*, 2298–2305. [CrossRef]

34. Zhang, Y.J.; Zhou, X.; Zhong, J.Z.; Tan, L.; Liu, C.M. Effect of pH on emulsification performance of a new functional protein from jackfruit seeds. *Food Hydrocoll.* **2019**, *93*, 325–334. [CrossRef]

35. Zhao, W.H.; Chi, C.F.; Zhao, Y.Q.; Wang, B. Preparation, physicochemical and antioxidant properties of acid-and pepsin-soluble collagens from the swim bladders of miiuy croaker (*Miichthys miiuy*). *Mar. Drugs* **2018**, *16*, 61. [CrossRef]

36. Wang, S.S.; Yu, Y.; Sun, Y.; Liu, N.; Zhou, D.Q. Comparison of physicochemical characteristics and fibril formation ability of collagens extracted from the skin of farmed river puffer (*Takifugu obscurus*) and tiger puffer (*Takifugu rubripes*). *Mar. drugs* **2019**, *17*, 462. [CrossRef]

37. Tang, Y.P.; Jin, S.J.; Li, X.Y.; Li, X.J.; Hu, X.Y.; Chen, Y.; Huang, F.F.; Yang, Z.S.; Yu, F.M.; Ding, G.F. Physicochemical properties and biocompatibility evaluation of collagen from the skin of giant croaker (*Nibea japonica*). *Mar. drugs* **2018**, *16*, 222. [CrossRef]

38. Li, Z.R.; Wang, B.; Chi, C.F.; Zhang, Q.H.; Gong, Y.D.; Tang, J.J.; Luo, H.Y.; Ding, G.F. Isolation and characterization of acid soluble collagens and pepsin soluble collagens from the skin and bone of Spanish mackerel (*Scomberomorous niphonius*). *Food Hydrocoll.* **2013**, *31*, 103–113. [CrossRef]

39. Zou, Y.; Wang, L.; Cai, P.P.; Li, P.P.; Zhang, M.H.; Sun, Z.L.; Sun, C.; Xu, W.M.; Wang, D.Y. Effect of ultrasound assisted extraction on the physicochemical and functional properties of collagen from soft-shelled turtle calipash. *Int. J. Biol. Macromol.* **2017**, *105*, 1602–1610. [CrossRef]

40. Hadian, M.; Corcoran, B.M.; Bradshaw, J.P. Molecular changes in fibrillar collagen in myxomatous mitral valve disease. *Cardiovasc. Pathol.* **2010**, *19*, 141–148. [CrossRef]

41. Kim, H.W.; Yeo, I.J.; Hwang, K.E.; Song, D.H.; Kim, Y.J.; Ham, Y.K.; Jeong, T.J.; Choi, Y.S.; Kim, C.J. Isolation and characterization of pepsin-soluble collagens from bones, skins, and tendons in duck feet. *Korean J. Food Sci. An.* **2016**, *36*, 665–670. [CrossRef]

42. Yu, W.J.; Xu, D.; Li, D.D.; Guo, L.N.; Su, X.Q.; Zhang, Y.; Wu, F.F.; Xu, X.M. Effect of pigskin-originated gelatin on properties of wheat flour dough and bread. *Food Hydrocoll.* **2019**, *94*, 183–190. [CrossRef]

43. Sai, K.P.; Babu, M. Studies on rana tigerina skin collagen. *Comp. Biochem. Phys B.* **2001**, *128*, 81–90.
44. Ju, H.Y.; Liu, M.; Dan, W.H.; Hu, Y.; Lin, H.; Dan, N.H. Dynamic rheological property of type I collagen fibrils. *J. Mech. Med. Biol.* **2013**, *13*, 1340015. [CrossRef]
45. Han, J.R.; Tang, Y.; Li, Y.; Shang, W.H.; Yan, J.N.; Du, Y.N.; Wu, H.T.; Zhu, B.W.; Xiong, Y.L.L. Physiochemical properties and functional characteristics of protein isolates from the scallop (*Patinopecten yessoensis*) gonad. *J. Food Sci.* **2019**, *84*, 1023–1034. [CrossRef]
46. Pham, T.T.; Tran, T.T.T.; Ton, N.M.N.; Le, V.V.M. Effects of pH and salt concentration on functional properties of pumpkin seed protein fractions. *J. Food Process. Pres.* **2016**, *41*, e13073. [CrossRef]
47. Ren, Z.Y.; Chen, Z.Z.; Zhang, Y.Y.; Zhao, T.; Ye, X.G.; Gao, X.; Lin, X.R.; Li, B. Functional properties and structural profiles of water-insoluble proteins from three types of tea residues. *LWT-Food Sci. Technol.* **2019**, *110*, 324–331. [CrossRef]
48. Wu, W.; Hua, Y.F.; Lin, Q.L.; Xiao, H.X. Effects of oxidative modification on thernal aggergation and gel properties of soy protein by peroxyl radicals. *Int. J. Food Sci. Tech.* **2011**, *46*, 1891–1897. [CrossRef]
49. Jitngarmkusol, S.; Hongsuwankul, J.; Tananuwong, K. Chemical compositions functional properties and microstructure of defatted macadamia flours. *Food Chem.* **2008**, *110*, 23–30. [CrossRef]
50. Maruyama, N.; Katsube, T.; Wada, Y.; Oh, M.H.; De La Rosa, A.P.B.; Okuda, E.; Nakagawa, S.; Utsumi, S. The roles of the N - Linked glycans and extension regions of soybean β - conglycinin in folding, assembly and structural features. *Eur. J. Biochem.* **1998**, *258*, 854–862. [CrossRef]
51. Chandi, G.K.; Sogi, D.S. Functional properties of rice bran protein concentrates. *J. Food Eng.* **2007**, *79*, 592–597. [CrossRef]
52. Deng, Y.J.; Huang, L.X.; Zhang, C.H.; Xie, P.J.; Cheng, J.; Wang, X.J.; Li, S.H. Physicochemical and functional properties of Chinese quince seed protein isolate. *Food Chem.* **2019**, *283*, 539–548. [CrossRef]
53. Zhang, H.Y.; Yang, Y.L.; Pan, J.F.; Long, H.; Huang, L.S.; Zhang, X.K. Compare study between icephobicity and superhydrophobicity. *Physica B.* **2019**, *556*, 118–130. [CrossRef]
54. Yi, X.Z.; Zheng, Q.H.; Pan, M.-H.; Chiou, Y.-S.; Li, Z.S.; Li, L.; Chen, Y.; Hu, J.; Duan, S.Z.; Wei, S.D.; et al. Liposomal vesicles-protein interaction: Influences of iron liposomes on emulsifying properties of whey protein. *Food Hydrocoll.* **2019**, *89*, 602–612. [CrossRef]
55. Xue, S.W.; Yu, X.B.; Lin, X.; Zhao, X.; Han, M.Y.; Xu, X.L.; Zhou, G.H. Structural changes and emulsion properties of goose liver proteins obtained by isoelectric solubilization/precipitation processes. *LWT-Food Sci. Technol.* **2019**, *102*, 190–196. [CrossRef]
56. Di Benedetto, C.; Barbaglio, A.; Martinello, T.; Alongi, V.; Fassini, D.; Cullorà, E.; Patruno, M.; Bonasoro, F.; Barbosa, M.A.; Carnevali, M.D.C.; et al. Production, characterization and biocompatibility of marine collagen matrices from an alternative and sustainable source: The sea urchin *Paracentrotus lividus*. *Mar. drugs* **2014**, *12*, 4912–4933. [CrossRef]
57. Laemmli, U.K. Cleavage of structural proteins during the assembly of the head of bacteriophage T4. *Nature* **1970**, *227*, 680–685. [CrossRef]
58. Veeruraj, A.; Arumugam, M.; Ajithkumar, T.; Balasubramanian, T. Isolation and characterization of collagen from the outer skin of squid (*Doryteuthis singhalensis*). *Food Hydrocoll.* **2015**, *43*, 708–716. [CrossRef]
59. Zhang, J.Y.; Jeevithan, E.; Bao, B.; Wang, S.J.; Gao, K.L.; Zhang, C.Y.; Wu, W.H. Structural characterization, in-vivo acute systemic toxicity assessment and in-vitro intestinal absorption properties of tilapia (*Oreochromis niloticus*) skin acid and pepsin soublilized type I collagen. *Process Biochem.* **2016**, *51*, 2017–2025. [CrossRef]
60. Deng, L.L.; Zhang, X.; Yang, L.; Oue, F.; Kang, X.F.; Liu, Y.Y.; Feng, F.Q. Characterization of gelatin/zein nanofibers by hybrid electrospinning. *Food Hydrocoll.* **2018**, *75*, 72–80. [CrossRef]
61. Huang, C.Y.; Kuo, J.M.; Wu, S.J.; Tsai, H.T. Isolation and characterization of fish scale collagen from tilapia (*Oreochromis Sp.*) by a novel extrusion-hydro-extraction process. *Food Chem.* **2016**, *190*, 997–1006. [CrossRef]
62. Wang, W.H.; Wang, X.; Zhao, W.P.; Gao, G.X.; Zhang, X.W.; Wang, Y.B.; Wang, Y.N. Impact of pork collagen superfine powder on rheological and texture properties of Harbin red sausage. *J. Texture Stud.* **2018**, *49*, 300–308. [CrossRef]
63. Wani, I.A.; Sogi, D.S.; Gill, B.S. Physico-chemical and functional properties of native and hydrolysed protein isolates from Indian black gram (*Phaseolus mungo* L.) cultivars. *LWT-Food Sci. Technol.* **2015**, *60*, 848–854. [CrossRef]

64. Celik, M.; Güzel, M.; Yildirim, M. Effect of pH on protein extraction from sour cherry kernels and functional properties of resulting protein concentrate and functional properties of resulting protein concentrate. *J. Food Sci. Tech.* **2019**, *56*, 3023–3032. [CrossRef]

 *marine drugs*

Article

The Effect of Depth on the Morphology, Bacterial Clearance, and Respiration of the Mediterranean Sponge *Chondrosia reniformis* (Nardo, 1847)

Mert Gökalp [1,*], Tjitske Kooistra [1], Miguel Soares Rocha [2,3], Tiago H. Silva [2,3], Ronald Osinga [1], AlberTinka J. Murk [1] and Tim Wijgerde [1]

1   Marine Animal Ecology Group, Wageningen University and Research, P.O. Box 338, 6700 AH Wageningen, The Netherlands; tjitske.kooistra@wur.nl (T.K.); ronald.osinga@wur.nl (R.O.); tinka.murk@wur.nl (A.J.M.); tim.wijgerde@wur.nl (T.W.)
2   3B's Research Group, I3Bs–Research Institute on Biomaterials, Biodegradables and Biomimetics, University of Minho, Headquarters of the European Institute of Excellence on Tissue Engineering and Regenerative Medicine, AvePark, Parque de Ciência e Tecnologia, Zona Industrial da Gandra, 4805-017 Barco, Guimarães, Portugal; miguelsoaresrocha@gmail.com (M.S.R.); tiago.silva@i3bs.uminho.pt (T.H.S.)
3   ICVS/3B's–PT Government Associate Laboratory, 4806-909 Braga/Guimarães, Portugal
*   Correspondence: mert.gokalp@wur.nl or mert.gokalp@gmail.com; Tel.: +90-5377086534

Received: 5 June 2020; Accepted: 3 July 2020; Published: 10 July 2020

Abstract: To support the successful application of sponges for water purification and collagen production, we evaluated the effect of depth on sponge morphology, growth, physiology, and functioning. Specimens of Eastern Mediterranean populations of the sponge *Chondrosia reniformis* (Nardo, 1847) (Demospongiae, Chondrosiida, Chondrosiidae) were reciprocally transplanted between 5 and 20 m depth within the Kaş-Kekova Marine Reserve Area. Control sponges at 5 m had fewer but larger oscula than their conspecifics at 20 m, and a significant inverse relationship between the osculum density and size was found in *C. reniformis* specimens growing along a natural depth gradient. Sponges transplanted from 20 to 5 m altered their morphology to match the 5 m control sponges, producing fewer but larger oscula, whereas explants transplanted from 5 to 20 m did not show a reciprocal morphological plasticity. Despite the changes in morphology, the clearance, respiration, and growth rates were comparable among all the experimental groups. This indicates that depth-induced morphological changes do not affect the overall performance of the sponges. Hence, the potential for the growth and bioremediation of *C. reniformis* in mariculture is not likely to change with varying culture depth. The collagen content, however, was higher in shallow water *C. reniformis* compared to deeper-growing sponges, which requires further study to optimize collagen production.

Keywords: sponge; osculum size; respiration; clearance rate; depth; *Chondrosia reniformis*; collagen; integrated multitrophic aquaculture

---

1. Introduction

Sponges are found at all latitudes, living in a wide array of ecosystems varying in temperature and depth [1,2]. They are filter-feeding organisms often dominating the benthos in terms of abundance and biomass [3–6]. Sponges process huge amounts of water daily at up to 900 times their body volume of water per hour and filter 50,000 liters of seawater per liter of sponge volume per day [7–10], which is comparable to well-established suspension feeders such as mussels [11–13]. Sponges have a high efficiency and capacity for particle retention [3,4,14], preferably small particles (<10 µm) such as bacteria [4,15–18], phytoplankton [3,19], viruses [20], and dissolved organic matter [21–24].

This efficient and versatile filtration makes sponges key drivers of the uptake, retention, and transfer of energy and nutrients within benthic ecosystems [25–28] and makes them interesting candidate species for the bioremediation of organic pollution, such as waste streams from aquaculture [29–31].

Around sea-based fish cultures, the elevated densities of bacteria and reduced oxygen levels caused by the excreta of cultured fish and uneaten fish feed negatively impact the seabed and add substantial pressure on the surrounding environment that could even affect the aquaculture and other ecosystem services [18,32–35]. However, the excess amount of bacteria in/around such an organically loaded culture system generates a potential food source for sponges [29]. In our recent study, we discovered that *Chondrosia reniformis* (Nardo, 1847) benefited from mariculture-sourced organic pollution and showed better a growth performance in polluted waters compared to a pristine environment (170% versus 79% in 13 months [31]). *C. reniformis* has been reported as a rich source of biomedically interesting types of collagen [36–38], which could be used in 3D printing inks for medical applications (i.e., scaffolds for bone and cartilage tissue regeneration [39,40]). This result suggests the potential for the utilization of this sponge in an integrated multi-trophic aquaculture (IMTA) system where it not only improves water quality but also produces collagen.

To optimize the combined use of sponges as producers of natural products and bioremediators, we need to assess the biomass production and associated natural product content and link these to the actual in situ filtration capacity of the model species. In the literature, different terms are used that relate to the filtration activity by sponges. In this paper, we will refer to the terms pumping, filtration, and clearance as follows: pumping is the volumetric water flow rate through the sponge body generated by the aquiferous system of the sponge; filtration and clearance are defined in the same way and refer to the amount of water per unit of time that is entirely cleared of a specific type of component, e.g., bacteria, microalgae, or dissolved organic matter (DOM). We use the term clearance usually in association with the measurements that are applied to determine the clearance/filtration rates. Pumping and clearance/filtration relate to each other through the retention efficiency for the specific component, which means that, for example, clearance rates for bacteria can be different from clearance rates for microalgae. Under in situ conditions in open water, the actual particle uptake rate for a sponge can be calculated by multiplying the ambient concentration by the clearance rate. In this case, a constant ambient particle concentration can be assumed because seawater around the sponge is replenished by water movement before the exhalant water can be inhaled for a second time.

To fully comprehend the role of sponges in the transfer of nutrients and energy in natural and artificial ecosystems such as IMTAs, it is important to understand the physiological and environmental factors that control their ability to take up specific components of interest out of the environment. Some of these factors have already been identified. For example, sponges hosting large quantities of associated microbes (high microbial abundance, or HMA sponges) including *C. reniformis* usually have lower pumping rates than species with low numbers of associated microbes [8]. Additionally, the elevated amounts of suspended sediment in the surrounding water reduce the sponge pumping activity [41]. The ambient flow and sponge morphology may influence the pumping and filtration rates of a sponge by either reducing the amount of energy that is needed for active pumping [42,43] or by altering the efficiency of retaining the particulate matter from the filtered water. Conflicting results have been reported on the relationship between temperature and filtration by sponges [14,44,45], and information on the effects of seasonality on pumping and filtration by sponges is scarce. Very recently, [46] conducted a comprehensive study spanning two annual cycles to assess the factors that regulate the in situ sponge pumping of five Mediterranean sponges (including *C. reniformis*). Unexpectedly, these authors reported no significant effect of temperature and particulate organic matter on the sponge pumping rates and no clear trend of seasonality. Instead, sponge size was found to be the main predictor of volumetric pumping rate, with larger sponges exhibiting a relatively lower pumping rate per unit of body volume.

To date, no studies are available that investigate the role of water depth on sponge pumping and filtration. In maricultures, culture depth may vary between and within locations. The eventual effects of water depth on sponge pumping activity may influence choices for sponge culture sites and sponge culture methodology. In an earlier survey, we observed some physiological differences between *C. reniformis* populations from different depth zones [47]. The osculum density, osculum size, and the associated osculum outflow rate differed significantly between two depth groups (0–3 m and 20–25 m). The bacterial clearance, however, was not measured, so the functional consequences of the observed morphological differences for the filtration/bioremediation capacity of *C. reniformis* remain unknown [47]. Concurrent with potential effects on bacterial clearance and filtration efficiency, water depth may also affect other parameters that are relevant to sponge mariculture, such as sponge growth and the production of secondary metabolites and collagen. Therefore, the aim of this study was to investigate the effect of depth on the filtration capacity (measured as in situ bacterial clearance rates), metabolism (respiration rate as oxygen consumption), morphology (density and size of oscula), growth, and the collagen/biomass production of *C. reniformis*. Up to this date, there is no scientific work that combines the sponge growth performance and related biomaterial content with the natural pumping and filtration rates of sponges (except from 51, who focused on the effect of copper tolerance on the clearance rate). Thus, the current work represents a key step forward, both in terms of understanding the morphological plasticity and performance of sponges and in the application of *C. reniformis* in IMTA for combined bioremediation and collagen production.

2. Results

2.1. General Observations

The water temperatures and salinity values recorded at the site during the transplantation and incubation experiments ranged between 26 and 28 °C and 38.5 and 38.8 g L^{-1}, respectively. The *C. reniformis* explants exhibited a swift recovery. Open surfaces had completely healed five days after sampling and cutting (Figure S1). The visible distinctive inner layers of the sponges were observed to change as early as from day one. A white layer was formed in and around the open surfaces, the cut marks completely disappeared, and the large gaps observed in the mesohyl of some explants were filled completely after 5 days. The explants rapidly reshaped their aquiferous structure, reorganizing the collagen-rich ectosome and mesohyl while regenerating the cut surfaces.

2.2. Natural Sponge Morphology

In sponges along a natural depth gradient, there was a significant decrease in the average osculum size with depth (Pearson's $r = -0.245$, $p = 0.044$; Figure 1a) and a significant increase in the OD (osculum density) with depth (Pearson's $r = 0.444$, $p < 0.001$; Figure 1c). No relationship between the depth and total OSA (osculum surface area) per sponge area was found for natural sponges (Spearman's rho $= 0.117$, $p = 0.335$, Figure 1e). Hence, on average deep water sponges have more yet smaller oscula than shallow water sponges, but the total osculum surface per SSA (sponge surface area) remains the same with increasing depth.

2.3. Effect of Depth and Transplantation on Morphology

For OSA, significant differences between the four experimental groups were found (Table 1). Sponges transplanted from 5 m to 20 m did not reduce their osculum size and had significantly larger oscula compared to the 20 m control explants (Table 1). In contrast, the sponges transplanted from 20 m to 5 m did increase their osculum size, resulting in no significant difference with the 5 m control group (Figure 1b and Table 1). For OD, a similar pattern was found, as sponges transplanted from 5 m to 20 m did not alter their morphology and thus had significantly different ODs from the 20 m control sponges (Figure 1d and Table 1). Sponges transplanted from 20 m to 5 m matched the OD of the 5 m control sponges, with no significant difference found between these groups (Figure 1d and Table 1). One observation to note is the effect of depth on the non-transplanted control sponges. Although not significant here (Table 1), there seems to be a considerable osculum size difference between the 5 and

20 m control groups (0.9 vs. 0.3 cm^2, respectively, Figure 1b). In addition, there is a highly significant difference (Table 1) in the OD between the groups (0.10 vs. 0.24 oscula cm^{-2}, respectively, Figure 1d). Thus, the osculum size is about three times larger at 5 m, whereas the OD is approximately two times lower, which matches the observations on the natural sponges (Figure 1a,c). For the total OSA per sponge area, no significant differences were found between the experimental groups, again in line with natural sponges (Figure 1e,f and Table 1).

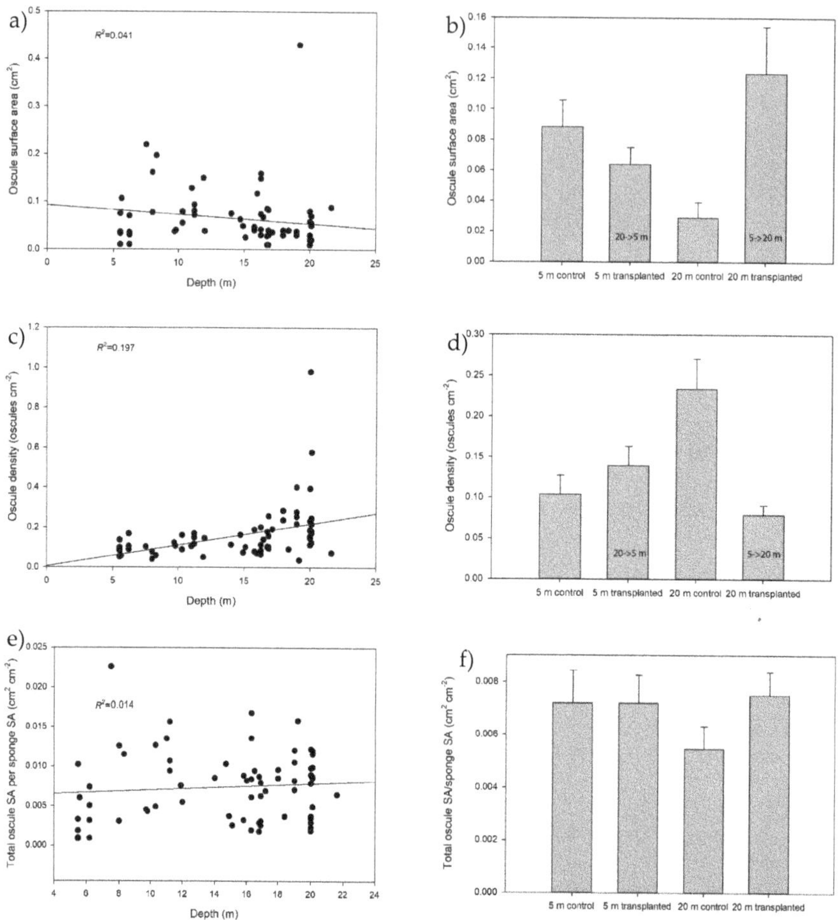

Figure 1. Natural sponges vs. Experimental Sponges. (**a**) Natural—significant decrease of average oscule surface area over depth (Pearson's $r = -0.245$, $p = 0.044$). Dots represent individual sponges. (**b**) Transplanted—oscule surface area (cm^2) measured 8 weeks after transplantation. Values are means + SEM, N = 8–10 per group (20 m control vs. 20 m transplanted F = 3.052, p = 0.005). (**c**) Natural—significant increase of oscule density over depth (Pearson's r =0.444, $p < 0.001$). Dots represent individual sponges. (**d**) Transplanted—oscule density measured 8 weeks after transplantation. Values are means + SEM, N = 8–10 per group (5 m control vs. 20 m control, F = 3.052, p = 0.001; 20 m control vs. 20 m transplanted, F = 3.052, p = 0.000). (**e**) Natural—total oscule surface area (OSA) per unit of sponge surface area (SSA) of natural sponges. (Spearman's rho = 0.117, p = 0.335). Dots represent individual sponges. (**f**) Transplanted—total oscule surface area (OSA) per unit of sponge surface (SSA) measured 8 weeks after transplantation. Values are means + SEM, N = 8–10 per group.

Table 1. One-way ANOVA testing all four experimental groups for differences in the oscule surface area, oscule density, % sponge growth in 68 days of culture, and clearance rate, along with one-way ANOVA testing for the effect of transplantation on the respiration rates (N = 8–10).

Variable	F/t	df	Error	p
Oscule surface area (OSA)				
	3.391	3	30	0.031 *
Planned contrasts with Bonferroni correction				
5 m control versus 20 m control	1.817	1	30	0.079
5 m control versus 5 m transplanted	0.811	1	30	0.424
20 m control versus 20 m transplanted	3.052	1	30	0.005 **
Oscule density (OD)				
	7.322	3	32	0.001 **
Planned contrasts with Bonferroni correction				
5 m control versus 20 m control	−3.557	1	32	0.001 **
5 m control versus 5 m transplanted	−1.012	1	32	0.319
20 m control versus 20 m transplanted	−4.464	1	32	0.000 ***
Total oscule surface area per unit of sponge surface area				
	0.821	3	31	0.492
Clearance rate				
	0.199	3	30	0.896
*Respiration rate *****				
	0.143	1	15	0.711
% growth				
	1.447	3	32	0.248

* p < 0.050, ** p < 0.010, *** p < 0.001, **** respiration rate measured at 5 m only.

2.4. Clearance and Respiration Rates

The mean clearance rates of the experimental sponges were 136.3 ± 21.09 mL cm^{-3} h^{-1} (Table 2). No significant effects of depth and transplantation on the clearance rates were found (Table 1). The respiration rates, only measured at 5 m due to technical limitations, were 0.07 ± 0.01 mg O$_2$ cm^{-3} h^{-1} (Table 2) and did not differ between the transplanted and control sponges (Table 1).

Table 2. Survival, growth, clearance, and respiration rates (%) of experimental sponges measured after 8 weeks in culture. Values are means + SEM.

Transplantation Experiment	Pooled	5 m Control N = 8	5 m Transplant. N = 10	20 m Control N = 10	20 m Transplant. N = 10
(8 weeks in culture)					
Survival Rates (%)	95	80	100	100	100
Growth Rates (%)	63.6 ± 5.4	83.0 ± 13.2	52.5 ± 8.3	60.3 ± 11.1	70.2 ± 10.4
Clearance Rate (mL cm^{-3} h^{-1})	136.3 ± 21.1	164.8 ± 43.9	120.8 ± 53.3	131.4 ± 32.9	136.5 ± 43.9
Respiration Rate (mg O$_2$ cm^{-3} h^{-1}) *	0.07 ± 0.00	0.06 ± 0.00	0.07 ± 0.01		

* Due to the cable reach only measured at 5 m depth.

2.5. Survival and Growth Rates

Both the control and transplanted sponges readily adapted to their new habitat on the incubation plates. Two sponges from the 5 m control group were lost; they disappeared due to unknown causes, but the rest of the explants ($N = 38$) remained at the center of the plates, where they were initially attached and a 95% survival was achieved (Table 2). The overall growth rate was 63.6 ± 5.4% after 8 weeks in culture (Table 2) without significant effects of depth and transplantation (Table 1).

2.6. Collagen Quantification

The collagen yields of the control and transplantation groups are shown in Table 3. The highest extraction yield was observed for the 5 m control group (35.5% of wet mass), whereas the 20→5 m transplanted group had the lowest collagen content (14.5%). The 20 m control and 5→20 m transplanted groups revealed similar collagen yields (18.4% and 21.6%, respectively).

Table 3. Total collagen content (% WM) of control and transplanted sponges after 64 weeks ($N = 1$ per group).

Treatment	Collagen Yield (%)
5 m control	35.5
5 m transplanted	14.5
20 m control	18.4
20 m transplanted	21.6

3. Discussion

The aim of this study was to investigate the effect of depth on the bacterial clearance, morphology, respiration, growth, and collagen production of *C. reniformis*. Our data show that the osculum morphology is depth-dependent, whereas the bacterial clearance rate, respiration, and growth are not. The collagen content also seems to be affected by depth, which will be further elaborated on below.

3.1. Morphology and Bacterial Clearance Rates in Relation to Depth

Along a natural depth transect, both OSA and OD significantly correlated with depth in a reciprocal way, with deeper living sponges having more but smaller oscula. These trends are in line with our earlier survey [47] and with measurements on non-transplanted experimental sponges in the current study. In the earlier survey, the average OSA of sponges in shallow water (0–3 m) was 0.71 cm^2 compared to 0.05 cm^2 in deeper water (20–25 m)—i.e., 14-fold larger. In the current study, the average OSA of the non-transplanted sponges was 0.09 cm^2 at a 5 m depth versus 0.03 cm^2 at a 20 m depth, and so was three-fold larger. The most prominent difference with our previous study was the almost eight times smaller OSA, which may relate to the difference in depth of the "shallow" sponges (0–3 m versus 5 m). We hypothesize that in more shallow water, effects of wave action and the concurrent resuspension of larger particles (e.g., course sand, small pebbles) are much greater than in deeper waters. In order to prevent clogging of their oscula by these particles, sponges in shallow waters need stronger pumping forces to remove these particles from their oscula, which is favoured by fewer but larger oscula. This effect may become most explicit in the surf zone (0–3 m), where wave forces are maximal.

In the current study, the total OSA:SSA ratio remained constant over depth, both in sponges growing along a natural depth gradient and in the non-transplanted experimental sponges at two depths. Hence, the sponges apparently compensate for the increase in OSA by making less oscula, thus maintaining a similar total pumping capacity. We conclude that *C. reniformis* shows morphological plasticity over depth that leads to a depth-independent capacity to filter bacteria from the surrounding water.

Interestingly, the morphological response of *C. reniformis* to transplantation differed between the two transplantation treatments. In agreement with sponges growing along a natural depth gradient, we found that transplantation of sponges from 20 to 5 m resulted in morphological changes, with an increase in OSA and a decrease in OD. Thus, deeper-growing sponges transplanted from 20 to 5 m adapted their morphology to match that of shallow-water individuals. However, the sponges transplanted from 5 to 20 m retained their original morphology. This suggests that the conditions in the wave action area immediately trigger the sponges to adapt their morphology, as described above. Additionally, the higher collagen content of the shallow sponges (Table 2), which makes them more robust, suggests an adaptation to the greater forces imposed on them. Another explanation for the lack of morphological adaption in the deeper sponges is that 8 weeks was insufficient time for the shallow-water sponges to reorganize their aquiferous system. Our own observations on fast healing after the cutting of explants, (Figure S1) and previous findings on morphological plasticity do not support this explanation. In *Dysidea avara*, the closure of oscula and formation of new ones occurred within days to weeks [43]. Moreover, many sponges, including *C. reniformis*, show very fast cell cycling [22] and rapid wound healing [48–50]. An alternative explanation for the lack of response of sponges transplanted to 20 m relates to the hypothesis outlined above, which suggests that the observed morphological plasticity relates to the wave action and corresponding resuspension of larger particles. Since sponges at 5 and 20 m depths showed a similar capacity to clear bacteria, there may be no immediate trigger for the replumbing of the aquiferous system in sponges transplanted from 5 to 20 m, whereas sponges transplanted from 20 to 5 m would have to modify their aquiferous system to cope with the higher coarse sediment loadings. It remains to be investigated whether the sponges transplanted to deeper waters will increase their OD and decrease their OSA over a longer period. To investigate whether and how the internal aquiferous system is affected by depth, histology is recommended for future experiments.

This study is the first to report on a combination of in situ bacterial clearance rates and respiration rates for *C. reniformis*. No earlier data on in situ respiration in this species are available in the literature. The mean respiration rate of 0.07 mg O_2 cm^{-3} h^{-1} (equal to 2.2 ± 0.1 µmol O_2 cm^{-3} h^{-1}) measured in our study falls in the middle of the range of respiration rates for various other sponge species [7,51]. The in situ clearance rates for *C. reniformis* have previously been reported [52] to range from 50 to 340 mL h^{-1}, which appears to be in good agreement with the mean value of 136.3 mL cm^{-3} h^{-1} reported here, but unfortunately [52] did not indicate whether the clearance rates had been normalized to a biomass parameter. Possibilities to compare our clearance data with the literature values for other species are limited due to the variation in the units in which clearance rates are expressed. We normalized the rates obtained in this study to both sponge volume (cm^3) and mg DM, using a conversion factor of 5.68 to convert the volume to DM (Table S1) to enable multiple comparisons. A comparison with the literature data (Table 4) shows that, similar to the respiration rates, the clearance rates measured for *C. reniformis* are also in the middle of the range reported for Mediterranean sponge species. *C. reniformis* is categorized as a High Microbial Abundant or HMA sponge [53]; these are species that host high numbers of bacteria in their bodies. HMA sponges tend to have lower pumping rates than Low Microbial Abundant (LMA) sponges [8]. The bacterial clearance rate reported here (136.3 mL cm^{-3} h^{-1}) is in the same order of magnitude as the average pumping rate of ~300 mL cm^{-3} h^{-1} reported for HMA sponge species (reviewed in [8]). Since clearance rates are always lower than volumetric pumping rates (the difference depending on the retention efficiency of the targeted particles), we conclude that our data are well in line with the literature values.

Table 4. Clearance rates for Mediterranean sponges in literature. DM = Dry Mass.

Species	Method (Incubation)	Sponge Size #	Clearance Rate	Reference
			Ex situ	
Crambe crambe	0.25 h	5 cm^2	506–790 mL g^{-1} DM h^{-1}	Turon et al. 1997, [54]
Dysida avara	0.25 h	5 cm^2	1380–3804 mL g^{-1} DM h^{-1}	Turon et al. 1997, [54]
Chondrilla nucula		25 cm^3	0.2 & 1.4 mL cm^{-3} h^{-1}	Milanese et al. 2003, [15]
Spongia officinalis	4 h, 1 L	91.4 cm^3	34–210 mL g^{-1} DM h^{-1}	Stabili et al. 2006, [17]
Corticium candelabrum	1 h, 1 L,	0.13–18.8 cm^2	1000–10.000 mL g^{-1} DM h^{-1}	de Caralt et al. 2008, [55]
			In situ	
Dysida avara	1 or 3 L, 4 h	25 cm^3	104–2046 mL g^{-1} DM h^{-1}	Ribes et al. 1999a, [14]
Chondrosia reniformis	4 L, 1 h		50–340	Cebrian et al. 2006, [49] *
Ircinia variabilis	1.5 h, 1 L	50 cm^3	15.96 mL cm^{-3} h^{-1}	Ledda et al. 2014, [56]
Agelas oroides	1.5 h, 1 L	35 cm^3	20.0 mL cm^{-3} h^{-1}	Ledda et al. 2014, [56]
Chondrosia reniformis	0.25 h, 6 L	9–49 cm^2	136.3 mL cm^{-3} h^{-1} 639 mL g^{-1} DM h^{-1}	This study

* Unit not specified in the study; # sponge size is expressed either as volume (m^3) or as projected surface area (m^2).

The *C. reniformis* clearance rates were considerably higher than those found for Mediterranean *C. nucula* [15], a closely related species. This may relate to the type of particles that were used for the incubation experiments. To detect clearance in *C. nucula*, cultures of *Escherichia coli* were used [15], which may not resemble the natural populations of bacteria in seawater. All the other studies referred to in Table 4 used natural food assemblages and show considerably higher but also highly variable rates. Sponges can exhibit differential clearance rates for different categories of picoplankton occurring within natural assemblages, and clearance rates can change over the season [55]. Comparative studies such as the current study on the effect of depth should therefore be performed using standardized methodology (i.e., always using the same target particles) within a short time frame.

3.2. Survival, Growth, and Collagen Content

Culturing *C. reniformis* explants has always been problematic as a result of the extreme motility of this species, making them "escape" from the intended location [29–31,57–59]. However, the custom design incubation plates with a protective PVC rim (poly vinyl chloride) preventing them from escaping and a protective chicken wire lid to prevent predation as applied in this study proved to be successful. The setup provided easy handling for clearance rate experiments and growth measurements. The main challenge was the pinning of explants onto the nails during outplanting due to the contraction of the sponges following initial disturbance (sampling and cutting), making their tissue very hard. When this succeeded, attachment was successful and the explants remained in place. The survival rate we achieved (95% after 8 weeks of culture) is considerably higher than the survival in our previous study, 39–86% after 56 weeks [31], or 55% in a study by [59]. In the study reported by [30], the protective stainless-steel cage around explants of *C. reniformis* resulted in comparable survival rates for explants cultured under pristine conditions.

In line with the similar pumping capacity, clearance rates, and respiration rates across the treatment groups, we also did not observe growth differences. This has implications from the perspective of integrated multitrophic aquaculture (IMTA) and bioremediation, as it does not seem to matter at what depth the *C. reniformis* explants are placed in terms of performance. The growth rates observed in this study (52–83% in two months) are in line with our previous results—70–114% in 6 months [31]. In spite of the variable growth rates and peculiarities (growing protrusions, dripping through meshes, etc.) reported by earlier studies [29,57,59], our current and previous studies show that *C. reniformis* remains a good candidate for mariculture [30,31]. Although the growth rate of this species is not as high as that reported for other species such as *Mycale hentcheli* (359–2437% year 1 [60]) or *Lissodendoryx* sp. (5000% month 1 [61]), the high survival rates and reproducible growth rates obtained when applying appropriate methods will enable the controlled production of *C. reniformis* biomass through sea-based aquaculture.

Clearly, *C. reniformis* is a potential source of collagen for biomedical applications in tissue engineering and regenerative medicine because of the unique physicochemical properties of the collagen and the high collagen content [62–65]. Our results are in line with earlier studies that quantified collagen in *C. reniformis*. Swatschek et al. [63] determined a collagen content of 30% of the freeze-dried mass of *C. reniformis*. Considering a wet mass to dry mass ratio of 5.68 (Table S1), this would translate into a collagen yield of 5.3% of the wet mass. More recently, Pozzolini et al. [37] found a similar value (355 mg collagen in 998 mg of dry weight, which equals 30%) using the methodology that was also applied in the current paper. Our yield of 14.5–35.5% wet mass (Table 3) is much higher. Possibly, the fact that we removed excess water during the homogenization of the sponge before weighing may have resulted in a relatively lower wet mass and thus a relatively higher wet mass-based collagen content. The results show, however, that the cultured materials and the methodology to extract collagen are suitable to collagen for pilot-scale processes. Additionally, the developed approach enables further collagen yield optimization, e.g., via the further investigation of genotype-specific yields and the effects of specific depths and other environmental variables on collagen content.

In conclusion, the filtration capacity, metabolism, and biomass production of *C. reniformis* are not affected by depth, in contrast to morphology and collagen content. Osculum morphology clearly is depth-dependent, where sponges transplanted from 5 m to 20 m reflect the conspecifics at their origin (do not adapt), whereas sponges transplanted from 20 m to 5 m reflect the conspecifics at their destination (adaptation). This adaptation to shallow water might relate to wave action and sediment loading. The sponges maintain their filtering capacity by presumably re-shaping their aquiferous system depending on their needs. This maintained filtering capacity is very promising for the future multifunctional application of *C. reniformis* to improve water quality by filtering out several types of organic pollution, including feces and unused feed from fish farms and pollution and microorganisms from urban sewage outlets.

4. Materials and Methods

4.1. Study Location and Seawater Parameters

The study was conducted over a period of 79 days during July–October 2018 at Pina Reef, a location within the Kaş-Kekova Special Environmental Protected Area (SEPA), Turkey (Figure 2). Pina Reef is located at the eastern side of Five-Islands, 4.3 km south of Kaş, and is exposed to wind and wave action originating from the open sea (west direction). The Pina reef wall is located right next to a 350 m-long wall structure located transversely Northwest–Southeast between 14 to 32 m depth. The Pina small reef, located 70 m south of the Pina reef wall, is a reef shoal with a depth ranging from 3.8 to 22 m, with adjacent Posidonia sea grass meadows in the south and east directions. The water temperature and salinity during the study period were measured by a multimeter during the clearance rate and respiration analysis (Multi 3620 IDS with TetraCon 959 and FDO 925 sensors, WTW, Weilheim, Germany).

4.2. Sponge Collection, Seeding, and Transplantation

The sponge specimens for the experiments were collected from 5 and 20 m (20 per depth, 40 in total), cut into pieces of 3–4 cm following the method described in Gökalp et al. [31] and fixed onto grey PVC plates with a rim with a radius of 17 cm and 2.7 cm high. Each PVC plate had a 15 cm-high cylindrical protective PVC rim (Figure 3) and was covered with 2 cm mesh size chicken wire in order to eliminate predation from larger marine life (e.g., sea turtle intrusion on the sponges in culture with boxes lacking protection; personal observation). The protective rim also prevented the sponges from migrating off the plate and prevented them from being carried away by occasional strong water currents [31].

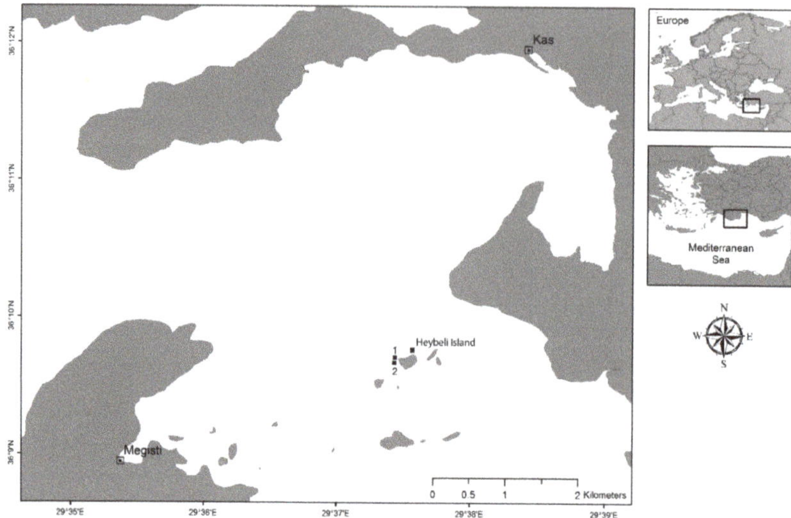

Figure 2. Map of Kaş-Kekova Special Environmental Protected Area. Pina Reef diving location is located at the eastern tip of Heybeli Island. The exact locations of experiments; (1) Pina reef wall (20 m depth), (2) Pina small reef (5 m depth) (36°09′42.4″ N 29°37′26.5″ E; 36°09′40.2″ N 29°37′26.3″ E; respectively).

Figure 3. Incubation chambers used in bacterial clearance and respiration experiments. The oxygen probe was connected to a surface multimeter with a Kevlar-protected 20 m cable (WTW, Germany). Four O-rings ensured a water-tight incubation chamber. The first one was located around the black engine and battery compartment, sealing the lid. The second one was in the oxygen sensor lid. A third one was located around the white lid, sealing the transparent acrylic chamber. The final one was located inside the grey base-rim, sealing the chamber. At the end of the black engine and battery compartment, a magnetic stirrer was located.

Each plate carried a chromium nail at the center with the sharp edge facing upwards (4 cm in height) to secure the specimens to the plate. After fixing the explants on pre-labelled PVC plates, they were left at their respective depth of origin (either 5 or 20 m) for 5 days to provide sufficient time to recover and attach to the PVC plates [49]. Subsequently, the PVC plates were transplanted by moving 10 plates from 5 to 20 m depth and 10 plates from 20 to 5 m depth (Figure 4; N = 10). Following transplantation, the sponges were left to acclimatize for seven days before the incubations started.

Figure 4. Overall view of the transplant and control groups: Group A—5 m control; Group B—5 m transplanted; Group C—20 m control; Group D—20 m transplanted. For the experiment, 10 specimens from 5 m moved to 20 m, and 10 specimens from 20 m moved to 5 m (50 m distance to each other). The other remaining sponges were kept at their depths as control, resulting in four groups of N = 10 sponges per group.

4.3. Determination of Sponge Size, Growth, and Morphological Characteristics

The size of the sponges and the number and size of the oscula were measured following the acclimatization period. Simultaneously, the sponges were photographed from the top and four sides with a ruler for scale using a Canon Eos 5D Mark IV camera and an Ikelite 5D housing setup. The sponge surface area (SSA) was measured from the projected surface area based on a picture taken from the top using the ImageJ software. The pictures from the sides were used to determine the average height of the sponge. Multiplying the measured average height with the SSA provided the approximate volume of the sponge (Vsponge), assuming a cylindrical shape. At the end of the experimental period, 5 sponges per group were weighed (wet mass, WM) and the volume was measured by water displacement to determine the accuracy of the volume calculation. To determine the wet mass (WM) to dry mass (DM) ratio, 6 individuals of the same species that were not used for the experiment were weighed, dried, and reweighed (Table S1). These size measurements were used (1) to determine the growth (using $\frac{t1-t0}{t0} \times 100\%$ to calculate the % increase in SSA over 8 weeks [31]) and (2) to normalize the metabolic rates to a biomass parameter (volume). The average osculum size (hereinafter referred to as the osculum surface area (OSA), sponge surface area (SSA), and osculum density (OD, the number of oscula per unit of sponge surface)) were determined for each of the 40 experimental sponges, hereby using the pictures taken for size measurements. From these parameters, the osculum number to sponge size ratio and the ratio of the oscular surface area to sponge surface area (OSA/SSA) were calculated. In addition, the multiplication of the OD with OSA provides the pumping potential of a sponge by expressing the total OSA per SSA. For comparison, the same parameters were determined for a series of *C. reniformis* specimens occurring along a natural depth gradient ranging from 2 to 25 m in depth with scuba diving at several sites within 1 km of the experimental locations.

4.4. Clearance and Respiration Rates

The circular rims on the square PVC plates (Figure 3) were designed to fit airtight to a transparent PVC chamber with a 6.75 L inner volume and were specifically developed for in situ clearance and respiration measurements (Wageningen University, Gelderland, The Netherlands). The upper side of the cylinder could be closed with a lid in which a magnetic stirrer with a battery pack (Jansen Tholen B.V., Tholen, The Netherlands) was mounted. The stirrer could create sufficient water circulation and the continuous mixing of seawater in order to equalize the oxygen distribution within the cylinder and prevent the particles from settling onto the bottom plate. Each lid also was fitted with two diaphragms for water sampling and an opening for an oxygen probe (WTW, Germany). Together, the PVC plate with rim, cylinder, and lid formed a water-tight incubation chamber (Figure 3). During the incubation experiments, the protective cylinder and lid of chicken wire were removed and replaced by the water-tight incubation chamber setup. The chambers were applied for the simultaneous in situ measurement of the bacterial clearance and respiration of the experimental sponges. Eight weeks after transplantation, all 40 specimens in the experiment were incubated to determine their bacterial clearance and respiration rate. A total 10 empty PVC plates were incubated to determine the background activity not associated with the sponges. Prior to the incubations, the chicken wire and the protective rim were carefully removed from the bottom plates, thus minimizing stress to the sponge specimens. The deposited sediment and live organisms accumulated on the O-rings and PVC plates were removed with a toothbrush and no organic matter was left on the PVC plate inside the base rim. Then, the cylinders were secured onto the PVC plates and left to acclimatize for 15 min before the incubations. After the acclimatization period, the chamber lid was closed, and using a syringe the first 10 mL water sample was taken directly (t0) and a second 10 mL water sample was taken 15 min later (t1). A total of 15 min was deemed sufficient time to obtain reliable results for both the oxygen depletion and bacterioplankton decrease, considering the size of the containers and the reported bacterial grazing rates and respiration rates from the literature [15,49,64]. The collected water samples were labelled right after surfacing from the dive, immediately fixed with 0.57 mL of 35% formaldehyde solution to a final concentration of 2–4%, and placed in ice containers. At the end of each incubation day, the fixed samples were filtered over white 25 mm diameter 0.2 μm polycarbonate membranes (GE Osmonics, Minneapolis, MN, USA), which were structurally supported by a 0.45 μm GF–F-type membrane (Whatman International Ltd., Maidstone, England, UK). Subsequently, the filters were air-dried in the laboratory for at least 15 min and stored at –20 °C in Eppendorf tubes until further use. To determine the total bacterial counts, the filters were placed on a microscopic slide and 10 μL of DAPI mix was added to stain the bacteria present, as described previously in de Goeij et al. (2007). The bacterial numbers were determined based on pictures taken under a fluorescence microscope (Leica DM6 B; Leica Microsystems, Wetzlar, Germany). Per filter, 10 images were taken using the Leica-LasX software (V3.3.0. Leica Microsystems. Wetzlar, Germany) and a DFC365 FC camera (Leica Microsystems, Germany). From the pictures, the average bacterial numbers were determined by counting 10 fields or up to a maximum number of 250 bacteria by using ImageJ. The clearance rates were calculated using the following formula (modified from [56]):

$$CR = V_{chamber} \cdot V_{sponge}^{-1} \cdot (rate\ sponge - rate\ blank), \tag{1}$$

in which $V_{chamber}$ stands for the volume of water in the incubation chamber (milliliter) and V_{sponge} is the volume of the sponge. The rate of bacterial concentration change for the sponge (rate sponge) and the blank chambers (rate blank) is calculated as:

$$\text{Cell change rate} = \frac{\ln(C0) - \ln(Ct)}{t},\tag{2}$$

in which $C0$ and Ct are the concentration of counted bacteria per milliliter at ($t0$) and ($t1$) and t is the total incubation time (15 min).

To determine the respiration rates, the oxygen concentrations in the chambers were measured during the incubations with a Multi 3620 IDS (WTW, Germany) connected with 20 m cables to two FDO925 probes (WTW, Germany). During the incubations, the oxygen values were logged every 10 s from a boat, which was anchored right on top of the divers to release a sufficient amount of cable underwater. The respiration rates were only measured for the sponges at 5 m depth due to the limited reach of the WTW cables and prolonged diving times. Linear graphs were fitted to the measurements, and the respiration rates were calculated with the slope of the graphs, which represent the oxygen decline in mg $O_2 \cdot L^{-1} \cdot min^{-1}$. The respiration rates (RR) were calculated with the following formula:

$$RR = (O_2 \text{ slope sponge} - O_2 \text{ slope blank}) \cdot V_{chamber} \cdot V_{sponge}^{-1}.\tag{3}$$

4.5. Collagen Extraction

After 8 weeks, 3 to 5 sponges from each of the four experimental groups were randomly collected and immediately frozen at −18 °C following the dives. The frozen samples were transported in dry ice containers to the facilities of University of Minho, Portugal, and kept at −20 °C until further analysis. All the procedures described below were performed separately with sponges pooled per experimental treatment to obtain enough material for the analysis. The samples were thawed and any exogenous materials on the sponges were removed by rinsing in dH$_2$O. All the steps for collagen extraction were carried out at 4 °C. Next, the wet weight was determined and the sponges were cut into small pieces of roughly 1 × 1 × 1 mm. Excess water was poured from the marine sponge samples, and 5 sponge volumes of disaggregating solution (50 mM Tris–HCl buffer pH 7.4, 1M NaCl, 50 mM EDTA and 100 mM 2-mercaptoethanol) were added and left under stirring for 4 days. The collagen solutions (CS) were filtered with a nylon mesh to remove any remaining undissolved fragments and the solution was extensively dialyzed in dialysis tubing cellulose membrane for 7 days with 2 dialyzing buffer changes every per day (CS/dialyzing buffer ratio 1:1000) against dH$_2$O to remove all traces of 2-mercaptoethanol. The suspensions were centrifuged for 10 min at 1200 g to further remove cell debris and sand particles. To collect the collagen from the suspension, another centrifugation followed for 30 min at 12,100 g, yielding pellets containing collagen which then were resuspended in dH$_2$O. The collagen re-extraction was performed by repeating the second centrifugation step. The collagen solutions were stored at 4 °C. The total collagen content was determined by freeze-drying and weighing (dry mass) the extracted material.

4.6. Collagen Quantification

Following the collagen extraction, the obtained collagen solutions were analyzed regarding the collagen content. To determine the total amount of collagen extracted, each collagen solution was freeze-dried and then weighted (dry mass). The wet collagen extraction yield was determined using Equation (4):

$$\text{Yield of collagen (wet) (\%)} = \frac{\text{Mass of collagen } (mg)}{\text{Wet mass C. reniformis } (mg)}) \times 100.\tag{4}$$

4.7. Data Analysis

All the data were tested for the normality and homogeneity of variances. A one-way ANOVA was performed to test for differences between the four treatment groups in terms of sponge morphology, clearance rate, and growth. The respiration rates of the transplanted and control sponges at 5 m depth were also compared with a one-way ANOVA. Planned contrasts with Bonferroni correction (i.e., using a corrected α for significance) were used to follow-up significant ANOVAs. Pearson's product-moment correlations were used to correlate the osculum size and density of the natural sponges with depth, whereas Spearman's rho for the total oscula surface area (SA) per unit of sponge surface area (SA) was used when the data were not normally distributed. A statistical analysis was performed using SPSS 25 (IBM SPSS Statistics for Windows, Version 25.0. Armonk, NY, USA: IBM Corp.); graphs were plotted with Sigmaplot 12 (Systat Software, San Jose, CA, USA).

Supplementary Materials: The following are available online at http://www.mdpi.com/1660-3397/18/7/358/s1: Figure S1: Healing phases of *C. reniformis* specimens following the sampling/cut; Table S1: Volumes, WM and DM from 6 randomLy collected *C. reniformis* specimen.

Author Contributions: Conceptualization, M.G.; Data curation, M.G., T.K. and T.W.; Formal analysis, M.G., T.K., M.S.R. and R.O.; Funding acquisition, T.H.S., R.O., A.J.M. and T.W.; Investigation, M.G. and T.K.; Methodology, M.G., T.K., M.S.R., T.H.S., R.O. and T.W.; Project administration, M.G., T.H.S., R.O. and A.J.M.; Resources, M.G., T.H.S., R.O. and T.W.; Software, M.G., T.K. and T.W.; Supervision, T.H.S., R.O., A.J.M. and T.W.; Validation, M.G., T.K. and T.W.; Visualization, M.G., T.K. and T.W.; Writing—original draft, M.G.; Writing—review and editing, M.G., T.K., M.S.R., R.O., A.J.M. and T.W. All authors have read and agreed to the published version of the manuscript.

Funding: This research was executed within the Connected Circularity program, financed by strategic funding of Wageningen University and Research and the knowledge base of the Ministry of Agriculture, Nature, and Food Quality (KB40), and was part of the ERA-NET project Biogenink (project 4195), funded by the European Commission in conjunction with the Dutch Science Foundation NWO and the Portuguese Foundation for Science and Technology (FCT) (project M-ERA-NET-2/0022/2016).

Acknowledgments: Special thanks to Marretje Adriaanse, Marlin Ter Huurne, Anne Top, Efecan Toker, Kemal Akçor, and Ellen van Marrewijk for scientific diving support during the incubation experiments; to KASAD, Dragoman Diving and Outdoor, and Kaş Adventure Diving for logistics and diving support; to Ozan Atabilen, Bora Kolbay, Mertcan Kırgız, Okan Avcı, and Melis Uman for technical diving support; to Serdar Taşkan, Orhan Batuhan Özyurt, and Uğur Gökberk Aytuğ for boating support; to Murat Draman, Tuba Atabilen, Çağatay Arıcan, Bora Ömeroğulları, Murat Kabaş, Murat Baykara, Orhun Can Varol, Çağla Çorumluoğlu, Namık Dikbaş, Aleyna Su Büyüktepe, Kenan Verbakel, and Çağla Karaalifor diving logistics; and to the Turkish Coast Guard Command and District Governorate of Kaş for the necessary permissions and security.

Conflicts of Interest: The authors declare no conflict of interest. The funders had no role in the design of the study; in the collection, analysis, or interpretation of data; in the writing of the manuscript; or in the decision to publish the results.

References

1. Hooper, J.; van Soest, R. *Systema Porifera: A Guide to the Classification of Sponges*; Springer: Boston, MA, USA, 2002. [CrossRef]

2. Van Soest, R.W.M.; Boury-Esnault, N.; Vacelet, J.; Dohrmann, M.; Erpenbeck, D.; De Voogd, N.J. Global diversity of sponges (Porifera). *PLoS ONE* **2012**, *7*, e35105. [CrossRef] [PubMed]

3. Reiswig, H.M. In situ pumping activities of tropical Demospongiae. *Mar. Biol.* **1971**, *9*, 38–50. [CrossRef]

4. Reiswig, H.M. The aquiferous systems of three marine Demospongiae. *J. Morphol.* **1975**, *145*, 493–502. [CrossRef] [PubMed]

5. Bell, J. Functional roles of sponges. *Estuar. Estuar. Coast. Shelf Sci.* **2008**, *79*, 341–353. [CrossRef]

6. Ribes, M.; Jiménez, E.; Yahel, G.; López-Sendino, P.; Diez, B.; Massana, R.; Sharp, J.H.; Coma, R. Functional convergence of microbes associated with temperate marine sponges. *Environ. Microbiol.* **2012**, *14*, 1224–1239. [CrossRef] [PubMed]

7. Osinga, R.; Tramper, J.; Wijffels, R. Cultivation of Marine Sponges. *Mar. Biotechnol.* **1999**, *1*, 509–532. [CrossRef]

8. Weisz, J.B.; Lindquist, N.; Martens, C.S. Do associated microbial abundances impact marine demosponge pumping rates and tissue densities? *Oecologia* **2008**, *155*, 367–376. [CrossRef]

9. Lavy, A.; Keren, R.; Yahel, G.; Ilan, M. Intermittent hypoxia and prolonged suboxia measured in situ in a marine sponge. *Front. Mar. Sci.* **2016**, *3*, 263. [CrossRef]

10. Ludeman, D.A.; Reidenbach, M.A.; Leys, S.P. The energetic cost of filtration by demosponges and their behavioural response to ambient currents. *J. Exp. Biol.* **2017**, *220*, 4743–4744. [CrossRef]

11. Jørgensen, C.B. Feeding-rates of sponges, lamellibranchs and ascidians. *Nature* **1949**, *163*, 912. [CrossRef]

12. Jørgensen, C.B. Quantitative aspects of filter feeding in invertebrates. *Biol. Rev.* **1955**, *30*, 391–453. [CrossRef]

13. Riisgård, H.U.; Larsen, P. Comparative ecophysiology of active zoobenthic filter feeding, essence of current knowledge. *J. Sea Res.* **2000**, *44*, 169–193. [CrossRef]

14. Ribes, M.; Coma, R.; Gili, J. Natural diet and grazing rate of the temperate sponge *Dysidea avara* (Demospongiae, Dendroceratida) throughout an annual cycle. *Mar. Ecol. Prog. Ser.* **1999**, *176*, 179–190. [CrossRef]

15. Milanese, M.; Chelossi, E.; Manconi, R.; Sara, A.; Sidri, M.; Pronzato, R. The marine sponge *Chondrilla nucula* Schmidt, 1862 as an elective candidate for bioremediation in integrated aquaculture. *Biomol. Eng.* **2003**, *20*, 363–368. [CrossRef]

16. Fu, W.T.; Sun, L.M.; Zhang, X.C.; Zhang, W. Potential of the marine sponge *Hymeniacidon perlevis* as a bioremediator of pathogenic bacteria in integrated aquaculture ecosystems. *Biotechnol. Bioeng.* **2006**, *93*, 1112–1122. [CrossRef]

17. Stabili, L.; Mercurio, M.; Marzano, C.N.; Corriero, G. Filtering activity of *Spongia officinalis* var. *adriatica* (Schmidt) (Porifera, Demospongiae) on bacterioplankton: Implications for bioremediation of polluted seawater. *Water Res.* **2006**, *40*, 3083–3090. [CrossRef] [PubMed]

18. Zhang, X.; Zhang, W.; Xue, L.; Zhang, B.; Jin, M.; Fu, W. Bioremediation of bacteria pollution using the marine sponge *Hymeniacidon perlevis* in the intensive mariculture water system of turbot *Scophthalmus maximus*. *Biotechnol. Bioeng.* **2010**, *105*, 59–68. [CrossRef] [PubMed]

19. Pile, A.; Patterson, M.; Witman, J. In situ grazing on plankton 10 μm by the boreal sponge Mycale lingua. *Mar. Ecol. Prog. Ser.* **1996**, *141*, 95–102. [CrossRef]

20. Hadas, E.; Marie, D.; Shpigel, M.; Ilan, M. Virus predation by sponges is a new nutrient-flow pathway in coral reef food webs. *Limnol. Oceanogr.* **2006**, *51*, 1548–1550. [CrossRef]

21. Yahel, G.; Sharp, J.H.; Marie, D. In situ feeding and element removal in the symbiont–bearing sponge *Theonella swinhoei*: Bulk DOC is the major source for carbon. *Limnol. Oceanogr.* **2003**, *48*, 141–149. [CrossRef]

22. De Goeij, J.M.; van Duyl, F.C. Coral cavities are sinks of dissolved organic carbon (DOC). *Limnol. Oceanogr.* **2007**, *52*, 2608–2617. [CrossRef]

23. Alexander, B.E.; Liebrand, K.; Osinga, R.; van der Geest, H.G.; Admiraal, W.; Cleutjens, J.P.M. Cell turnover and detritus production in marine sponges from tropical and temperate benthic ecosystems. *PLoS ONE* **2014**, *9*, e109486. [CrossRef]

24. Mueller, B.; de Goeij, J.M.; Vermeij, M.J.A.; Mulders, Y.; van der Ent, E.; Ribes, M. Natural diet of coral–excavating sponges consists mainly of dissolved organic carbon (DOC). *PLoS ONE* **2014**, *9*, e90152. [CrossRef] [PubMed]

25. Gili, J.M.; Coma, R. Benthic suspension feeders: Their paramount role in littoral marine food webs. *Trends Ecol. Evol.* **1998**, *13*, 316–321. [CrossRef]

26. Perez-Blázquez, A.; Davy, S.K.; Bell, J.J. Estimates of particulate organic carbon flowing from the pelagic environment to the benthos through sponge assemblages. *PLoS ONE* **2012**, *7*, e29569. [CrossRef]

27. Maldonado, M.; Ribes, M.; van Duyl, F.C. Nutrient fluxes through sponges: Biology, budgets, and ecological implications. *Adv. Mar. Biol.* **2012**, *62*, 113–182. [CrossRef] [PubMed]

28. De Goeij, J.M.; van Oevelen, D.; Vermeij, M.J.A.; Osinga, R.; Middelburg, J.J.; de Goeij, A.F.P.M.; Admiraal, W. Surviving in a marine desert: The sponge loop retains resources within coral reefs. *Science* **2013**, *342*, 108–110. [CrossRef]

29. Pronzato, R.; Bavestrello, G.; Cerrano, C.; Magnino, G.; Manconi, R.; Pantelis, J.; Sarà, A.; Sidri, M. Sponge farming in the Mediterranean Sea: New perspectives. *Mem. Qld. Mus.* **1999**, *44*, 485–491.

30. Osinga, R.; Sidri, M.; Cerig, E.; Gokalp, S.Z.; Gokalp, M. Sponge Aquaculture Trials in the East–Mediterranean Sea: New Approaches to Earlier Ideas. *Open Mar. Biol. J.* **2010**, *4*, 74–81. [CrossRef]

31. Gökalp, M.; Wijgerde, T.; Sarà, A.; de Goeij, J.M.; Osinga, R. Development of an Integrated Mariculture for the Collagen–Rich Sponge *Chondrosia reniformis*. *Mar. Drugs* **2019**, *17*, 29. [CrossRef]

32. Naylor, R.; Goldburg, R.; Primavera, J. Effect of aquaculture on world fish supplies. *Nature* **2000**, *405*, 1017–1024. [CrossRef]

33. Sarà, G.; Scilipoti, D.; Mazzola, A.; Modica, A. Effects of fish farming waste to sedimentary and particulate organic matter in a southern Mediterranean area (Gulf of Castellammare, Sicily): A multiple stable isotope study (??13C and ??15N). *Aquaculture* **2004**, *234*, 199–213. [CrossRef]

34. Nimmo, F.; Cappell, R.; Huntington, T.; Grant, A. Does fish farming impact on tourism in Scotland? *Aquac. Res.* **2011**, *42*, 132–141. [CrossRef]

35. Aguilar-Manjarrez, J.; Soto, D.; Brummett, R. *Aquaculture Zoning, Site Selection and Area Management under the Ecosystem Approach to Aquaculture*; A handbook; Report ACS18071; Rome, FAO, and World Bank Group: Washington, DC, USA, 2017.

36. Silva, J.; Barros, A.; Aroso, I.; Fassini, D.; Silva, T.H.; Reis, R.L.; Duarte, A. Extraction of Collagen/Gelatin from the Marine Demosponge *Chondrosia reniformis* (Nardo, 1847) Using Water Acidified with Carbon Dioxide—Process Optimization. *Ind. Eng. Chem. Res.* **2016**, *55*, 6922–6930. [CrossRef]

37. Pozzolini, M.; Scarfi, S.; Gallus, L.; Castellano, M.; Vicini, S.; Cortese, K.; Gagliani, M.C.; Bertolino, M.; Costa, G.; Giovine, M. Production, Characterization and Biocompatibility Evaluation of Collagen Membranes Derived from Marine Sponge *Chondrosia reniformis* Nardo, 1847. *Mar. Drugs* **2018**, *16*, 111. [CrossRef]

38. Fassini, D.; Duarte, A.R.C.; Reis, R.L.; Silva, T.H. Bioinspiring *Chondrosia reniformis* (Nardo, 1847) Collagen-Based Hydrogel: A New Extraction Method to Obtain a Sticky and Self-Healing Collagenous Material. *Mar. Drugs.* **2017**, *15*, 380. [CrossRef]

39. Silva, T.H.; Moreira-Silva, J.; Marques, A.L.; Domingues, A.; Bayon, Y.; Reis, R.L. Marine origin collagens and its potential applications. *Mar. Drugs.* **2014**, *12*, 5881–5901. [CrossRef]

40. Chattopadhyay, S.; Raines, R.T. Review collagen-based biomaterials for wound healing. *Biopolymers* **2014**, *101*, 821–833. [CrossRef]

41. Gerrodette, T.; Flechsig, A.O. Sediment–induced reduction in the pumping rate of the tropical sponge *Verongia lacunosa*. *Mar. Biol.* **1979**, *55*, 103–110. [CrossRef]

42. Vogel, S. Current-Induced Flow through the Sponge, Halichondria. *Biol. Bull.* **1974**, *147*, 443–456. [CrossRef]

43. Mendola, D.; De Caralt, S.; Uriz, M.; Van den End, F.; Van Leeuwen, J.L.; Wijffels, R.H. Environmental Flow Regimes for *Dysidea avara* Sponges. *Mar. Biotechnol.* **2008**, *10*, 622–630. [CrossRef] [PubMed]

44. Frost, T.M. Clearance rate determinations for the fresh–water sponge *Spongilla–Lacustris* effects of temperature, particle type and concentration and sponge size. *Arch. Hydrobiol.* **1978**, *90*, 330–356.

45. Riisgård, H.U.; Thomassen, S.; Jakobsen, H.; Weeks, J.M.; Larsen, P.S. Suspension feeding in marine sponges *Halichondria panicea* and *Haliclona urceolu*: Effects of temperature on filtration rate and energy cost of pumping. *Mar. Ecol. Prog. Ser.* **1993**, *96*, 177–188. [CrossRef]

46. Morganti, T.M.; Ribes, M.; Yahel, G.; Coma, R. Size Is the Major Determinant of Pumping Rates in Marine Sponges. *Front. Physiol.* **2019**, *10*, 1474. [CrossRef]

47. Gökalp, M.; Kuehnhold, H.; De Goeij, J.M.; Osinga, R. Depth and turbidity affect in situ pumping activity of the Mediterranean sponge *Chondrosia reniformis* (Nardo, 1847). *BioRxiv* **2020**, *2020*, 009290. [CrossRef]

48. Alexander, B.E.; Achlatis, M.; Osinga, R.; van der Geest, H.G.; Cleutjens, J.P.M.; Schutte, B.; de Goeij, J.M. Cell kinetics during regeneration in the sponge *Halisarca caerulea*: How local is the response to tissue damage? *PeerJ* **2015**, *3*, e820. [CrossRef]

49. Nickel, M.; Brümmer, F. In vitro sponge fragment culture of Chondrosia reniformis (Nardo, 1847). *J. Biotechnol.* **2003**, *100*, 147–159. [CrossRef]

50. Pozzolini, M.; Gallus, L.; Ghignone, S. Insights into the evolution of metazoan regenerative mechanisms: Roles of TGF superfamily members in tissue regeneration of the marine sponge *Chondrosia reniformis*. *J. Exp. Biol.* **2019**, *222*, jeb207894. [CrossRef]

51. Morganti, T.M. In Situ Direct Study of filtration and Respiration Rate of Mediterranean Sponges. Ph.D. Thesis, Universidad Politécnica de Cataluña, Barcelona, Spain, 2015. Available online: https://digital.csic.es/bitstream/10261/141995/1/Morganti_Tesis_2016.pdf (accessed on 27 July 2016).

52. Cebrian, E.; Agell, G.; Martí, R.; Uriz, M.J. Response of the Mediterranean sponge *Chondrosia reniformis* Nardo to copper pollution. *Environ. Pollut.* **2006**, *141*, 452–458. [CrossRef]

53. Erwin, P.M.; Coma, R.; López-Sendino, P.; Serrano, E.; Ribes, M. Stable symbionts across the HMA-LMA dichotomy: Low seasonal and interannual variation in sponge-associated bacteria from taxonomically diverse hosts. *FEMS Microbiol. Ecol.* **2015**, *91*, 10. [CrossRef]

54. Turon, X.; Galera, J.; Uriz, M.J. Clearance rates and aquiferous systems in two sponges with contrasting life-history strategies. *J. Exp. Zool.* **1997**, *278*, 22–36. [CrossRef]

55. Caralt, S.D.; Uriz, M.J.; Wijffels, R.H. Grazing, differential size-class dynamics and survival of the Mediterranean sponge *Corticium candelabrum*. *Mar. Ecol. Prog. Ser.* **2008**, *360*, 97–106. [CrossRef]

56. Ledda, F.D.; Pronzato, R.; Manconi, R. Mariculture for bacterial and organic waste removal: A field study of sponge filtering activity in experimental farming. *Aquac. Res.* **2014**, *45*, 1389–1401. [CrossRef]

57. Wilkinson, C.; Vacelet, J. Transplantation of marine sponges to different conditions of light and current. *J. Exp. Mar. Biol. Ecol.* **1979**, *37*, 91–104. [CrossRef]

58. Garrabou, J.; Zabala, M. Growth dynamics in four Mediterranean Demosponges. *Estuar. Coast. Shelf Sci.* **2001**, *52*, 293–303. [CrossRef]

59. Van Treeck, P.; Eisinger, M.; Muller, J.; Paster, M.; Schuhmacher, H. Mariculture trials with Mediterranean sponge species—The exploitation of an old natural resource with sustainable and novel methods. *Aquaculture* **2003**, *218*, 439–455. [CrossRef]

60. Page, M.J.; Handley, S.J.; Northcote, P.T.; Cairney, D.; Willan, R.C. Successes and pitfalls of the aquaculture of the sponge *Mycale hentscheli*. *Aquaculture* **2011**, *312*, 52–61. [CrossRef]

61. Battershill, C.N.; Page, M.J. Sponge aquaculture for drug production. *Aquac. Update.* **1996**, 5–6.

62. Garrone, R.; Huc, A.; Junqua, S. Fine structure and physicochemical studies on the collagen of the marine sponge *Chondrosia reniformis* Nardo. *J. Ultrastruct. Res.* **1975**, *52*, 261–275. [CrossRef]

63. Swatschek, D.; Schatton, W.; Kellerman, J.; Muller, W.E.G.; Kreuter, J. Marine sponge collagen: Isolation, characterization and effects on the skin parameters surface–pH, moisture and sebum. *Eur. J. Pharm. Biopharm.* **2002**, *53*, 107–113. [CrossRef]

64. Wilkie, I.C.; Parma, L.; Bonasoro, F.; Bavestrello, G.; Cerrano, C.; Candia Carnevali, M.D. Mechanical adaptability of a sponge extracellular matrix: Evidence for cellular control of mesohyl stiffness in *Chondrosia reniformis* Nardo. *J. Exp. Biol.* **2006**, *209*, 4436–4443. [CrossRef] [PubMed]

65. Fassini, D.; Parma, L.; Wilkie, L.; Bavestrello, G.; Bonasoro, F.; Candia, D. Ecophysiology of mesohyl creep in the demosponge *Chondrosia reniformis* (Porifera: Chondrosida). *J. Exp. Mar. Biol. Ecol.* **2012**, *428*, 24–31. [CrossRef]

*marine drugs*

Article

Electrochemical Approach for Isolation of Chitin from the Skeleton of the Black Coral *Cirrhipathes* sp. (Antipatharia)

Krzysztof Nowacki [1,]*, Izabela Stępniak [1,]*, Enrico Langer [2], Mikhail Tsurkan [3], Marcin Wysokowski [4,5], Iaroslav Petrenko [5], Yuliya Khrunyk [6,7], Andriy Fursov [5], Marzia Bo [8], Giorgio Bavestrello [8], Yvonne Joseph [5] and Hermann Ehrlich [5,9,]*

1   Institute of Chemistry and Technical Electrochemistry, Faculty of Chemical Technology, Poznan University of Technology, ul. Berdychowo 4, 60965 Poznan, Poland
2   Institute of Semiconductors and Microsystems, TU Dresden, 01062 Dresden, Germany; enrico.langer@tu-dresden.de
3   Leibniz Institute of Polymer Research Dresden, 01069 Dresden, Germany; tsurkan@ipfdd.de
4   Institute of Chemical Technology and Engineering, Faculty of Chemical Technology, Poznan University of Technology, Berdychowo 4, 60965 Poznan, Poland; marcin.wysokowski@put.poznan.pl
5   Institute of Electronics and Sensor Materials, TU Bergakademie Freiberg, Gustav-Zeuner str. 3, 09599 Freiberg, Germany; iaroslavpetrenko@gmail.com (I.P.); andriyfur@gmail.com (A.F.); yvonne.joseph@esm.tu-freiberg.de (Y.J.)
6   Department of Heat Treatment and Physics of Metal, Ural Federal University, Mira Str. 19, 620002 Ekaterinburg, Russia; juliakhrunyk@yahoo.co.uk
7   The Institute of High Temperature Electrochemistry of the Ural Branch of the Russian Academy of Sciences, Akademicheskaya Str. 20, 620990 Ekaterinburg, Russia
8   Dipartimento di Scienze della Terra, dell'Ambiente e della Vita, Università degli Studi di Genova, Corso Europa 26, 16132 Genova, Italy; marzia.bo@unige.it (M.B.); giorgio.bavestrello@unige.it (G.B.)
9   Center for Advanced Technology, Adam Mickiewicz University, 61614 Poznan, Poland
*   Correspondence: krzysztof.j.nowacki@doctorate.put.poznan.pl (K.N.); izabela.stepniak@put.poznan.pl (I.S.); hermann.ehrlich@esm.tu-freiberg.de (H.E.)

Received: 9 May 2020; Accepted: 29 May 2020; Published: 2 June 2020

Abstract: The development of novel and effective methods for the isolation of chitin, which remains one of the fundamental aminopolysaccharides within skeletal structures of diverse marine invertebrates, is still relevant. In contrast to numerous studies on chitin extraction from crustaceans, mollusks and sponges, there are only a few reports concerning its isolation from corals, and especially black corals (Antipatharia). In this work, we report the stepwise isolation and identification of chitin from *Cirrhipathes* sp. (Antipatharia, Antipathidae) for the first time. The proposed method, aiming at the extraction of the chitinous scaffold from the skeleton of black coral species, combined a well-known chemical treatment with in situ electrolysis, using a concentrated Na_2SO_4 aqueous solution as the electrolyte. This novel method allows the isolation of α-chitin in the form of a microporous membrane-like material. Moreover, the extracted chitinous scaffold, with a well-preserved, unique pore distribution, has been extracted in an astoundingly short time (12 h) compared to the earlier reported attempts at chitin isolation from Antipatharia corals.

Keywords: chitin; biological materials; electrolysis; Antipatharia; black corals; *Cirrhipathes* sp.

---

1. Introduction

Chitin is composed of β-(1,4)-*N*-acetyl-ᴅ-glucosamine units, and plays a crucial role in the formation of skeletal structures in invertebrate organisms, where rigidity and strength are required [1].

This most abundant aminopolysaccharide has been isolated and identified in skeletal structures of diverse species of fungi, algae and invertebrates (i.e., sponges, hydrozoans, mollusks, worms, insects, spiders and crustaceans) [2–12]. Here, chitin is present in the form of biocomposites, being chemically bound to proteins, pigments and other polysaccharides, as well as mineral phases [13,14]. Consequently, its extraction from such biocomposites is fraught with a number of methodological difficulties that must be overcome using different approaches.

Nowadays, chitin of marine invertebrate origin is commonly isolated via two types of extraction process: chemical or biological [15,16]. In brief, the chemical treatment requires three main steps, namely deproteinization, demineralization and depigmentation. The deproteinization is generally carried out as the first stage. During this part of the process, the chitin-based structure is treated with alkaline solutions, which causes the dissolution of most of the proteins. This step is highly important in terms of medical and technological applications, because it determines the purity of the obtained product as well as the deacetylation degree and the possible hydrolysis of the chitin polymeric chain, depending on temperature conditions used [17]. The demineralization step usually follows hydrolysis of the proteins of the chitinous structure, and involves its treatment with acid solution (i.e., CH_3COOH or HCl). This step is conducted to treat highly mineralized biomaterials; it ensures the elimination of calcium carbonates via decomposition of these insoluble compounds into water-soluble calcium salts, along with the release of carbon dioxide [13]. The third step, called depigmentation, is a treatment optimally carried out by adding highly reactive oxidizing agents, such as hydrogen peroxide. However, all three steps of the chemical treatment rely on extraction agents that have to be used in great excess, thus generating effluents that are hazardous to the environment. Moreover, a relatively long treatment time, along with the increased temperature of the deproteinization process, can cause uncontrolled degradation of the chitinous polymeric chain [18–20]. The biological isolation of chitin, being an alternative method to the chemical treatment, uses microorganisms which produce diluted organic acids and enzymes to fulfill the role of the chemical extraction agents. Despite the longer time of treatment, chitin obtained via the biologically catalyzed process possesses a better-preserved spatial structure than that of an industrial source [21]. Regardless, in order to increase efficiency and reduce the environmental impact of chemical the process, novel and modified methods have been developed [22–24]. The recently reported assisted methods are mostly focused on the use of microwave irradiation as the accelerating factor [9]. Indeed, the application of this approach leads to a significant reduction of treatment time (from days to a few hours) [15].

Among the assisted methods for the isolation of chitin, electrolysis has been relatively poorly investigated. Only one proposed approach that includes the electrochemical treatment of chitin-containing crustacean exoskeletons has been described to date [25–27]. The principle of this method is based on the electrolysis of a diluted NaCl aqueous solution to ensure the acidic and alkali treatment of the crustacean *Gammarus pulex*'s (Linnaeus, 1758) biomass. However, being based on the electrolysis of a low-concentration NaCl solution, this method is characterized by significant treatment time (13–19 h) [28]. The evolution of chlorine gas on the anode surface, which is a highly corrosive compound, is another drawback of the method [29]. Thus, electrolysis-assisted isolation of chitin has not received much attention over traditionally used extraction procedures. As we recently showed with experiments on chitin from the skeleton of the marine sponge *Aplysina aerophoba* (Nardo, 1833), the electrolysis process can be very flexible in terms of electrochemical conditions [16]. Thus, there is possible scope for the improvement of this overlooked approach—for example, by the application of different electrolytes.

In this study, we present the results of electrochemical isolation of chitin from the black coral *Cirrhipathes* sp. (Antipatharia, Antipathidae) (Figures 1 and 2) for the first time. The intense research on these marine invertebrates has been focused mostly on the isolation and characterization of the special, dark-pigmented biopolymer antipathin, and the accompanying diphenol compounds [30–32]. Unfortunately, *Cirrhipathes* species are poorly described as a source of chitinous scaffolds. Previously, we have shown that the isolation of chitin from selected black corals is mostly performed via the chemical method [33]. It involves long, alternating alkaline and acidic extraction steps, and hence the duration of treatment often exceeds 7 days [33]. Therefore, this method can be modified in order to reduce the treatment time and amount of chemicals used. For this purpose, we decided to modify the known electrolysis method for the isolation of chitin from crustaceans, as reported earlier [34]. Since the described process is based on similar steps to the chemical process, it was essential to ensure a highly alkaline environment in order to achieve a successful result. Therefore, in this study, a concentrated Na_2SO_4 aqueous solution was utilized as the electrolyte, and a novel electrochemical method (which combined a well-known chemical treatment with in situ electrolysis) was investigated in terms of its usefulness for the extraction of chitinous scaffold from *Cirrhipathes* sp. skeletons.

Figure 1. Overview of the *Cirrhipathes* sp. coral fragments used in the study. (**A**) Central portion of the unbranched, unpinnulated stem of the colony. (**B**) Close-up view of the skeletal surface showing the multiple longitudinal rows of spines. (**C**) Basal plates of the spines after erosion. (**D**) Close-up view of one spine basal plate showing the concentric layers of skeleton.

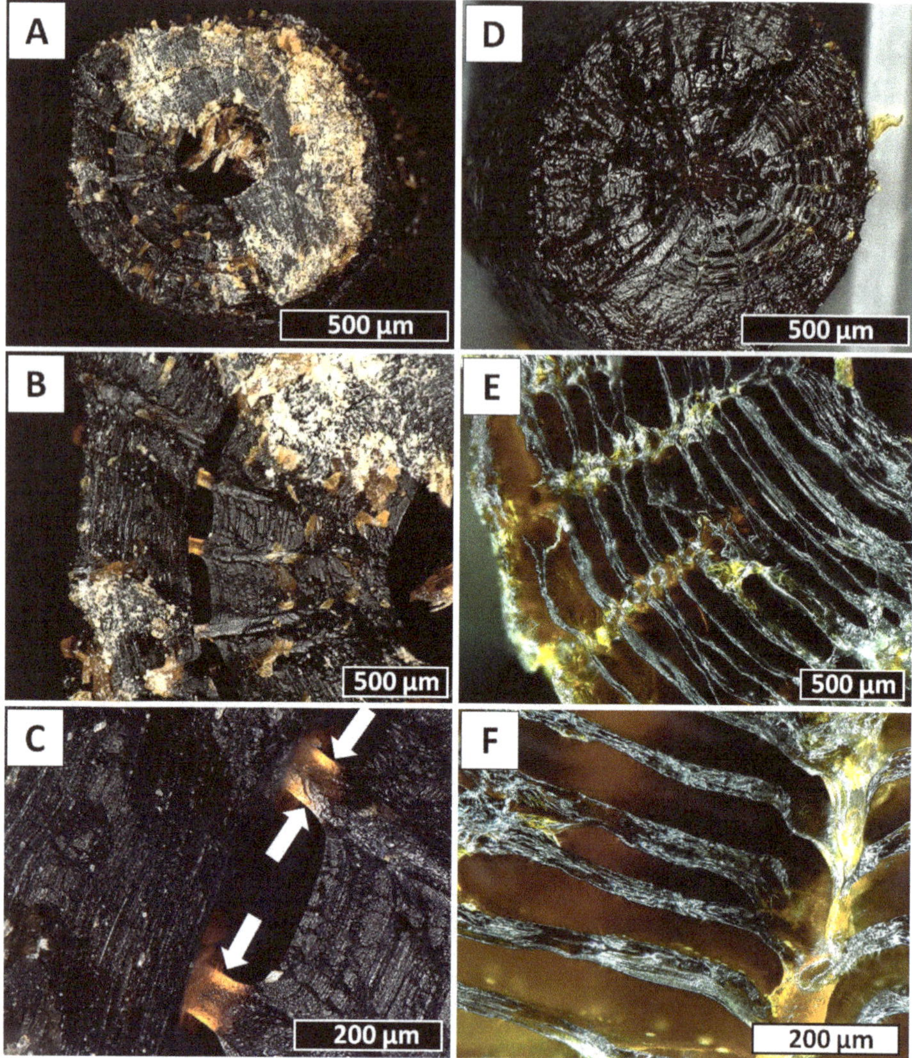

Figure 2. Insights into the inner structure of *Cirrhipathes* sp. skeleton. (**A**) Transversal section of the stem, showing a clearly hollow central canal surrounded by concentric layers of skeleton. The outer surface is covered in small triangular spines. (**B,C**) Spines' roots visible between the skeletal concentric layers. (**D**) Transversal section of the stem with a central canal partially closed by a skeletal septum. (**E,F**) Clusters of concentric skeletal layers intersecting perpendicularly with the spines' roots, connecting vertically the outer surface with the internal channel.

2. Results

The water electrolysis process is a well-known electrochemical phenomenon that has to be thermodynamically forced via the flow of a direct electric current from an external source [35]. In order to pass the current between two electrodes, a specific electrolytic cell (electrolyzer) must be constructed. Briefly, the modern electrolyzer is composed of two symmetrical, polarizable electrodes, made of electrically and chemically inert materials with high active surface areas. Usually, both electrodes are dipped in an electrically conductive solution (electrolyte) and separated with an ion exchange membrane (cation, anion or bipolar), forming two compartments [36–41]. The chamber with the anode contains the electrolyte solution, called the anolyte, whereas the chamber with the cathode is filled with the catholyte. Aqueous solutions of low molecular salts, such as NaCl or Na_2SO_4, which are generated as by-products in a wide variety of chemical processes, serve as perfect substrates for the production of alkalis and acids by electrolysis. The splitting of the Na_2SO_4 aqueous solution into NaOH and H_2SO_4 solutions, which occurs in the cation exchange membrane (CEM) of the electrolyzer, is one of the most popular ways to utilize the overproduction of this salt [42]. Figure 3 shows the basic principle of this process [43]. During Na_2SO_4 aqueous solution electrolysis (decomposition of water particles), fundamental electrochemical reactions take place on the electrodes' surfaces. These redox reactions result in an excess of H^+ and OH^- ions in the anolyte and catholyte, respectively. Simultaneously, sodium ions from the anolyte migrate through the CEM towards the cathode, where they are reduced to a sodium metal, which immediately reacts with the water to form NaOH [44]. Thanks to this phenomenon, it is possible to establish and change the pH in each part of the electrolyzer by applying a specific potential. This feature can be useful in terms of the extraction process that is responsive to the pH parameter. Thus, we have applied the fundamental feature of chitin, i.e., strong resistance to alkaline solutions up to temperatures between 70 °C and 80 °C, where its de-acetylation and transformation into chitosan occurs [45]. However, in our method, the alkaline environment has been achieved via electrolysis.

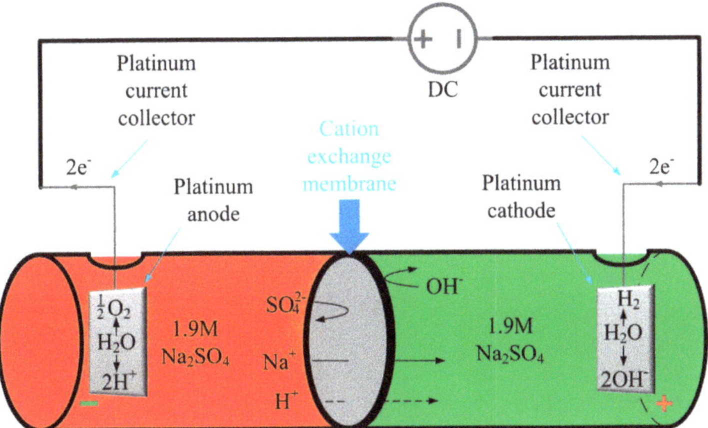

Figure 3. Schematic illustration of the electrolysis cell assembled in this study, and a general principle of Na_2SO_4 aqueous solution electrolysis [44].

The changes in the surface morphology of the selected fragments of *Cirrhipathes* sp., occurring during the electrolysis within the electrolytic chamber, were monitored using digital microscopy. Nearly three hours after the electrolysis treatment was started, the rejection of the upper layer, in the form of a membrane-like film, became observable (Figure 4A,B). At this step, no structural changes on the surface of this film could be observed at the microlevel (Figure 5), however, the fact of the appearance of such an alkali-resistant structure implies the destruction of proteins that secured it in the coral's skeleton.

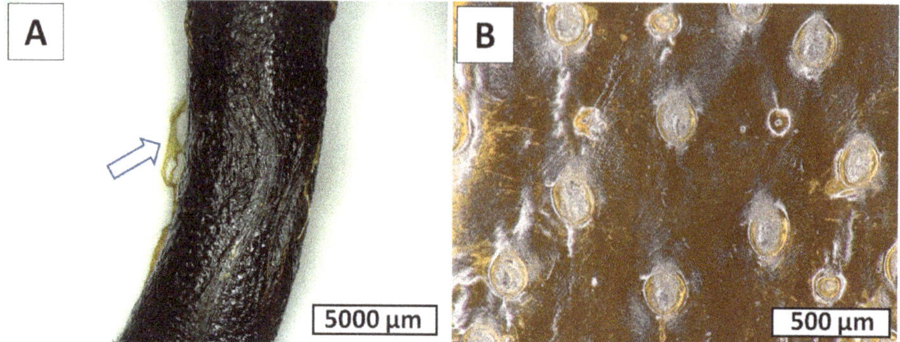

Figure 4. Rejection of the membranous film-like structure following 3 h of catholyte treatment on the *Cirrhipathes* sp. black coral surface (**A**) becomes well visible. This biological material was still pigmented and kept regular spine formations (**B**) on its surface (See also Figure 5).

Figure 5. Regular longitudinal rows of spines (**A**,**B**), which are characteristic of the surface of *Cirrhipathes* sp. skeleton, remain without visible changes (**C**) also on the surface of the rejected film-like membrane after 3 h of electrolysis. The surface ornamentation of small, sparse papillae is visible.

Following 6 h of electrolysis, both digital (Figure 6A,B,D,E) and scanning electron microscopy (Figure 6C,F) images showed defined structural changes. Spines partially disappeared, and only spongy and nanoporous structures (Figure 6C) became visible at the sites of their previous localization, though at some sites these structures also disappeared (Figure 6F). We suggest that these spines are composed of proteins and not of chitin, which is well recognized as a biological material with high resistance to alkaline treatment [4–6]. After 12 h of electrolysis, the nanoporous structures disappeared completely (Figure 7). Thus, we obtained a membranous organic matrix, with regular pores up to

100 µm large. Calcofluor white staining of this matter (Figure 7B) allows us to assume, with a high probability, the chitinous nature of the matrix, taking into account previously published results on chitin identification using this broadly applied technique [3–5,9,10,13,16,46–58].

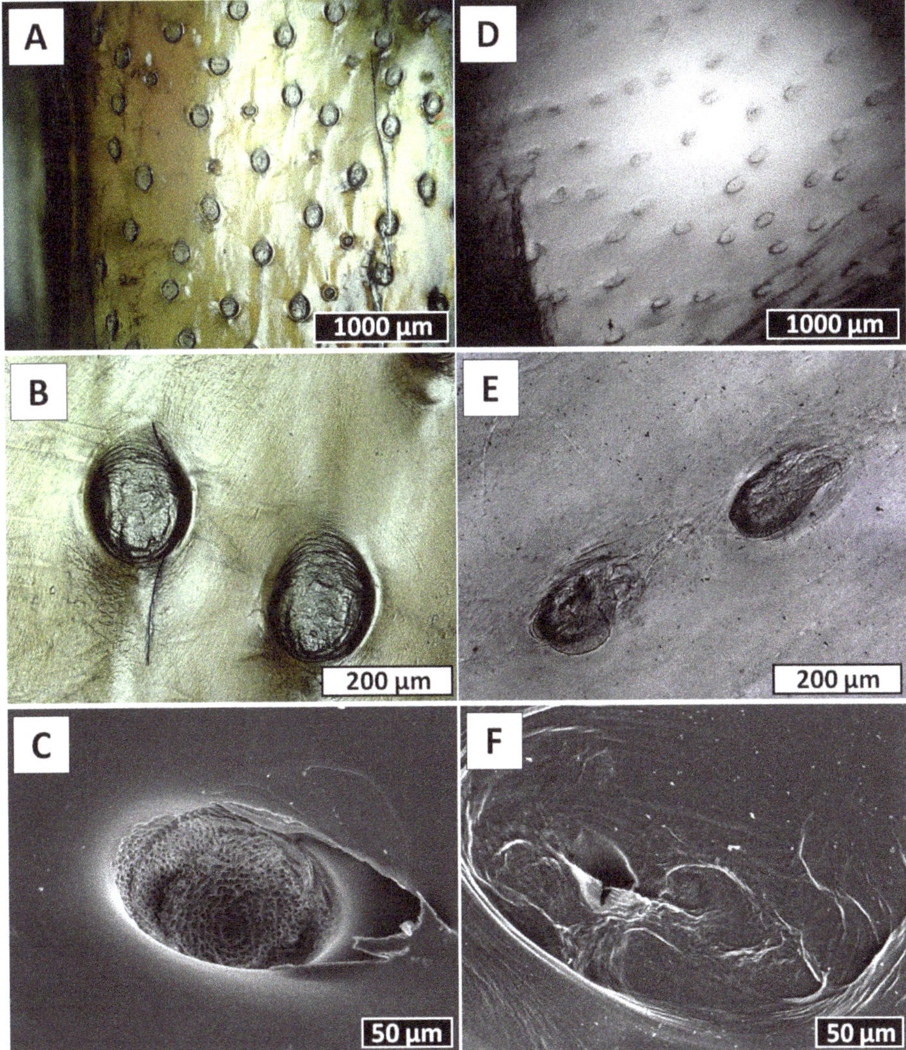

Figure 6. Surface morphology of the film-like, membranous structure, that was rejected from the *Cirrhipathes* sp. coral stem after 3 h of electrolysis (see Figures 4 and 5), continues to be structurally changed following 6 h of electrolysis. Disappearance of spines becomes well visible using digital (**A,B,D,E**) as well as SEM (**C,F**) microscopy (see also Figure 7).

For indisputable identification of chitin, we used infrared spectroscopy (ATR FT-IR), the chitinase digestion test, and ESI-MS-based analytics as represented below.

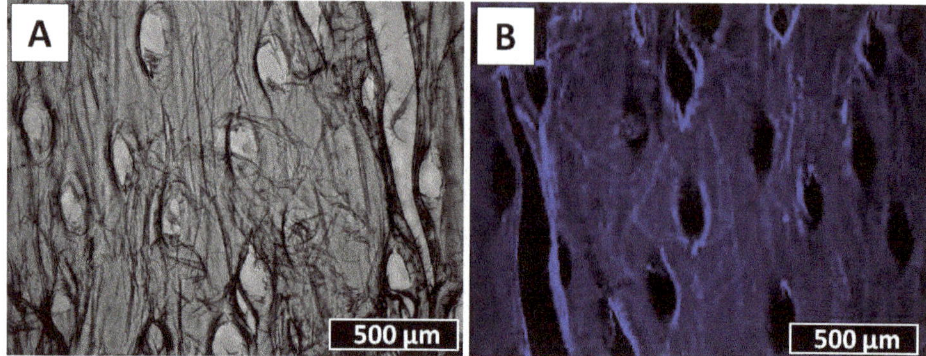

Figure 7. *Cirrhipathes* sp. sample after 12 h of electrolysis (**A**) and after Calcofluor white staining for preliminary chitin identification (**B**) (Light exposure time 1/500 s).

ATR-FTIR spectroscopy was used to identify the functional groups typical for chitin in the investigated samples. Spectra, obtained for the naturally occurring rod of *Cirrhipathes* sp., the α-chitin standard, and a chitinous membrane-like scaffold (Figure 7A) which was isolated after full electrolysis treatment, are shown in Figure 8. The spectrum of the analyzed coral skeleton fragment (Figure 1B) shows bands which are similar to those previously reported for antipatharians by other authors [30,31]. Since the chemical composition of the *Cirrhipathes* sp.'s skeleton is highly diverse and complex (scleroproteins, lipids, diphenols and polysaccharides), most of the characteristic bands within the spectrum are overlapped by each other. Despite this, the characteristic bands for α-chitin, such as amide I (carbonyl stretching vibrations of *N*-acetyl groups) at 1645 cm^{-1}, amide II (νN–H and νC–N) at 1531–1510 cm^{-1} and amide III (νC–N and δN–H) at 1306 cm^{-1}, visible in the analyzed spectrum (Figure 8, black line), are sufficient to confirm the presence of chitin [6,59] within the sample under study. IR analysis of the isolated chitinous scaffold spectrum (Figure 8, red line) indicated that the characteristic band at 895 cm^{-1} (C–H deformation of the β-glycosidic bond, as well as the C–O–C bridge) suggests the occurrence of α-chitin in the sample (for comparison, 890 cm^{-1} for β-chitin) [4,6,33,45–48]. Moreover, the wavelengths of all other characteristic bands in this specimen are nearly identical to the α-chitin standard spectrum (Figure 8, green line), which additionally proves the presence of chitin in the α form in the electrochemically isolated organic matter. No presence of chitosan, as a possible product of the chitin de-acetylation under electrolysis conditions used here, has been confirmed using infrared spectroscopy.

Determination of *N*-acetylglucosamine (GlcNAc) is a key step for chitin identification in biological materials of unknown origin. To quantify chitin in the specimens of the membranaceous matter, isolated from *Cirrhipathes* sp. after 12 h of electrolysis, we measured the quantity of GlcNAc, released by chitinases using a classical Morgan–Elson colorimetric assay, which, owing to its specificity, is recognized as the most reliable method for the identification of alkali-insoluble chitin [4–6,47,48]. We detected 875.3 + 0.5 µg of *N*-acetylglucosamine per mg of depigmented skeleton of *Cirrhipathes* sp. Furthermore, the chitinase digestion test, based on the observation of the enzymatic dissolution of the purified, pigments- and proteins-free black coral organic matrix by light microscopy, confirmed the presence of pure chitin (Figure 9). Previously, we have shown that chemically impure chitin cannot be digested in chitinase solution [3,13].

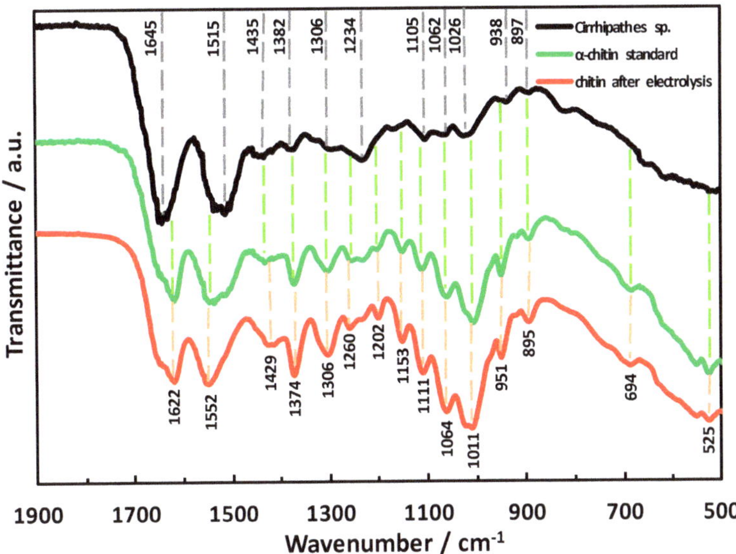

Figure 8. ATR-FTIR spectra of *Cirrhipathes* sp. sample (black line), α-chitin standard (green line) and electrochemically isolated chitinous scaffold (red line) in the region of 1900–500 cm^{-1}.

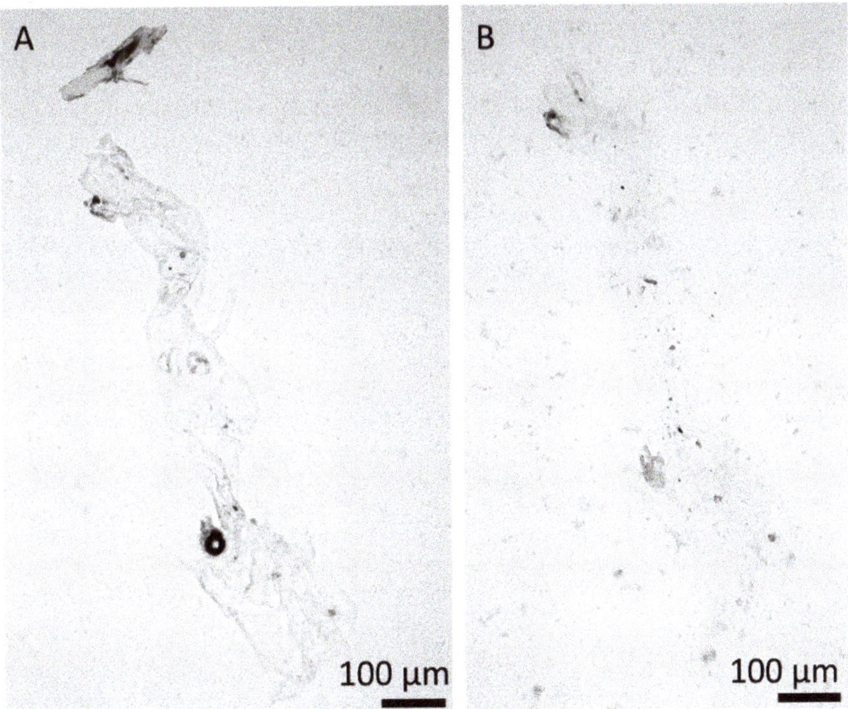

Figure 9. Chitinase digestion of electrochemically extracted chitin isolated from *Cirrhipathes* sp. at room temperature (light microscopic images). (**A**) Initial stage; (**B**) after 12 h of chitinase treatment.

It is well recognized that the D-glucosamine (GlcN) signals in the mass spectrum of hydrolyzed (6M HCl) biological samples reveal the presence of chitin. This method was utilized previously for chitin identification in diverse chitin-producing organisms [9,13,47,48], including heavy mineralized fossil specimens [46]. The acid hydrolysis of the *Cirrhipathes* sp. sample (see Figure 7A) revealed five main signals (Figure 10). The signals with m/z = 162 and 180 correspond to the [M − H$_2$O + H$^+$], [M + H$^+$], while species with signal m/z = 359 correspond to proton-bound non covalent GlcN dimmer [2M + H$^+$], which is common for ESI-MS spectra of amino monosaccharides. The signals with m/z = 202 and 381 correspond to the same species in which the hydrogen ion is substituted on sodium [M + Na$^+$] and [2M + Na$^+$], respectively, which is very common for the spectra of natural samples. Together, these results prove that the membrane-like, alkali-resistant biological material isolated from *Cirrhipathes* sp. contains chitin biopolymer.

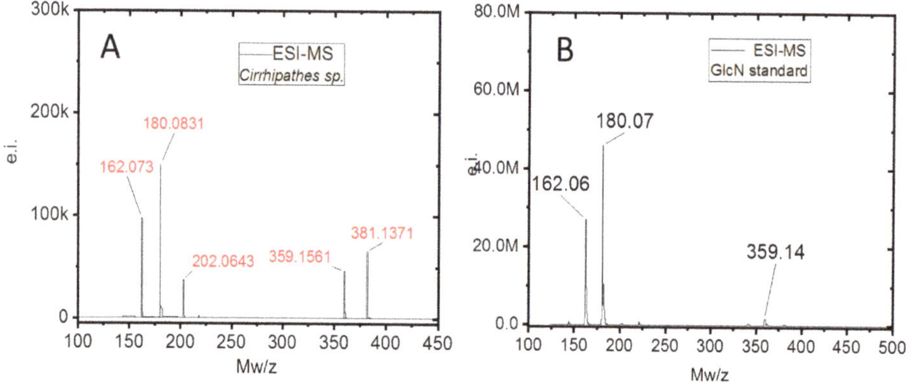

Figure 10. Electrospray-ionization mass spectroscopy (ESI-MS) characterization. (**A**) *Cirrhipathes* sp. after acid hydrolysis. (**B**) ESI-MS spectrum of GlcN standard reference.

Chitin was not the only product that was isolated from *Cirrhipathes* sp. using the electrochemical procedure represented in our study. Electrochemically mediated alkalization of the medium leads to the extraction of black pigment from the chitinous matter. Figure 11 shows the UV-visible spectra of the extracts obtained from the cathode chamber, after 6 h (extract I) and 12 h (extract II) of electrolytic treatment. According to data in the literature [60], both spectra show bands characteristic for polyphenols, most likely catechol derivatives, due to the peaks at 212 nm and 293 nm (Figure 11; extract I—green line) and 216 nm and 290 nm (extract II—red line). The diphenol trace within the extract samples could have resulted from the decomposition of the antipathin–chitin structural complex, as well as the reduction of 3,4-dihydroxybenzaldehyde (DOBAL) and 3-(3,4-dihydroxyphenyl)-L-alanine (DOPA) compounds [32,61]. We suggest that the presence of catechol-related compounds in the skeleton of the black coral *Cirrhipathes* sp. affects its fine, solid structure, ultimately resulting in the easier erosion of this biocomposite-based construct by NaOH. Similar results were reported for chitin isolated from beetle larva, where derivatives of catechol in its cuticle inhibited chitin crystallinity and led to the amorphous chitin structure [62].

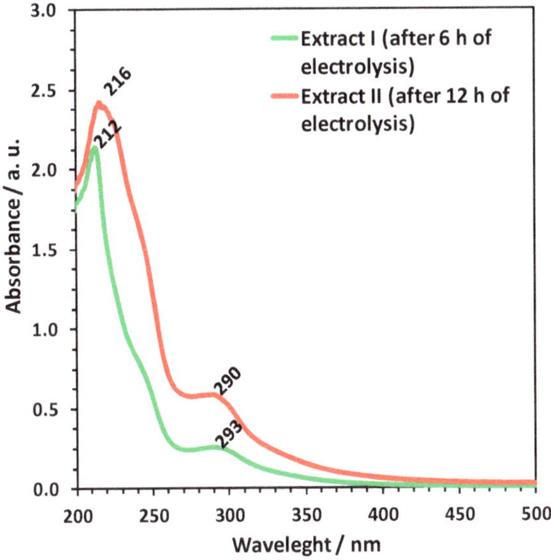

Figure 11. UV-visible absorption spectra of catholyte extracts isolated from the specimens of *Cirrhipathes* sp., following electrolysis for 6 h (green line) and 12 h (red line), suggest the catechol-like nature of the pigments.

3. Discussion

Black corals (Hexacorallia: Antipatharia) are a taxonomic group that appear worldwide in all the oceans and exhibit a wide depth distribution, ranging between coastal shallows and abyssal depths [63–65]. To date, the order Antipatharia comprises about 247 valid species, and they are most abundant in tropical and subtropical seas, in deep waters (≥50 m) beneath the photic zone [63,66,67]. For this reason, since most species are found below the depth limits of conventional SCUBA diving, very little is known about the basic biology and ecology of black corals. Despite this, all these species are characterized as having exclusively colonial habitus, with generally slow growth rates depending on the environment and high longevity, varying from decades to millennia [68–72]. Antipatharian colonies have various morphologies, either unbranched in the form of a whip or branched into a bush or a fan, with maximum sizes among species ranging from a few centimeters to many meters [73,74]. Moreover, to ensure stiffness and high mechanical strength, which are necessary to withstand high hydrostatic pressure and turbulent oceanic currents, the skeletons of antipatharians (Figure 1) usually have a layered scleroproteinaceous structure, strengthened by an inner chitinous scaffold [33,75]. The general morphology of their skeletons has been described as a laminated composite (Figure 2), where chitinous fibrils play a crucial role in terms of the growth support and elasticity of the entire colony. In addition, the internal structure of spines is believed to represent a significant reinforcing effect in the architecture of black coral skeletons, where the forces of the torsion imposed by currents are released [32]. This is particularly important in whip black corals, such as those belonging to the genera *Cirrhipathes*, *Stichopathes* and *Pseudocirrhipathes*, forming dense forests in the sites with the highest currents [76]. However, the dominant fraction within black coral skeletons is represented by a non-fibrillar formation, composed mostly of a halogen-containing scleroprotein and chitin [77]. This compound, known as antipathin, is exclusive to this taxon, and has unequalled thermal and mechanical stability which ensures the stiffness of the coral skeleton. Moreover, the chitin–antipathin based composite shows a unique combination of flexibility and hardness that provides better resistance to stress factors in a marine environment than inorganic structural materials [14]. Beside this, the chemical arrangement of the black coral skeleton also contains proteins, lipids and diphenols, and the chitin content within is estimated to constitute between 6% and 18% of the total organism mass (which is a considerable amount for marine invertebrates). Therefore, black corals can be considered as a potential source of chitinous scaffolds [78].

The finding of chitin within the skeletal structures of black coral *Cirrhipathes* sp. is important for gaining a better understanding of the structural biology of these organisms. Of course, black coral chitin cannot be used for practical application on a large scale, as is the case with crustacean or sea sponge chitin. However, this type of chitin is an interesting biomaterial in terms of its inspirational potential for biomimetics and material science. For example, the development of new chitin–catechol composite materials has an intriguing potential in biomedicine and technology. The first attempts in this direction have been recently made and patented (EP2778179A2 Chitosan and/or chitin composite having reinforced physical properties and use thereof. 2015). Definitively, more detailed studies on the chemistry and biosynthesis of naturally occurring chitin–polyphenol composites should be carried out in the near future. Chitin matrices of this type, with regularly located micropores, can serve as model systems for biomimetic studies into the creation of chitin-based membranes. An equally promising direction may be the creation of chitin membranes modified with polyphenolic compounds, which have antibacterial properties. The study of such membranes for the treatment of burns and other wounds seems to be in demand.

4. Materials and Methods

4.1. Biological Samples and Chemicals

Cirrhipathes sp. dry sample was purchased from INTIB GmbH (Freiberg, Germany). Sodium sulfate (Na_2SO_4, ≥99.7%), purchased from VWR (Darmstadt, Germany), was used for the preparation of aqueous electrolyte solution. Sodium hydroxide (NaOH, ≥99.0%), purchased from VWR (Darmstadt, Germany), was utilized as substrate to prepare an extracting solution. Distilled water was used to prepare all aqueous solutions.

4.2. Electrolytic Cell Setup

The schematic illustration of the experimental system for the electrochemically-assisted isolation of chitin is shown in Figure 12. The CEM (cation exchange membrane) electrolyzer consisted of two cylindrical poly(propylene) chambers (50 mL each) separated by a cellulose membrane made from filter paper (75 g cm^{-2}, ChemLand, Poland) and sealed with parafilm (Bemis Company Inc., Neenah, WI, USA). Electrodes (cathode and anode) were made of platinum sheets (effective area: 2.2 cm^2). Distance between both electrodes was kept at about 10.0 cm and they were connected with DC power supply VoltCraft PS2043D (Conrad Electronic International GmbH & Co., Wels, Austria) by platinum wire current collectors. 1.9 M sodium sulfate aqueous solution with an initial temperature of 40 °C was utilized as anolyte as well as catholyte.

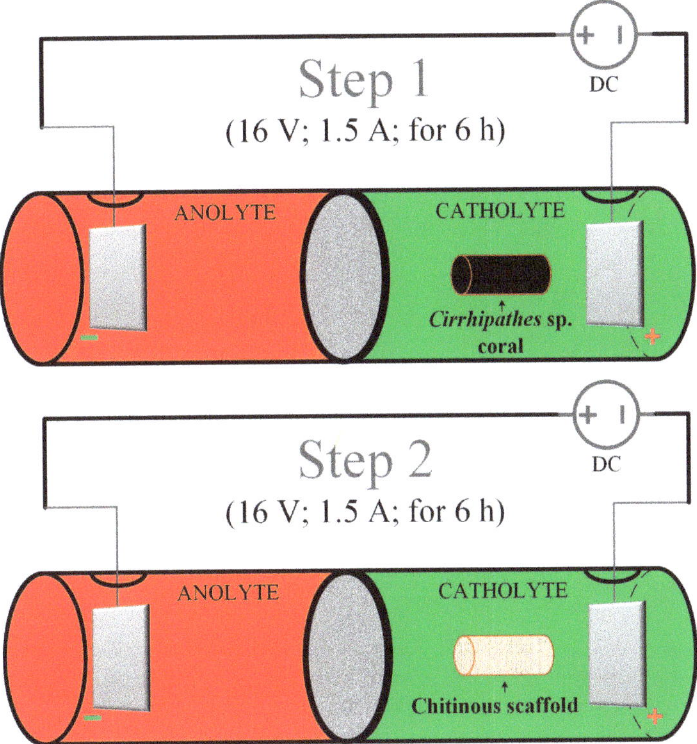

Figure 12. Schematic illustration of the experimental setup for all steps of the electrochemically-assisted isolation of *Cirrhipathes* sp. chitin [43].

4.3. Electrochemically-Assisted Isolation of Chitin

A novel electrochemically-assisted method of chitinous scaffold isolation from *Cirrhipathes* sp. was performed in two main steps (see Figure 12). In both stages, the sample was treated in the catholyte solution, and anolyte treatment (low pH) was not necessary to remove proteins, lipids and pigments. It should be noted that the initial concentration of the electrolyte (Na_2SO_4) for every step was 1.9 mol L^{-1} and the starting temperature for both anolyte and catholyte was 40 °C.

Pretreatment: A 0.3 g piece was cut from the *Cirrhipathes* sp. sample and rinsed repeatedly with distilled water (25 °C) in order to get rid of the major solid impurities and water-soluble salts of marine origin.

Step 1: The first part of the electro-alkali treatment was performed in the cathode chamber for 6 h (16 V, 1.5 A, 70 °C). High pH (up to 12.5) of the catholyte caused complete lysis of corals cells and degradation of lipids and proteins, which resulted in the removal of soft tissues from interlayer spaces of the chitinous skeleton. Moreover, this effect was followed by partial depigmentation and possible desilicification of the sample. Remaining chitinous skeleton was in the form of a light brown cell-free layered tube.

Step 2: In order to complete the depigmentation and deproteinization, the exchange of the electrolyte was required. With fresh 1.9 M Na_2SO_4 solution, the process was further carried out in the cathode chamber for 6 h (16 V, 1.5 A, 70 °C). Free access of the catholyte solution to the coral skeleton, along with high pH (up to 12.5), resulted in complete dissolution of pigments and residual proteins. After treatment, the remaining sample in the form of a colorless scaffold was extensively rinsed using distilled water up to neutral pH, and stored in ethanol absolute (4 °C).

4.4. Calcofluor White Staining

The Calcofluor white staining (CFW) (Fluorescent Brightener M2R, Sigma-Aldrich, St. Louis, MO, USA) was used to confirm the presence of chitin in the sponge skeleton several times (see [5–8]). For the staining process, 30 μL of a solution containing 10 g glycerin and 10 g NaOH in 90 mL of water was applied. After a minute, the CFW was added, and the investigated material was incubated in staining solution for 6 h without light at 25 °C. Then, the sample was washed with distilled water to eliminate the unattached stain, dried at 25 °C and analyzed using fluorescent microscopy. On binding to polysaccharides containing β-glycosidic bonds (such as chitin), this fluorochrome secretes bright blue light under UV excitation even with a very short light exposure time.

4.5. Chitinase Digestion Test

The fragment of isolated chitinous scaffold from *Cirrhipathes* sp. was treated with Yatalase enzyme solution (pH 6.5) [58]. The treatment was carried out for 6 h at 37 °C. The progress of digestion was observed under light microscopy using BZ-9000 microscope (Keyence, Osaka, Japan).

4.6. Attenuated Total Reflectance Fourier Transform Infrared Spectroscopy

Attenuated Total Reflectance Fourier Transform Infrared Spectroscopy (ATR-FTIR) was used for the qualitative characterization and identification of the isolated materials. The samples were analyzed by Nicolet 210c spectrometer (Thermo Fisher Scientific, Waltham, MA, USA).

4.7. Estimation of N-acetyl-D-glucosamine (NAG) Contents

The Morgan–Elson assay was used to quantify the *N*-acetyl-D-glucosamine released after chitinase treatment, as described previously [79]. Purified and dried *Cirrhipathes* sp. samples (6 mg) were pulverized to fine powder in an agate mortar. The samples were suspended in 400 mL of 0.2 M phosphate buffer at pH 6.5. A positive control was prepared by solubilizing 0.3% colloidal chitin (INTIB GmbH, Freiberg, Germany) in the same buffer. Equal amounts of 1 mg/mL from three chitinases (EC 3.2.1.14 and EC 3.2.1.30)—*N*-acetyl-D-glucosaminidase from *Trichoderma viride* (Sigma, No. C-8241), and two poly (1,4-α-[2-acetamido-2-deoxy-D-glucoside]) glycanohydrolases from *Serratia marcescens* (Sigma, No. C-7809), and *Streptomyces griseus* (Sigma, No. C-6137)—were suspended in 100 mM sodium phosphate buffer at pH 6.0. Digestion was initiated by mixing 400 mL of the sample and 400 mL of the chitinase mix. Incubation was performed at 37 °C and stopped after 114 h by adding 400 mL of 1% NaOH, followed by boiling for 5 min. The vessels were centrifuged at 7000 rpm for 5 min and the purified reducing sugars were used for 3,5-dinitrosalicylic acid assay (DNS) [47,48]. For this purpose, 250 mL of the supernatants and 250 mL of 1% DNS were dissolved in a solution containing 30% sodium potassium tartrate in 0.4 M NaOH. The reagents were mixed and incubated for 5 min in a boiling water bath. Thereafter, the absorbance at 540 nm was recorded using a Tecan Spectrafluor Plus Instrument (Mannedorf/Zurich, Switzerland). Data were interpolated using a standard curve prepared with a series of dilutions (0–3.0 mM) of *N*-acetyl-D-glucosamine (Sigma, No. A-8625) and DNS. A sample, which contained chitinase solution without substrate, was used as a control.

4.8. Electrospray Ionization Mass Spectrometry (ESI-MS)

Specimens obtained after electrochemical isolation in the final step (Figure 7A) were hydrolyzed in 6 M HCl for 24 h at 50 °C. Following the HCl hydrolysis, the samples were filtrated with 0.4 μm filter and freeze-dried in order to remove excess HCl. The remaining solid was dissolved in water for ESI-MS analysis. As standard, D-glucosamine was purchased from Sigma-Aldrich (Taufkirchen, Germany). The ESI-MS analytical measurements were performed using Agilent Technologies 6230 TOF LC/MS spectrometer (Applied Biosystems, Santa Clara, CA, USA). Nitrogen was used as the nebulizing and desolvation gas. Graphs were generated using Origin 8.5 for PC (Originlab Corporation, Northampton, MA, USA).

4.9. UV-VIS Spectroscopy

To conduct UV-VIS Spectroscopy, 0.5 mg of pigments, electrochemically isolated from *Cirrhipathes* sp. (Figure 1), were dissolved in 1 mL of 0.1 M KOH. The spectra were measured by JASCO V-750 spectrometer, in the wavelength range of 200 to 800 nm, which was operated at a resolution of 5 nm using a quartz cuvette with path length of 1 cm (quartz suprasil, Hellma Analytics, Müllheim, Germany).

4.10. Scanning Electron Microscopy (SEM)

The specimens were fixed on an aluminum sample holder with conductive carbon adhesive tabs and were sputtered with platinum for 15 s at a distance of 30 mm by an Edwards S150B sputter coater. The scanning electron micrographs were observed using a high-resolution Hitachi S-4700-II (Hitachi, Ltd., Tokyo, Japan) equipped with a cold field emission gun.

5. Conclusions

In the present work, we utilized the in-situ electrolysis of a 1.9 M Na_2SO_4 aqueous solution in the CEM of an electrolyzer as the method for isolating chitinous scaffolds from *Cirrhipathes* sp. black coral. The final results of the electrochemically-assisted isolation of chitin were a colorless, membrane-like film, and catechol-based extracts. The digital light and scanning electron microscopy investigations of this final product revealed that, despite the highly alkaline environment of the catholyte and the destruction of proteins within the coral's skeleton, the general spatial structure of the sample preserved its original membranous formation, with regular pores up to 100 μm large. Further characterization of the isolated sample with various techniques proved that a pure chitinous scaffold can be obtained via the application of the described method. Moreover, as the ATR-FTIR spectroscopy analysis showed, the electrochemically-supported isolation process does not cause a chitin–chitosan transformation, and the obtained scaffold was fully α-chitin. All these features, boosted additionally by the advantages of the electrolysis method (i.e., reduction of time treatment and amount of chemicals used), show that our method can be considered as an alternative to the standard chemical chitin extraction process. Thus, without doubt further development of the electrochemical isolation of chitin from marine sources should be carried out in the near future.

Author Contributions: H.E., K.N., M.W., M.T. designed the study protocol and wrote the manuscript; M.B., G.B. collected the materials and carried out species identification; K.N., I.P. carried out the electrochemical isolation of chitin; E.L., Y.K., A.F., Y.J., M.T. prepared samples and performed detailed physicochemical characterization of obtained chitin; I.S. and H.E. edited the manuscript. All authors have read and agreed to the published version of the manuscript.

Funding: This work was performed with the financial support of Poznan University of Technology, Poland (Grant No. 03/31/SBAD/0395/2020). K.N. was supported by the Erasmus Plus program (2019). This research was also funded by DFG Project HE 394/3 and SMWK Project no. 02010311 (Germany) and a subsidy from the Ministry of Science and Higher Education, Poland, to PUT: project no. 03/32/SBAD/0906. M.W. was financially supported by Polish National Agency for Academic Exchange (PPN/BEK/2018/1/00071).

Acknowledgments: We thank T. Machałowski and P. Machałowska for technical support during UV/VIS measurements.

Conflicts of Interest: We declare no conflicts of interest.

References

1. Rinaudo, M. Chitin and chitosan: Properties and applications. *Prog. Polym. Sci.* **2006**, *31*, 603–632. [CrossRef]
2. Rahman, A.; Halfar, J. First evidence of chitin in calcified coralline algae: New insights into the calcification process of Clathromorphum compactum. *Sci. Rep.* **2014**, *4*, 6162. [CrossRef] [PubMed]
3. Brunner, E.; Ehrlich, H.; Schupp, P.; Hedrich, R.; Hunoldt, S.; Kammer, M.; Machill, S.; Paasch, S.; Bazhenov, V.; Kurek, D.; et al. Chitin-based scaffolds are an integral part of the skeleton of the marine demosponge Ianthella basta. *J. Struct. Biol.* **2009**, *168*, 539–547. [CrossRef] [PubMed]

4. Ehrlich, H.; Maldonado, M.; Spindler, K.-D.; Eckert, C.; Hanke, T.; Born, R.; Goebel, C.; Simon, P.; Heinemann, S.; Worch, H. First evidence of chitin as a component of the skeletal fibers of marine sponges. Part I. Verongidae (demospongia: Porifera). *J. Exp. Zoöl. Part B: Mol. Dev. Evol.* **2007**, *308*, 347–356. [CrossRef] [PubMed]

5. Ehrlich, H.; Krautter, M.; Hanke, T.; Simon, P.; Knieb, C.; Heinemann, S.; Worch, H. First evidence of the presence of chitin in skeletons of marine sponges. Part II. Glass sponges (Hexactinellida: Porifera). *J. Exp. Zoöl. Part B: Mol. Dev. Evol.* **2007**, *308*, 473–483. [CrossRef] [PubMed]

6. Ehrlich, H.; Ilan, M.; Maldonado, M.; Muricy, G.; Bavestrello, G.; Kljajic, Z.; Carballo, J.L.; Schiaparelli, S.; Ereskovsky, A.; Schupp, P.; et al. Three-dimensional chitin-based scaffolds from Verongida sponges (Demospongiae: Porifera). Part I. Isolation and identification of chitin. *Int. J. Biol. Macromol.* **2010**, *47*, 132–140.

7. Liu, S.; Sun, J.; Yu, L.; Zhang, C.; Bi, J.; Zhu, F.; Qu, M.; Jiang, C.; Yang, Q. Extraction and Characterization of Chitin from the Beetle Holotrichia parallela Motschulsky. *Molecules* **2012**, *17*, 4604–4611. [CrossRef]

8. Kaya, M.; Seyyar, O.; Baran, T.; Erdogan, S.; Kar, M. A physicochemical characterization of fully acetylated chitin structure isolated from two spider species: With new surface morphology. *Int. J. Biol. Macromol.* **2014**, *65*, 553–558. [CrossRef]

9. Machałowski, T.; Wysokowski, M.; Tsurkan, M.; Galli, R.; Schimpf, C.; Rafaja, D.; Brendler, E.; Viehweger, C.; Żółtowska-Aksamitowska, S.; Petrenko, I.; et al. Spider Chitin: An Ultrafast Microwave-Assisted Method for Chitin Isolation from Caribena versicolor Spider Molt Cuticle. *Molecules* **2019**, *24*, 3736. [CrossRef]

10. Machałowski, T.; Wysokowski, M.; Żółtowska-Aksamitowska, S.; Bechmann, N.; Binnewerg, B.; Schubert, M.; Guan, K.; Bornstein, S.R.; Czaczyk, K.; Pokrovsky, O.; et al. Spider Chitin. The biomimetic potential and applications of Caribena versicolor tubular chitin. *Carbohydr. Polym.* **2019**, *226*, 115301. [CrossRef]

11. Tolesa, L.D.; Gupta, B.S.; Lee, M.-J. Chitin and chitosan production from shrimp shells using ammonium-based ionic liquids. *Int. J. Biol. Macromol.* **2019**, *130*, 818–826. [CrossRef] [PubMed]

12. Mohan, K.; Ravichandran, S.; Muralisankar, T.; Uthayakumar, V.; Chandirasekar, R.; Rajeevgandhi, C.; Rajan, D.K.; Seedevi, P. Extraction and characterization of chitin from sea snail Conus inscriptus (Reeve, 1843). *Int. J. Biol. Macromol.* **2019**, *126*, 555–560. [CrossRef] [PubMed]

13. Ehrlich, H.; Bazhenov, V.; Debitus, C.; De Voogd, N.; Galli, R.; Tsurkan, M.; Wysokowski, M.; Meissner, H.; Bulut, E.; Kaya, M.; et al. Isolation and identification of chitin from heavy mineralized skeleton of Suberea clavata (Verongida: Demospongiae: Porifera) marine demosponge. *Int. J. Biol. Macromol.* **2017**, *104*, 1706–1712. [CrossRef] [PubMed]

14. Ehrlich, H. *Marine Biological Materials of Invertebrate Origin*; Springer Science and Business Media LLC: Berlin/Heidelberg, Germany, 2019.

15. Klinger, C.; Żółtowska-Aksamitowska, S.; Wysokowski, M.; Tsurkan, M.; Galli, R.; Petrenko, I.; Machałowski, T.; Ereskovsky, A.V.; Martinovic, R.; Muzychka, L.; et al. Express Method for Isolation of Ready-to-Use 3D Chitin Scaffolds from Aplysina archeri (Aplysineidae: Verongiida) Demosponge. *Mar. Drugs* **2019**, *17*, 131. [CrossRef]

16. Nowacki, K.; Stępniak, I.; Machałowski, T.; Wysokowski, M.; Petrenko, I.; Schimpf, C.; Rafaja, D.; Langer, E.; Richter, A.; Ziętek, J.; et al. Electrochemical method for isolation of chitinous 3D scaffolds from cultivated Aplysina aerophoba marine demosponge and its biomimetic application. *Appl. Phys. A* **2020**, *126*, 1–16. [CrossRef]

17. Soon, C.Y.; Tee, Y.B.; Tan, C.H.; Rosnita, A.T.; Khalina, A. Extraction and physicochemical characterization of chitin and chitosan from Zophobas morio larvae in varying sodium hydroxide concentration. *Int. J. Biol. Macromol.* **2018**, *108*, 135–142. [CrossRef]

18. Younes, I.; Rinaudo, M. Chitin and Chitosan Preparation from Marine Sources. Structure, Properties and Applications. *Mar. Drugs* **2015**, *13*, 1133–1174. [CrossRef]

19. Ehrlich, H.; Shaala, L.A.; Youssef, D.T.A.; Żółtowska-Aksamitowska, S.; Tsurkan, M.; Galli, R.; Meissner, H.; Wysokowski, M.; Petrenko, I.; Tabachnick, K.R.; et al. Discovery of chitin in skeletons of non-verongiid Red Sea demosponges. *PLoS ONE* **2018**, *13*, e0195803. [CrossRef]

20. Percot, A.; Viton, C.; Domard, A. Optimization of Chitin Extraction from Shrimp Shells. *Biomacromolecules* **2003**, *4*, 12–18. [CrossRef]

21. Khanafari, A.; Marandi, R.; Sanatei, S. Recovery of chitin and chitosan from shrimp waste by chemical and microbal methods. *Iran. J. Environ. Health Sci. Eng.* **2008**, *5*, 19–24.

22. Knidri, H.; Dahmani, J.; Addaou, A.; Laajeb, A.; Lahsini, A. Rapid and efficient extraction of chitin and chitosan for scale-up production: Effect of process parameters on deacetylation degree and molecular weight. *Int. J. Biol. Macromol.* **2019**, *139*, 1092–1102. [CrossRef] [PubMed]

23. Kuprina, E.E.; Maslova, G.V.; Bachische, E.V. Electrochemical method for obtaining water-soluble oligomers of chitin in the presence of NaCl. In Proceedings of the IXth International Conference: Modern Perspectives in Chitin and Chitosan Studies, Stavropol, Russia, 13–17 October 2008; pp. 30–33.

24. Feng, M.; Lu, X.; Zhang, J.; Li, Y.; Shi, C.; Lu, L.; Zhang, S. Direct conversion of shrimp shells to O-acylated chitin with antibacterial and anti-tumor effects by natural deep eutectic solvents. *Green Chem.* **2019**, *21*, 87–98. [CrossRef]

25. Kuprina, E.E.; Vodolazhskaya, S.V.; Nyanikova, G.G.; Timofeeva, K.G. Development of technology for obtaining biologically active chitin sorbents based on the electrochemical conversion of crustaceans. In Proceedings of the VIth International Conference: New Achievements in Study of Chitin and Chitosan, Shchelkovo, Russia, 22–24 October 2001; pp. 31–34.

26. Kuprina, E.E.; Timofeeva, K.G.; Kozlova, I.; Pimenov, A. Electrochemical method extracting sorbitol from chitin-containing raw material with strengthened anitimicrobal properties. In Proceedings of the VIIth International Conference: Modern Perspectives in Chitin and Chitosan Studies, St. Petersburg, Russia, 15–18 September 2003; pp. 19–22.

27. Kuprina, E.E.; Timofeeva, K.G.; Krasavtsev, V.E.; Boykov, I.O.A. Experimental producing unit for getting chitin-mineral complex "chizitel" by electrochemical method. In Proceedings of the VIIIth International Conference: Modern Perspectives in Chitin and Chitosan Studies, Kazan, Russia, 13–17 June 2006; pp. 34–37.

28. Kuprina, E.É.; Timofeeva, K.G.; Vodolazhskaya, S.V. Electrochemical Preparation of Chitin Materials. *Russ. J. Appl. Chem.* **2002**, *75*, 822–828. [CrossRef]

29. Tennakone, K. Hydrogen from brine electrolysis: A new approach. *Int. J. Hydrogen Energy* **1989**, *14*, 681–682. [CrossRef]

30. La Rosa, B.J.-D.; Ardisson, P.-L.; Azamar-Barrios, J.; Quintana, P.; Alvarado-Gil, J.J. Optical, thermal, and structural characterization of the sclerotized skeleton of two antipatharian coral species. *Mater. Sci. Eng. C* **2007**, *27*, 880–885. [CrossRef]

31. Nowak, D.; Florek, M.; Nowak, J.; Kwiatek, W.M.; Lekki, J.; Chevallier, P.; Hacura, A.; Wrzalik, R.; Ben-Nissan, B.; Van Grieken, R.; et al. Morphology and the chemical make-up of the inorganic components of black corals. *Mater. Sci. Eng. C* **2009**, *29*, 1029–1038. [CrossRef]

32. Kim, K.; Goldberg, W.M.; Taylor, G.T. Architectural and Mechanical Properties of the Black Coral Skeleton (Coelenterata: Antipatharia): A Comparison of Two Species. *Biol. Bull.* **1992**, *182*, 195–209. [CrossRef]

33. Bo, M.; Bavestrello, G.; Kurek, D.; Paasch, S.; Brunner, E.; Born, R.; Galli, R.; Stelling, A.L.; Sivkov, V.N.; Petrova, O.V.; et al. Isolation and identification of chitin in black coral Paranthipates larix (Anthozoa: Cnidaria). *Int. J. Biol. Macromol.* **2012**, *51*, 129–137. [CrossRef]

34. Kuprina, E.E.; Krasavtsev, V.; Kozlova, I.; Vodolazhskaya, S.; Bogeruk, A.; Ezhov, V. Electrochemical method of chitinous products with enhanced ecology rehabilitation ability. In Proceedings of the Vth International Conference: New Prospects in Study of Chitin and Chitosan, Shchelkovo, Russia, 25–27 May 1999; pp. 42–44.

35. Pletcher, D.; Walsh, F.C. *Industrial Electrochemistry*; Springer: Dordrecht, The Netherlands, 1993.

36. Strathmann, H. Ion-Exchange Membrane Separation Processes. *Membr. Sci. Tech.* **2004**, *9*, 1–22.

37. Savari, S.; Sachdeva, S.; Kumar, A. Electrolysis of sodium chloride using composite poly(styrene-co-divinylbenzene) cation exchange membranes. *J. Membr. Sci.* **2008**, *310*, 246–261. [CrossRef]

38. Zeppilli, M.; Lai, A.; Villano, M.; Majone, M. Anion vs. cation exchange membrane strongly affect mechanisms and yield of CO_2 fixation in a microbial electrolysis cell. *Chem. Eng. J.* **2016**, *304*, 10–19. [CrossRef]

39. Holze, R.; Ahn, J. Advances in the use of perfluorinated cation exchange membranes in integrated water electrolysis and hydrogen/oxygen fuel cell systems. *J. Membr. Sci.* **1992**, *73*, 87–97. [CrossRef]

40. Park, J.E.; Kang, S.Y.; Oh, S.-H.; Kim, J.K.; Lim, M.S.; Ahn, C.-Y.; Cho, Y.-H.; Sung, Y.-E. High-performance anion-exchange membrane water electrolysis. *Electrochim. Acta* **2019**, *295*, 99–106. [CrossRef]

41. Salvatore, D.A.; Weekes, D.M.; He, J.; Dettelbach, K.E.; Li, Y.C.; Mallouk, T.E.; Berlinguette, C.P. Electrolysis of Gaseous CO_2 to CO in a Flow Cell with a Bipolar Membrane. *ACS Energy Lett.* **2017**, *3*, 149–154. [CrossRef]

42. Pisarska, B.; Wicher, I.; Dylewski, R. Studies on the parameters for membrane-electrolysis conversion of sodium sulfate solutions. *Przemysł Chem.* **2004**, *83*, 186–190.

43. Holze, S.; Jörissen, J.; Fischer, C.; Kalvelage, H. Hydrogen consuming anodes for energy saving in sodium sulphate electrolysis. *Chem. Eng. Technol.* **1994**, *17*, 382–389. [CrossRef]

44. Jörissen, J.; Simmrock, K.H. The behavior of ion exchange membranes in electrolysis and electrodialysis of sodium sulfate. *J. Appl. Electrochem.* **1991**, *21*, 869–876. [CrossRef]

45. Kumirska, J.; Czerwicka, M.T.; Kaczynski, Z.; Bychowska, A.; Brzozowski, K.; Thöming, J.; Stepnowski, P. Application of Spectroscopic Methods for Structural Analysis of Chitin and Chitosan. *Mar. Drugs* **2010**, *8*, 1567–1636. [CrossRef]

46. Ehrlich, H.; Rigby, J.K.; Botting, J.P.; Tsurkan, M.; Werner, C.; Schwille, P.; Petrásek, Z.; Pisera, A.; Simon, P.; Sivkov, V.N.; et al. Discovery of 505-million-year old chitin in the basal demosponge Vauxia gracilenta. *Sci. Rep.* **2013**, *3*, 3497. [CrossRef]

47. Ehrlich, H.; Kaluzhnaya, O.V.; Tsurkan, M.; Ereskovsky, A.V.; Tabachnick, K.R.; Ilan, M.; Stelling, A.; Galli, R.; Petrova, O.V.; Nekipelov, S.V.; et al. First report on chitinous holdfast in sponges (Porifera). *Proc. R. Soc. B* **2013**, *280*, 20130339. [CrossRef]

48. Ehrlich, H.; Kaluzhnaya, O.V.; Brunner, E.; Tsurkan, M.; Ereskovsky, A.V.; Ilan, M.; Tabachnick, K.R.; Bazhenov, V.; Paasch, S.; Kammer, M.; et al. Identification and first insights into the structure and biosynthesis of chitin from the freshwater sponge Spongilla lacustris. *J. Struct. Biol.* **2013**, *183*, 474–483. [CrossRef] [PubMed]

49. Henriques, B.S.; Garcia, E.S.; Azambuja, P.; Genta, F.A. Determination of Chitin Content in Insects: An Alternate Method Based on Calcofluor Staining. *Front. Physiol.* **2020**, *11*, 117. [CrossRef] [PubMed]

50. Denny, G.; Khanna, R.; Hornstra, I.; Kwatra, S.G.; Grossberg, A.L. Rapid detection of fungal elements using calcofluor white and handheld ultraviolet illumination. *J. Am. Acad. Dermatol.* **2020**, *82*, 1000–1001. [CrossRef] [PubMed]

51. Connors, M.J.; Ehrlich, H.; Hog, M.; Godeffroy, C.; Araya, S.; Kallai, I.; Gazit, D.; Boyce, M.; Ortiz, C. Three-dimensional structure of the shell plate assembly of the chiton Tonicella marmorea and its biomechanical consequences. *J. Struct. Biol.* **2012**, *177*, 314–328. [CrossRef] [PubMed]

52. Wysokowski, M.; Motylenko, M.; Walter, J.; Lota, G.; Wojciechowski, J.; Stöcker, H.; Galli, R.; Stelling, A.L.; Himcinschi, C.; Niederschlag, E.; et al. Synthesis of nanostructured chitin–hematite composites under extreme biomimetic conditions. *RSC Adv.* **2014**, *4*, 61743–61752. [CrossRef]

53. Żółtowska-Aksamitowska, S.; Tsurkan, M.; Lim, S.; Meissner, H.; Tabachnick, K.; Shaala, L.A.; Youssef, D.T.; Ivanenko, V.N.; Petrenko, I.; Wysokowski, M.; et al. The demosponge Pseudoceratina purpurea as a new source of fibrous chitin. *Int. J. Biol. Macromol.* **2018**, *112*, 1021–1028. [CrossRef]

54. Żółtowska-Aksamitowska, S.; Shaala, L.A.; Youssef, D.T.A.; Elhady, S.S.; Tsurkan, M.; Petrenko, I.; Wysokowski, M.; Tabachnick, K.; Meißner, H.; Ivanenko, V.N.; et al. First Report on Chitin in a Non-Verongiid Marine Demosponge: The Mycale euplectellioides Case. *Mar. Drugs* **2018**, *16*, 68. [CrossRef]

55. Fromont, J.; Zoltowska-Aksamitowska, S.; Galli, R.; Meissner, H.; Erpenbeck, D.; Vacelet, J.; Diaz, C.; Tsurkan, M.V.; Petrenko, I.; Youssef, D.; et al. New family and genus of a Dendrilla-like sponge with characters of Verongiida. Part II. Discovery of chitin in the skeleton of Ernstilla lacunosa. *Zoologischer Anzeiger* **2019**, *280*, 21–29. [CrossRef]

56. Shaala, L.A.; Asfour, H.; Youssef, D.T.A.; Żółtowska-Aksamitowska, S.; Wysokowski, M.; Tsurkan, M.; Galli, R.; Meißner, H.; Petrenko, I.; Tabachnick, K.; et al. New Source of 3D Chitin Scaffolds: The Red Sea Demosponge Pseudoceratina arabica (Pseudoceratinidae, Verongiida). *Mar. Drugs* **2019**, *17*, 92. [CrossRef]

57. Schubert, M.; Binnewerg, B.; Voronkina, A.; Muzychka, L.; Wysokowski, M.; Petrenko, I.; Kovalchuk, V.; Tsurkan, M.; Martinovic, R.; Bechmann, N.; et al. Naturally Prefabricated Marine Biomaterials: Isolation and Applications of Flat Chitinous 3D Scaffolds from Ianthella labyrinthus (Demospongiae: Verongiida). *Int. J. Mol. Sci.* **2019**, *20*, 5105. [CrossRef]

58. Kovalchuk, V.; Voronkina, A.; Binnewerg, B.; Schubert, M.; Muzychka, L.; Wysokowski, M.; Tsurkan, M.; Bechmann, N.; Petrenko, I.; Fursov, A.; et al. Naturally Drug-Loaded Chitin: Isolation and Applications. *Mar. Drugs* **2019**, *17*, 574. [CrossRef]

59. Kaya, M.; Mujtaba, M.; Ehrlich, H.; Salaberria, A.M.; Baran, T.; Amemiya, C.T.; Galli, R.; Akyuz, L.; Sargin, I.; Labidi, J. On chemistry of γ-chitin. *Carbohydr. Polym.* **2017**, *176*, 177–186. [CrossRef]

60. Pillar, E.A.; Zhou, R.; Guzman, M. Heterogeneous Oxidation of Catechol. *J. Phys. Chem. A* **2015**, *119*, 10349–10359. [CrossRef]

61. Holl, S.M.; Schaefer, J.; Goldberg, W.M.; Kramer, K.J.; Morgan, T.D.; Hopkins, T.L. Comparison of black coral skeleton and insect cuticle by a combination of carbon-13 NMR and chemical analyses. *Arch. Biochem. Biophys.* **1992**, *292*, 107–111. [CrossRef]

62. Zhang, M.; Haga, A.; Sekiguchi, H.; Hirano, S. Structure of insect chitin isolated from beetle larva cuticle and silkworm (*Bombyx mori*) pupa exuvia. *Int. J. Biol. Macromol.* **2000**, *27*, 99–105. [CrossRef]

63. Cairns, S.D. Deep-water corals: An overview with special reference to diversity and distribution of deep-water scleractinian corals. *Bull. Mar. Sci.* **2007**, *81*, 311–322.

64. Farfan, G.A.; Cordes, E.E.; Waller, R.G.; Decarlo, T.M.; Hansel, C.M. Mineralogy of Deep-Sea Coral Aragonites as a Function of Aragonite Saturation State. *Front. Mar. Sci.* **2018**, *5*, 473. [CrossRef]

65. Brugler, M.R.; Opresko, D.M.; France, S. The evolutionary history of the order Antipatharia (Cnidaria: Anthozoa: Hexacorallia) as inferred from mitochondrial and nuclear DNA: Implications for black coral taxonomy and systematics. *Zool. J. Linn. Soc.* **2013**, *169*, 312–361. [CrossRef]

66. Molodtsova, T.N.; Opresko, D.M. Black corals (Anthozoa: Antipatharia) of the Clarion-Clipperton Fracture Zone. *Mar. Biodivers.* **2017**, *47*, 349–365. [CrossRef]

67. Daly, M.; Brugler, M.R.; Cartwright, P.; Collins, A.G.; Dawson, M.N.; Fautin, D.G.; France, S.; McFadden, C.S.; Opresko, D.M.; Rodriguez, E.; et al. The phylum Cnidaria: A review of phylogenetic patterns and diversity 300 years after Linnaeus*. *Zootaxa* **2007**, *1668*, 127–182. [CrossRef]

68. Bo, M.; Di Camillo, C.; Addamo, A.M.; Valisano, L.; Bavestrello, G. Growth strategies of whip black corals (Cnidaria: Antipatharia) in the Bunaken Marine Park (Celebes Sea, Indonesia). *Mar. Biodivers. Rec.* **2009**, *2*, 1–6. [CrossRef]

69. Bo, M.; Bavestrello, G.; Angiolillo, M.; Calcagnile, L.; Canese, S.; Cannas, R.; Cau, A.; D'Elia, M.; D'Oriano, F.; Follesa, M.C.; et al. Persistence of Pristine Deep-Sea Coral Gardens in the Mediterranean Sea (SW Sardinia). *PLoS ONE* **2015**, *10*, e0119393. [CrossRef] [PubMed]

70. Roark, E.B.; Guilderson, T.; Dunbar, R.; Ingram, B. Radiocarbon-based ages and growth rates of Hawaiian deep-sea corals. *Mar. Ecol. Prog. Ser.* **2006**, *327*, 1–14. [CrossRef]

71. Prouty, N.; Roark, E.B.; Buster, N.; Ross, S. Growth rate and age distribution of deep-sea black corals in the Gulf of Mexico. *Mar. Ecol. Prog. Ser.* **2011**, *423*, 101–115. [CrossRef]

72. Wagner, D.; Luck, D.G.; Toonen, R.J. The Biology and Ecology of Black Corals (Cnidaria: Anthozoa: Hexacorallia: Antipatharia). *Adv. Mar. Biol.* **2012**, *63*, 67–132. [CrossRef]

73. Wagner, D.; Shuler, A. The black coral fauna (Cnidaria: Antipatharia) of Bermuda with new records. *Zootaxa* **2017**, *4344*, 367. [CrossRef]

74. Gąsiorek, P.; Cordeiro, R.T.S.; Perez, C.D. Black Corals (Anthozoa: Antipatharia) from the Southwestern Atlantic. *Zootaxa* **2019**, *4692*, 1–67. [CrossRef]

75. Goldberg, W.M. Chemical changes accompanying maturation of the connective tissue skeletons of gorgonian and antipatharian corals. *Mar. Biol.* **1978**, *49*, 203–210. [CrossRef]

76. Tazioli, S.; Bo, M.; Boyer, M.; Rotinsulu, H.; Bavestrello, G. Ecological observations of some common antipatharian corals in the marine park of Bunaken (North Sulawesi, Indonesia). *Zool. Stud.* **2007**, *46*, 227–241.

77. Goldberg, W.M.; Hopkins, T.L.; Holl, S.M.; Schaefer, J.; Kramer, K.J.; Morgan, T.D.; Kim, K. Chemical composition of the sclerotized black coral skeleton (Coelenterata: Antipatharia): A comparison of two species. *Comp. Biochem. Physiol. Part B: Comp. Biochem.* **1994**, *107*, 633–643. [CrossRef]

78. Goldberg, W.M. Chemistry and structure of skeletal growth rings in the black coral Antipathes fiordensis (Cnidaria, Antipatharia). *Hydrobiologia* **1991**, *216*, 403–409. [CrossRef]

79. Boden, N.; Sommer, U.; Spindler, K.-D. Demonstration and characterization of chitinases in the Drosophila-K-cell Line. *Insect Biochem.* **1985**, *15*, 19–23. [CrossRef]

 *marine drugs*

Article

Physicochemical and Biological Properties of Gelatin Extracted from Marine Snail *Rapana venosa*

Alexandra Gaspar-Pintiliescu [1], Laura Mihaela Stefan [1,*], Elena Daniela Anton [1], Daniela Berger [2], Cristian Matei [2], Ticuta Negreanu-Pirjol [3] and Lucia Moldovan [1]

[1] Departament of Cellular and Molecular Biology, National Institute of R&D for Biological Sciences, 296 Splaiul Independentei, 060031 Bucharest, Romania; alex.gaspar@yahoo.com (A.G.-P.); danielaanton89@gmail.com (E.D.A.); moldovanlc@yahoo.com (L.M.)

[2] Faculty of Applied Chemistry and Material Science, University "Politehnica" of Bucharest, 1-7 Gheorghe Polizu street, 011061 Bucharest, Romania; danaberger01@yahoo.com (D.B.); cristi_matei@yahoo.com (C.M.)

[3] Faculty of Pharmacy, University "Ovidius" of Constanta, 1 Aleea Universitatii, 900470 Constanta, Romania; ticuta_np@yahoo.com

* Correspondence: lauramihaelastefan@yahoo.com; Tel./Fax: +40-21-2200882

Received: 26 July 2019; Accepted: 15 October 2019; Published: 17 October 2019

Abstract: In this study, we aimed to obtain gelatin from the marine snail *Rapana venosa* using acidic and enzymatic extraction methods and to characterize these natural products for cosmetic and pharmaceutical applications. Marine gelatins presented protein values and hydroxyproline content similar to those of commercial mammalian gelatin, but with higher melting temperatures. Their electrophoretic profile and Fourier transform infrared (FTIR) spectra revealed protein and absorption bands situated in the amide region, specific for gelatin molecule. Scanning electron microscopy (SEM) analysis showed significant differences in the structure of the lyophilized samples, depending on the type of gelatin. In vitro studies performed on human keratinocytes showed no cytotoxic effect of acid-extracted gelatin at all tested concentrations and moderate cytotoxicity of enzymatic extracted gelatin at concentrations higher than 0.5 mg/mL. Also, both marine gelatins favored keratinocyte cell adhesion. No irritant potential was recorded as the level of IL-1α and IL-6 proinflammatory cytokines released by HaCaT cells cultivated in the presence of marine gelatins was significantly reduced. Together, these data suggest that marine snails are an alternative source of gelatins with potential use in pharmaceutical and skincare products.

Keywords: gelatin; marine gastropod; Black Sea; acidic and enzymatic extraction; biocompatibility; cytokines

1. Introduction

Gelatin is a protein obtained by thermal denaturation of collagen, the main constituent of connective tissue. Being a derivate product of collagen, gelatin has similar structural features and properties [1]. The primary structure of collagen type I consists of two α1-chains and one α2-chain containing the repeating amino acid sequence Gly–X–Y, where X and Y are mainly proline and hydroxyproline [2]. The specific primary structure leads to left-handed helices (secondary structure) and three alpha chains organize into a right-handed triple helix, forming a collagen molecule of 300 nm in length and less than 2 nm in diameter [3]. The denaturation process of collagen implies partial destruction of its tertiary, secondary and, to some extent, its primary structure, resulting in gelatin as a mixture of proteins and polypeptides [4].

Gelatin is extensively used as a natural biomaterial in tissue engineering due to its high biocompatibility, biodegradability, low antigenicity, and ability to stimulate cellular attachment and growth [5]. So far, commercial gelatin is conventionally obtained from mammalian tissues, like skin and bones from bovine, porcine, or caprine species. Recently, there has been great interest in obtaining gelatin from other sources, especially marine organisms, in order to avoid transmitting bovine spongiform encephalopathy or swine flu, as well as for religious reasons [6]. Moreover, byproducts of different fish species [7,8] and other marine sources have been used for gelatin extraction, such as sponges [9–11], jellyfishes [12,13], squids [14,15], or snails [16,17].

Rapana venosa is a marine snail belonging to the Muricidae family that is rich in proteins, amino acids, sterols, and vitamins [18–20]. It is a predatory marine snail that quickly expanded in the Black Sea, having a negative effect on the ecosystem, especially inducing the decline of different species of mussels and mollusks [21]. Besides its nutritional value [18,22], *R. venosa* is also appreciated as a potential source for biotechnological applications. Previous studies have reported that amino acids and lipids extracted from *R. venosa* exhibited wound-healing properties on rat skin burns [18,23]. Recently, Luo et al. [24] obtained protein hydrolysates from *R. venosa* with significant antioxidant activity.

The aim of this study was to extract and characterize gelatin from the soft tissue of *R. venosa* using acidic and enzymatic methods, in order to assess their use in the pharmaceutical and cosmetic fields. Its physicochemical and ultrastructural properties were analyzed and compared to those of commercial pig skin gelatin. In addition, marine gelatins were tested on human keratinocyte cells for their cytocompatibility, cell adhesion capacity, and irritant potential.

2. Results and Discussion

2.1. Physicochemical and Structural Properties of Marine Gelatins

2.1.1. Yield Extraction and Gelatin Characteristics

In order to obtain gelatin, insoluble native collagen is subjected to thermal hydrolysis, using chemical and enzymatic methods. Both methods are intended to break the inter- and intramolecular crosslinks without cleavage of the peptide bonds, so that the polypeptide chains remain intact [25]. In our study, after the pretreatment of the marine snail soft tissue with NaOH, chemical hydrolysis was performed using acetic acid, an organic acid capable to induce higher solubility of the tissue during the extraction process [25], resulting in an acid-solubilized gelatin (ASG) solution. The pepsin-solubilized gelatin (PSG) solution was obtained by enzymatic hydrolysis with pepsin, which cleaves bonds in the telopeptide region of the collagen structure [26]. Applying these two methods at 60 °C, we have obtained a higher extraction yield for the acidic treatment (9.71%), compared to the enzymatic one (8.65%) (Table 1).

Table 1. Yield and characteristics of acid-solubilized gelatin (ASG) and pepsin-solubilized gelatin (PSG) from *R. venosa* and commercial pig skin gelatin (CG). The results are expressed as mean $\pm$ SD ($n = 3$). * $p < 0.05$, compared to CG sample.

Gelatin Type	Extraction Yield (%)	Protein Content (%)	Hyp Content (%)	Melting Temperature (°C)
ASG	9.71 ± 0.38	91.48 ± 4.61	10.62 ± 0.37	35.30 ± 1.56
PSG	8.65 ± 0.42	83.12 ± 3.30	9.39 ± 0.51 *	33.20 ± 1.38
CG	-	86.12 ± 3.23	11.17 ± 0.21	28.80 ± 1.93

The yield values were comparable to those of gelatin extracted from squids (7.5%), but were lower than those obtained from jellyfish (11.8%–12%) and several fish species (ranging between 11.3% and 20.27%) [14,27,28]. A previous study reported a yield of 8.69%, 10.57%, and 6.54% of gelatin extracted from the body, foot, and viscera of *Ficus variegata* gastropod [16]. The gelatin extraction yields are affected by the tissue used for extraction, the quantity of soluble components in the source, or the collagen content [17]. According to Jamilah and Harvinder [29], the yield extraction can also be influenced by other factors, such as temperature, time of extraction, concentration of NaOH or acetic acid, tissue/enzyme ratio, and pH. The temperatures ranging between 45 and 60 °C are considered optimal regarding the extraction yield of gelatin [30,31]. Higher temperature and longer time of extraction could result in an increased yield, but loss of functional properties of gelatin [30].

The protein content of both marine gelatins was high (91.48% for ASG and 83.12% for PSG) and comparable to that of the commercial pig skin gelatin (CG), indicating the efficiency of the used extraction methods (Table 1).

ASG and CG samples showed similar Hyp values (10.62% and 11.17%, respectively), while PSG exhibited a slightly lower Hyp content (9.39%). Glycine, proline, and hydroxyproline are the most abundant amino acids found in variable amounts in gelatin compositions depending on the source. The total amount of proline and hydroxyproline in fish gelatin is about 16%–20% [32]. Our data were comparable to those reported for gelatin extracted from ribbon jellyfish (*Chrysaora* sp.) and tilapia fish skin, with a Hyp content of 8.2% and 10.31%, respectively [33,34].

The melting temperature, determined by differential scanning calorimetry, was higher for ASG and PSG samples (35.3 and 33.2 °C, respectively) compared to CG (28.8 °C) (Table 1, Supplementary Figure S1). However, the melting temperature values of ASG and PSG extracted from *R. venosa* were close to those reported for gelatin derived from other marine sources, such as *Chondrosia reniformis* (30.48 °C) and *Thymosia guernei* (31.02 °C) marine sponges [35]. In our study, the gelatin samples obtained from the marine snail *R. venosa* showed thermal stability, indicating the possibility of using these components for the development of new biomaterials that require heat resistance.

2.1.2. SDS-Polyacrylamide Gel Electrophoresis (SDS-PAGE)

Figure 1 shows the marine gelatins from *R. venosa* evaluated by electrophoresis in polyacrylamide gel and compared to that of CG from pig skin.

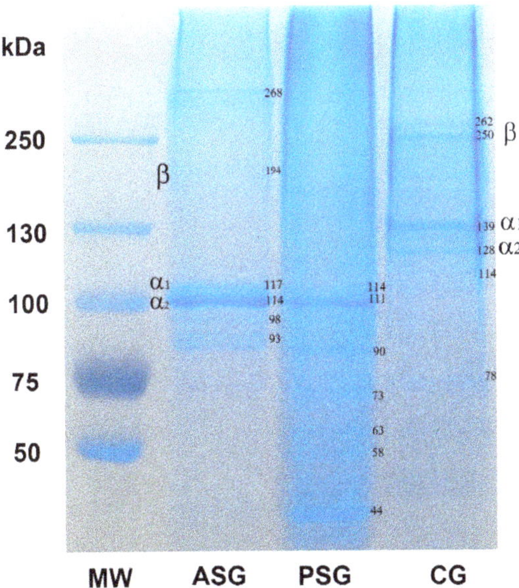

Figure 1. SDS-polyacrylamide gel electrophoresis (SDS-PAGE) showing ASG and PSG marine gelatins from *R. venosa* and CG from pig skin; MW—molecular weight marker. Numbers represent the molecular weight of different protein bands identified in marine and commercial gelatins.

The ASG sample presented α- and β-chains, as major protein constituents, corresponding to the following molecular weights: ~114–117 and ~194 kDa, respectively. For ASG, a higher molecular weight protein of ~268 kDa and two lower molecular weight proteins of ~93–98 kDa were also observed.

The presence of β-dimer and the protein of ~268 kDa indicated that ASG contains intermolecular crosslinks which have not been hydrolyzed during the extraction. The electrophoretic pattern of ASG was similar with those reported for gelatin extracted from marine snails *Hexaplex trunculus* and *Ficus variegate* [16,17]. The PSG did not display β-chains, but revealed the presence of α-chains at ~111–114 kDa and several protein bands with low molecular mass of ~90, 73, 63, 58, and 44 kDa. This mixture of polypeptides is probably the result of a high degree of hydrolysis due to the pepsin treatment for 24 h and the additional heat treatment at 98 °C. Previous studies showed that gelatin extracted from shark skin and African catfish exhibited an increase in shorter chain fragments and a decrease in the intensity of high molecular weight chains, findings which were consistent with our results [36,37]. Both gelatin samples obtained in this study exhibited bands corresponding to α-chains with a lower molecular mass (~110–117 kDa) than that of CG (123–139 kDa). The CG presented the typical α- (~123–139 kDa) and β-chains (~250–262 kDa) corresponding to collagen type I and several bands of polypeptides with molecular weight of 114 and 78 kDa. Similar results were reported for collagen extracted from different fish species, which consisted of two α_1 and one α_2 chains with slightly lower molecular weight than collagen type I from calf skin [38]. Collagen from small-spotted catshark also exhibited α subunits lower than 110 kDa [39]. The differences in the α- and β-chains position between species is probably related to the number of amino acids, which differ between marine and mammalian collagens [7].

2.1.3. Fourier Transform Infrared (FTIR) Spectroscopy

In order to study the secondary structure of gelatin samples isolated from marine snail, we have used FTIR spectroscopy analysis. The FTIR spectra of the tested samples are shown in Figure 2.

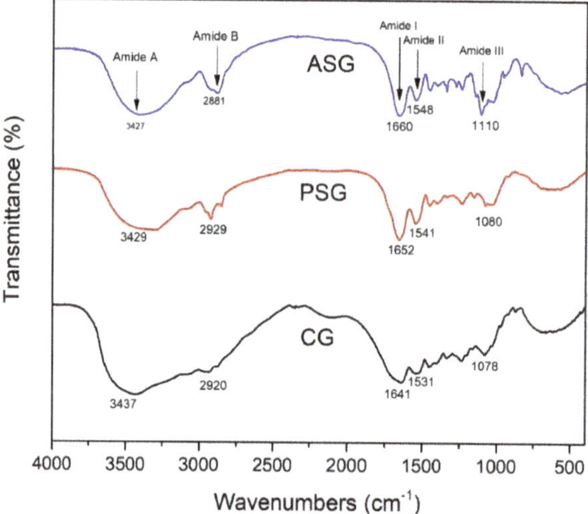

Figure 2. Fourier transform infrared (FTIR) spectra of ASG and PSG marine gelatins from *R. venosa* compared to CG from pig skin.

R. venosa gelatins exhibited main absorption bands, specific for the peptide bonds in the amide band regions. Thus, the amide A large band, associated with N–H stretching vibration, depending on the conformation of gelatin, shifted to lower wavenumbers for ASG (3427 cm^{-1}) and PSG (3429 cm^{-1}) samples compared to CG (3437 cm^{-1}), which demonstrated a lower structural order of polypeptide chains. The amide B bands, corresponding to symmetric and asymmetric vibrations of C–H bonds, were observed at 2881 cm^{-1} for ASG. The amide I band corresponding to the C=O stretching vibration from amide group shifted towards higher wavenumbers, at 1660 and 1652 cm^{-1} for ASG and PSG, respectively, when compared to CG (1641 cm^{-1}), due to the weaker H-bond formation in snail gelatins.

The amide II band resulted from the overlap of amide N–H bending and C–N stretching vibrations appeared at 1548 cm^{-1} for ASG sample and shifted at 1541 cm^{-1} for PSG sample, probably as a result of higher content of imide bonds formation. The amide II band of CG from 1531 cm^{-1} was larger and shifted towards lower wavenumbers than that of marine gelatins, which was consistent with a more disordered helical structure with more imide bonds formation. In the FTIR spectra, the amide III band was assigned to NH bending at 1230 cm^{-1}, while C–O stretching vibrations superimposed with C–N stretching vibrations were identified at 1110, 1080, and 1078 cm^{-1} for ASG, PSG and CG, respectively, suggesting a degree of glycosylation [40].

Overall, these results indicated a comparable structure and chemical composition of the gelatins obtained from *R. venosa* using acidic or enzymatic extraction, but a structure less associated by hydrogen bonding in the case of ASG. In addition, slight differences between marine and pig skin gelatins were observed probably due to variations in the sequence of amino acids. Similar FTIR spectra were reported for collagen/gelatin extracted from *C. reniformis* marine sponge [10].

2.1.4. Scanning Electron Microscopy (SEM) Observations

The ultrastructure of the obtained samples was observed by SEM. The lyophilized samples of ASG exhibited a rough, multilayered appearance, in contrast to PSG and CG samples, which showed a fibril-like network forming a microporous structure with uneven sized pores (Figure 3). A previous study has reported a nonfibrillar form of collagen/gelatin extract obtained from *C. reniformis* marine sponge, which presented a nodular structure with a rough appearance [10].

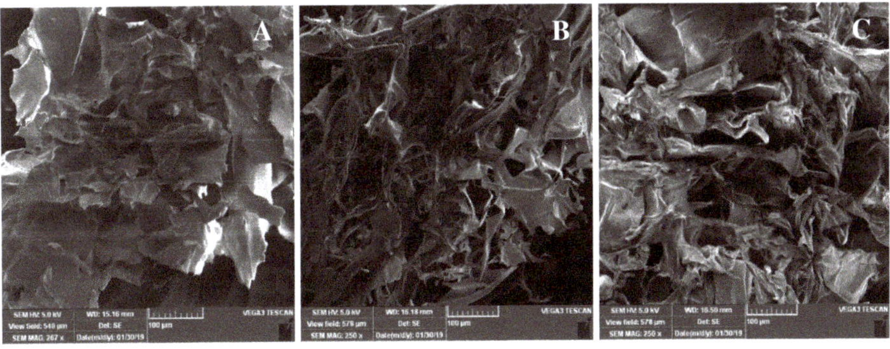

Figure 3. Scanning electron micrographs showing the surface of freeze-dried (**A**) ASG and (**B**) PSG from *R. venosa* and (**C**) CG from pig skin.

2.2. *In Vitro Biocompatibility of Marine Gelatins*

2.2.1. Evaluation of Cell Viability

The percentage of cell viability after the treatment with different concentrations of ASG, PSG, and CG was assessed by MTT assay, which evaluates the activity of mitochondrial dehydrogenases. ASG sample showed a good biocompatibility at all tested concentrations (0.05–1.5 mg/mL) and at both exposure times (24 and 48 h) (Figure 4). All cell viability values were above 80% (noncytotoxic effect), ranging between 90.49% and 97.59% after 24 h of treatment, and between 80.05% and 97.47% after 48 h. On the other hand, PSG sample showed no cytotoxic activity at all tested concentrations after 24 h of exposure (viability values between 82.55% and 98.68%), whereas CG exhibited viability values above 80% only within the concentration range of 0.05–0.5 mg/mL. After 48 h, PSG maintained a good biocompatibility up to the concentration of 0.5 mg/mL (viability values between 81.93% and 96.63%), but cell viability decreased below 75% at higher concentrations. The CG sample showed a similar profile with that observed after 24 h of treatment, with viability values below 80% at concentrations higher than 0.5 mg/mL. Overall, the ASG exhibited no cytotoxic effect at all tested concentrations, whereas the PSG and CG showed a similar biocompatibility, with cell viability values decreasing below 80% at concentrations higher than 0.5 mg/mL.

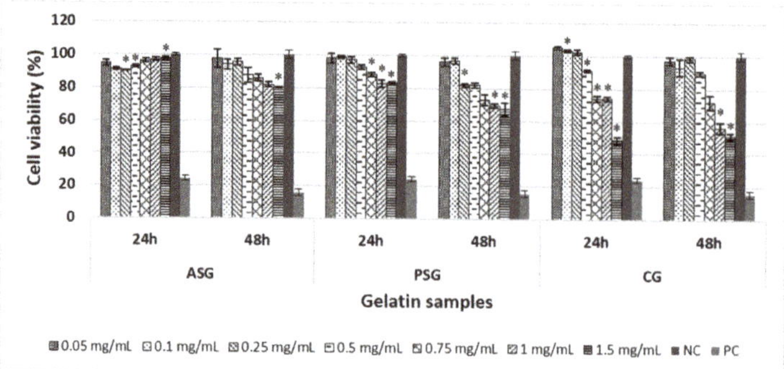

Figure 4. Cell viability of HaCaT cells exposed to increasing concentrations of ASG, PSG, and CG samples for 24 and 48 h, as evaluated by MTT assay. The negative control (NC) was represented by untreated cells and the positive control (PC) was represented by 100 μM H_2O_2. All samples were normalized to the NC considered to be 100% viable. Data were presented as mean ± SD (n = 3). * $p < 0.05$ compared to the NC.

Previous studies have reported no cytotoxic effects of collagen/gelatin obtained from different marine sources on various cell lines. For example, collagen from codfish skin did not affect lung fibroblast metabolism at concentrations ranging between 0.01 and 0.05 mg/mL, but exhibited cytotoxic activity at concentrations higher than 0.1 mg/mL [41]. Collagen obtained from the starfish *Asterias amurensis* also promoted growth and viability of human dermal fibroblasts at concentrations ranging between 0.01 and 1 mg/mL [42], whereas squid gelatin peptides (0.025–0.1 mg/mL) exhibited a dose-dependent increase of cell viability in oxidation-induced human lung fibroblasts [43].

2.2.2. Morphological Examination

Live/dead staining was used to evaluate the cell morphology and viability after the treatment with marine gelatin samples. Thus, HaCaT cells maintained their viability after 48 h of cultivation in the presence of marine and commercial gelatins, while few dead cells could be observed (Figure 5).

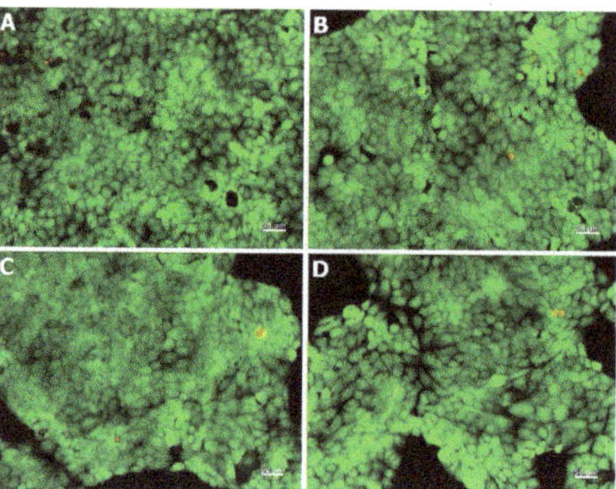

Figure 5. Live/dead staining with calcein-AM (green) and ethidium homodimer-1 (red) of HaCaT cells untreated control; (**A**) and treated with ASG (**B**), PSG (**C**), and CG (**D**) at the concentration of 0.25 mg/mL.

Furthermore, cells treated with marine and commercial gelatins showed no significant morphological changes, maintaining their normal phenotype when compared to that of the untreated cells. In addition, quantitative analysis of cell density performed with ImageJ software showed that more than 99% of cells were viable for all samples. ASG, PSG, and CG promoted growth and viability of human keratinocytes, results which correlated well with those obtained by the quantitative MTT assay.

2.3. Biological Properties of Marine Gelatins

2.3.1. Cell Adhesion Capacity

The adhesion of HaCaT cells on gelatin coatings was evaluated using phalloidin TRITC staining, which highlighted the actin filaments. The distribution of actin revealed the morphological changes at the cytoskeleton level in HaCaT cells (Figure 6).

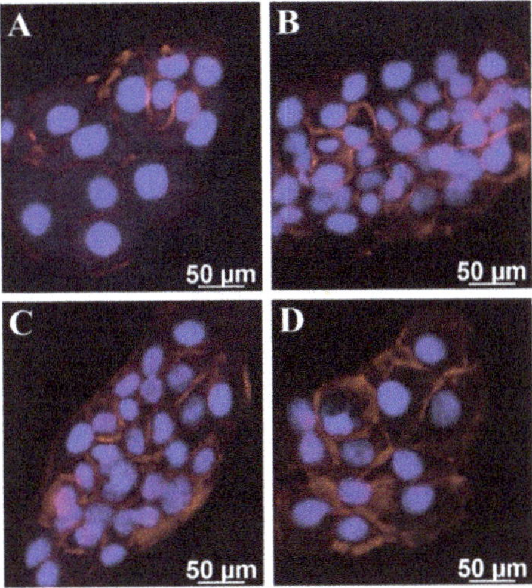

Figure 6. Distribution of actin filaments in HaCaT cells, assessed by fluorescence microscopy. HaCaT cells adhered to plastic (**A**) and to 0.25 mg/mL ASG (**B**), PSG (**C**), and CG coatings (**D**). Cells were stained for actin (red) and nuclei (blue).

In the control, cells exhibited a polygonal-shaped morphology typical of HaCaT cells, with actin filaments assembled into large radial bundles (Figure 6A). The cells cultivated on ASG- and PSG-coated coverslips presented a different actin distribution, with smaller filaments which were visualized close to the nucleus (Figure 6B,C). The cells that adhered to the CG coating presented morphological features similar to the control (Figure 6A,D).

2.3.2. Irritant Potential

According to the Organization for Economic Cooperation and Development (OECD) guidelines for testing of chemicals, a substance is considered to be an irritant if it has the ability to decrease cell viability below the defined threshold of 50%, when compared to the negative control (untreated cells). In our study, marine gelatin samples exhibited cell viability values higher than 50% at all tested concentrations, with values ranging between 73.33% and 108.33% and, therefore, can be considered non-irritants for skin. These results were also correlated with the low release of IL-6 and IL-1α proinflammatory cytokines as measured by ELISA assay. These interleukins are active and pleiotropic inflammatory cytokines that play a key role in the inflammatory process [44,45]. IL-1α is a known endpoint to predict skin irritation and the concentration of IL-1α released by keratinocytes in culture medium has been reported to increase after exposure to different irritants [46]. In our study, the cytokine content, expressed as pg/mL, varied with the sample concentration (Tables 2 and 3). Thus, in the case of ASG, the IL-1α content varied from 1.39 to 6.72, whereas slightly higher values were observed for PSG (Table 2). For the CG sample, the concentration of IL-1α was similar to that of the negative control, ranging between 0.37 and 1.06. However, all these values were significantly lower than that of 0.1% SDS (40.93) used as positive control (Table 2).

Table 2. IL-1α secretion levels expressed as pg/mL in the culture medium of HaCaT keratinocytes treated with increasing concentrations of marine gelatin samples.

Sample	Tested Concentrations			
	0.1 mg/mL	0.25 mg/mL	0.5 mg/mL	0.75 mg/mL
ASG	1.39 ± 0.27 *	1.10 ± 0.20 *	3.40 ± 0.55 *	6.72 ± 1.27 *
PSG	1.90 ± 0.87 *	2.82 ± 0.94 *	5.85 ± 1.83 *	7.35 ± 1.00 *
CG	0.37 ± 0.11 *	0.61 ± 0.18 *	0.77 ± 0.21 *	1.06 ± 0.13 *
NC	0.34 ± 0.07 *			
PC	40.93 ± 4.02			

NC—negative control (untreated cells), PC—positive control (SDS 0.1%). Values are expressed as mean ± SD (n = 3). * p < 0.05 compared to PC.

The levels of IL-6 increased from 54.08 at the concentration of 0.1 mg/mL to 279.65 at the concentration of 0.75 mg/mL for ASG, and ranged between 92.67 and 384.02 for PSG (Table 3). Although the concentration of IL-6 released in culture medium was higher for PSG compared to ASG, all values were significantly lower compared to the positive control (1094.09). For the CG sample, the IL-6 content was slightly higher than that of the negative control, ranging between 30.49 and 98.52 (Table 3). For TNF-α, no evidence of the release of this cytokine was detected in the cell culture medium after the treatment with gelatin samples for 18 h. Similar results were reported by Alves et al. [47] in the case of human keratinocyte treatment with codfish skin collagen, where no release of IL-6 and IL-18 was detected in the culture medium, highlighting the non-irritant effect and cosmetic potential of marine collagen.

Table 3. IL-6 secretion levels expressed as pg/mL in the culture medium of HaCaT keratinocytes treated with increasing concentrations of marine gelatin samples.

Sample	Tested Concentrations			
	0.1 mg/mL	0.25 mg/mL	0.5 mg/mL	0.75 mg/mL
ASG	54.08 ± 1.49 *	162.81 ± 3.03 *	244.24 ± 19.70	279.65 ± 41.15 *
PSG	92.67 ± 7.55 *	196.44 ± 6.79 *	311.89 ± 16.10 *	384.02 ± 15.64
CG	46.46 ± 3.10 *	30.49 ± 1.59 *	98.52 ± 1.78 *	42.53 ± 4.65 *
NC	6.45 ± 2.30 *			
PC	1094.88 ± 188.96			

NC—negative control (untreated cells), PC—positive control (SDS 0.1%). Values are expressed as mean ± SD (n = 3). * p < 0.05 compared to PC.

3. Experimental Section

3.1. Raw Materials

Marine snails were collected in August 2018 from the Romanian seacoast of the Black Sea between the 2 Mai and Vama Veche areas. The samples were washed with cold distilled water and stored at −20 °C until use. CG from pig skin and other chemicals were purchased from Sigma-Aldrich (Saint Louis, MO, USA) unless otherwise specified.

3.2. Gelatin Extraction

Snail soft tissue was removed from the hard shell, washed with distilled water for 30 min, and cut into small pieces (2–5 mm) using scissors. In order to remove noncollagenous proteins, the cleaned tissue was pretreated with 0.5 M NaOH solution in a ratio of 1:10 (*w/v*) at room temperature for 24 h. After centrifugation at 5000 *g* for 30 min, the obtained residue was washed with distilled water until neutral pH was achieved. Gelatin was extracted using acidic and enzymatic methods.

Acidic extraction was performed by gentle stirring of the pretreated tissue in 0.5 M acetic acid solution (1:10, *w/v*) at room temperature for 24 h. The sample was centrifuged at 8000 *g* for 40 min and heated at 60 °C in a shaking water bath (Witeg, Wertheim, Germany) for 20 h. Then, the ASG solution was filtered to remove the insoluble material, dialyzed against distilled water, and freeze-dried at −40 °C for 48 h.

For enzymatic extraction, the pretreated tissue was digested using pepsin from porcine gastric mucosa (2000 FIP-U/g, Carl Roth, Karlsruhe, Germany) in 0.5 M acetic acid solution at a pepsin/dry tissue ratio of 1:10 (*w/w*) and continuously stirred at room temperature for 24 h. The sample was centrifuged at 8000 *g* for 40 min and the resulting solution was subjected to thermal treatment at 98 °C for 1 min, in order to inactivate the enzyme. Then, the solution was heated at 60 °C in a shaking water bath for 20 h. Finally, the PSG solution was filtered, dialyzed against distilled water, and freeze-dried at −40 °C for 48 h. The obtained gelatin powders were stored at 4 °C until use.

3.3. Yield of Gelatin Extraction

The yield of gelatin extraction was calculated based on wet weight of fresh tissue using the following formula:

$$\text{Extraction yield (\%)} = \text{Gelatin dried weight/Fresh tissue wet weight} \times 100.$$

3.4. Protein Content

The total protein content was assessed using a bicinchoninic acid (BCA) protein assay kit, according to the manufacturer's instructions. The absorbance of the samples was read at 562 nm, using an UV/VIS spectrophotometer (Jasco, V650, Tokyo, Japan). Bovine serum albumin (BSA) was used as standard.

3.5. Hydroxyproline Content

Hydroxyproline (Hyp) content of gelatin samples was determined according to the method of Edwards and O'Brien Jr [48], with several modifications. Briefly, gelatin samples (0.05 g) were hydrolyzed in 5 mL perchloric acid 70 % at 120 °C for 8 h. The solutions were neutralized with 2.5 N NaOH at pH 6 and then an oxidant solution (a mixture of 1.41% chloramine T and acetate/citrate buffer, pH 6) was added. The mixtures were incubated at room temperature for 20 min and, finally, 26% perchloric acid and 10% 4-(dimethylamino)benzaldehyde (DMAB) dissolved in *n*-propanol were added. The solutions were heated at 60 °C for 20 min and the absorbance was then read at 560 nm using an UV/VIS spectrophotometer (Jasco, V650, Tokyo, Japan). A hydroxyproline standard curve was prepared from serial dilutions in the range of concentrations 1–10 µg/mL. The Hyp content was expressed as g/100 g dry weight.

3.6. Differential Scanning Calorimetry

The melting temperature of gelatins was assessed by differential scanning calorimetry (DSC) using a Mettler Toledo (Greifensee, Switzerland) equipment. Freeze-dried samples (2–3 mg) were mixed with 10 µL distilled water and incubated at temperatures ranging between 25 and 85 °C, with a heating rate of 2.5 °C/min.

3.7. SDS-PAGE Analysis

SDS-PAGE was conducted according to the method of Laemmli [49] with minor modifications. Samples were diluted in Laemmli buffer at a ratio of 1:2 (*v/v*), heated at 100 °C for 5 min, and loaded on a 5% stacking gel and a 7.5% resolving gel. Then, they were migrated in a vertical gel unit (Biometra, Analytik Jena, Jena, Germany) at a constant current of 10 mA for 4 h. After electrophoresis, the gels were stained using Roti Blue solution (Carl-Roth, Karlsruhe, Germany), destained in a solution of methanol/distilled water 1:3, and photographed. Commercial gelatin (CG) from pig skin was used as control and high molecular weight marker (55–250 kDa) was migrated in the same conditions. The molecular weight of protein bands was determined using a logarithmic regression analysis by plotting the log of molecular weight versus relative mobility.

3.8. FTIR Spectroscopy

Gelatin samples were mixed with potassium bromide (KBr) and ground into powder. FTIR spectra were performed in the range of wavelength between 4000 and 400 cm^{-1} with a resolution of 5 cm^{-1} using a Bruker Tensor 27 (Billerica, MA, USA) infrared spectrometer. A total of 50 scans was carried out for each sample.

3.9. Scanning Electron Microscopy

The structural morphology of gelatins was examined by SEM. Lyophilized samples were coated with a gold layer and visualized on TESCAN VEGA 3 LMH scanning microscope (Brno, Czech Republic) operated at 15 kV in low vacuum mode.

3.10. Cell Viability Evaluation

In vitro experiments were performed on the spontaneously immortalized human keratinocyte cell line HaCaT purchased from AddexBio, San Diego, CA, USA. Cells were grown in RPMI 1640 culture medium (Biochrom, Berlin, Germany) supplemented with 10% fetal bovine serum (FBS) and 1% antibiotics (penicillin, streptomycin, and neomycin) at 37 °C in a humidified atmosphere with 5% CO_2. In order to evaluate cell cytotoxicity, HaCaT cells were seeded in 96-well culture plates at a density of 5×10^4 cells/mL and incubated for 24 h to allow cell attachment. After this period, fresh medium containing different concentrations of ASG, PSG, and CG (0.05; 0.1; 0.25; 0.5; 0.75; 1 and 1.5 mg/mL) were added into each well and plates were incubated in standard conditions for 24 and 48 h, respectively. Untreated cells and cells cultivated in the presence of 100 µM H_2O_2 served as negative and positive controls, respectively. Cell metabolic activity was measured using an MTT assay [47]. Briefly, MTT working solution (0.25 mg/mL prepared in culture medium without FBS) was added to the cells and the plates were incubated at 37 °C for 3 h. The insoluble formazan crystals were dissolved using isopropanol and, after 15 min of incubation at room temperature with gentle stirring, the absorbance was read at 570 nm using a Mithras LB 940 microplate reader (Berthold Technologies, Bad Wildbad, Germany). The recorded value directly correlates to the number of metabolically active cells. The results of the MTT assay were calculated using the following equation: %cell viability = sample absorbance/negative control absorbance × 100. The negative control was considered 100% viable.

3.11. Live/Dead Assay

Cell morphology was assessed by fluorescence microscopy using the Live/Dead assay kit (Molecular Probes, Thermo Fisher Scientific, Eugene, OR, USA) according to the manufacturer's instructions. Briefly, after 48 h of cultivation in the presence of different gelatin samples, cells were stained with calcein-AM (2 µM) and ethidium homodimer-1 (4 µM) at room temperature for 30 min. Fluorescent images were acquired using an Axio Observer D1 microscope and analyzed using AxioVision 4.6 software (Carl Zeiss, Oberkochen, Germany). All images were processed using ImageJ 1.51 software (Bethesda, MD, USA) and quantitative analysis of cell density were performed counting calcein and ethidium homodimer-1 positive cells. The obtained values have been normalized to the control (100% viability).

3.12. Cell Adhesion Assay

For cell adhesion assay, solutions containing 0.25 mg/mL of ASG, PSG and CG were added on coverslips previously inserted in 24-well plates and allowed to dry at room temperature, overnight. After evaporation, coatings were sterilized by UV irradiation for 3 h. HaCaT cells were seeded at a density of 1×10^5 cells/mL on coverslips coated with gelatin solution. Coverslips without gelatin coatings were used as control. After 24 h of incubation at 37 °C in a humidified atmosphere with 5% CO_2, the culture medium was discarded, and cells were washed with PBS and fixed with 4% formaldehyde. Cells were permeated with 0.1% Triton X-100 solution at room temperature for 10 min and stained with phalloidin TRITC (50 µg/mL) for 40 min and then with DAPI (1 µg/mL) for 10 min. Fluorescent images were acquired using an Axio Observer D1 microscope and analyzed using AxioVision 4.6 software (Carl Zeiss, Oberkochen, Germany). All images were processed using ImageJ 1.51 software (Bethesda, MD, USA).

3.13. Irritant Potential Test

In order to assess the irritant potential of gelatins, HaCaT cells were seeded in 24-well culture plates at a density of 1×10^5 cells/mL and cultivated in RPMI 1640 medium supplemented with 10% FBS. After 24 h, gelatin samples (0.1; 0.25; 0.5, and 0.75 mg/mL), negative control (cells grown in culture medium) and positive control (cells treated with 0.1% SDS) were added to the cells and incubated in standard conditions for 18 h. The culture medium was collected and used for cytokine analysis. Levels of IL-6, IL-1α, and TNF-α proinflammatory cytokines were determined using commercial ELISA kits, according to the manufacturer's instructions (Invitrogen, Thermo Fisher Scientific, Vienna, Austria). The absorbance was recorded at 450 nm using a microplate reader Tecan Sunrise (Tecan, Grodig, Austria).

3.14. Statistical Analysis

All experiments were performed in triplicate, and the data are presented as mean ± standard deviation (SD). Statistical analysis was performed using Student *t*-test. A value of $p < 0.05$ was considered to be significant.

4. Conclusions

Gelatin was isolated for the first time from the marine snail *R. venosa* using acidic and enzymatic methods. The samples presented comparable values to commercial gelatin, in terms of protein, hydroxyproline content, and melting temperature. The structural features highlighted by FTIR spectra were slightly different for all tested gelatins. The electrophoretic profile showed a high degree of hydrolysis and peptides with low molecular weight for the enzymatic gelatin compared to acidic gelatin. Regarding the in vitro tests, gelatin obtained by chemical treatment showed a better cytocompatibility compared to the enzymatic extracted gelatin, with no signs of cytotoxicity and irritant potential. In addition, both marine gelatins promoted cell proliferation and cell-induced adhesion capacity when compared to mammalian gelatin. Overall, our results suggested that *R. venosa* marine snail could be a valuable alternative and safe source of gelatin, useful as an additive in biomedical and pharmaceutical skincare products.

Supplementary Materials: The following are available online at http://www.mdpi.com/1660-3397/17/10/589/s1, Figure S1: Thermal melting curves of ASG and PSG from *R. venosa* and CG from pig skin.

Author Contributions: A.G.-P., L.M.S., L.M. and T.N.-P. conceived and designed the study. A.G.-P., L.M., L.M.S., E.D.A., D.B. and C.M. performed the experiments. A.G.-P., L.M.S. and L.M. analyzed the data. A.G.-P., L.M.S., T.N.-P. and L.M. wrote the manuscript. All authors revised the manuscript and approved the final version.

Funding: This work was supported by the Romanian Ministry of Research an Innovation, UEFISCDI, project No. PN-III-P1-1.2-PCCDI-2017-0701, grant no. 85PCCDI/2018. APC was funded by the Romanian Ministry of Research an Innovation, Institutional Performance Projects for Excellence Financing in RDI, grant No. 22PFE/2018.

Acknowledgments: The authors thank Raul-Augustin Mitran from "Ilie Murgulescu" Institute of Physical Chemistry, Bucharest, for melting temperature experiments.

Conflicts of Interest: The authors declare no conflict of interest.

References

1. Gorgieva, S.; Kokol, V. Collagen- vs. gelatin-based biomaterials and their biocompatibility: Review and perspectives. In *Biomaterials Applications for Nanomedicine*; Rosario, P., Ed.; InTech: Rijeka, Croatia, 2011; pp. 17–52. ISBN 978-953-307-661-4.

2. Duconseille, A.; Astruc, T.; Quintana, N.; Meersman, F.; Sante-Lhoutellier, V. Gelatin structure and composition linked to hard capsule dissolution: A review. *Food Hydrocoll.* **2015**, *43*, 360–376. [CrossRef]

3. Shoulders, M.D.; Raines, R.T. Collagen structure and stability. *Ann. Rev. Biochem.* **2009**, *78*, 929–958. [CrossRef] [PubMed]

4. See, S.F.; Ghassem, M.; Mamot, S.; Babji, A.S. Effect of different pretreatments on functional properties of African catfish (Clarias gariepinus) skin gelatin. *Int. J. Food Sci. Technol.* **2015**, *52*, 753–762. [CrossRef] [PubMed]

5. Nikkhah, M.; Akbari, M.; Paul, A.; Memic, A.; Dolatshahi-Pirouz, A.; Khademhosseini, A. Gelatin-based biomaterials for tissue engineering and stem cell bioengineering. In *Biomaterials from Nature for Advanced Devices and Therapies*; Neves, N.M., Reis, R.L., Eds.; John Wiley & Sons, Inc.: Hoboken, NJ, USA, 2016; pp. 37–62. ISBN 9781119126218.

6. Karim, A.A.; Bhat, R. Fish gelatin: Properties, challenges, and prospects as an alternative to mammalian gelatins. *Food Hydrocoll.* **2009**, *23*, 563–576. [CrossRef]

7. Silva, T.; Moreira-Silva, J.; Marques, A.; Domingues, A.; Bayon, Y.; Reis, R. Marine origin collagens and its potential applications. *Mar. Drugs* **2014**, *12*, 5881–5901. [CrossRef]

8. Gómez-Guillén, M.C.; Giménez, B.; López-Caballero, M.A.; Montero, M.P. Functional and bioactive properties of collagen and gelatin from alternative sources: A review. *Food Hydrocoll.* **2011**, *25*, 1813–1827. [CrossRef]

9. Tziveleka, L.A.; Ioannou, E.; Tsiourvas, D.; Berillis, P.; Foufa, E.; Roussis, V. Collagen from the marine sponges Axinella cannabina and Suberites carnosus: Isolation and morphological, biochemical, and biophysical characterization. *Mar. Drugs* **2017**, *15*, 152. [CrossRef]

10. Silva, J.C.; Barros, A.A.; Aroso, I.M.; Fassini, D.; Silva, T.H.; Reis, R.L.; Duarte, A.R.C. Extraction of collagen/gelatin from the marine demosponge Chondrosia reniformis (Nardo, 1847) using water acidified with carbon dioxide–process optimization. *Ind. Eng. Chem. Res.* **2016**, *55*, 6922–6930. [CrossRef]

11. Barros, A.A.; Aroso, I.M.; Silva, T.H.; Mano, J.F.; Duarte, A.R.C.; Reis, R.L. Water and carbon dioxide: Green solvents for the extraction of collagen/gelatin from marine sponges. *ACS Sustain. Chem. Eng.* **2015**, *3*, 254–260. [CrossRef]

12. Chancharern, P.; Laohakunjit, N.; Kerdchoechuen, O.; Thumthanaruk, B. Extraction of type A and type B gelatin from jellyfish (Lobonema smithii). *Int. Food Res. J.* **2016**, *23*, 419.

13. Cho, S.; Ahn, J.R.; Koo, J.S.; Kim, S.B. Physicochemical properties of gelatin from jellyfish Rhopilema hispidum. *Fish. Aquat. Sci.* **2014**, *17*, 299–304. [CrossRef]

14. Chan-Higuera, J.E.; Robles-Sánchez, R.M.; Burgos-Hernández, A.; Márquez-Ríos, E.; Velázquez-Contreras, C.A.; Ezquerra-Brauer, J.M. Squid by-product gelatines: Effect on oxidative stress biomarkers in healthy rats. *Czech J. Food Sci.* **2016**, *34*, 105–110. [CrossRef]

15. Uriarte-Montoya, M.H.; Santacruz-Ortega, H.; Cinco-Moroyoqui, F.J.; Rouzaud-Sández, O.; Plascencia-Jatomea, M.; Ezquerra-Brauer, J.M. Giant squid skin gelatin: Chemical composition and biophysical characterization. *Food Res. Int.* **2011**, *44*, 3243–3249. [CrossRef]

16. Nazeer, R.A.; Suganya, U.S. Porous scaffolds of gelatin from the marine gastropod Ficus variegate with commercial cross linkers for biomedical applications. *Food Sci. Biotechnol.* **2014**, *23*, 327–335. [CrossRef]

17. Zarai, Z.; Balti, R.; Mejdoub, H.; Gargouri, Y.; Sayari, A. Process for extracting gelatin from marine snail (Hexaplex trunculus): Chemical composition and functional properties. *Process. Biochem.* **2012**, *47*, 1779–1784. [CrossRef]

18. Luo, F.; Xing, R.; Wang, X.; Peng, Q.; Li, P. Proximate composition, amino acid and fatty acid profiles of marine snail Rapana venosa meat, visceral mass and operculum. *J. Sci. Food Agric.* **2017**, *97*, 5361–5368. [CrossRef] [PubMed]

19. Badiu, D.L.; Balu, A.M.; Barbes, L.; Luque, R.; Nita, R.; Radu, M.; Rosoiu, N. Physico-chemical characterisation of lipids from Mytilus galloprovincialis (L.) and Rapana venosa and their healing properties on skin burns. *Lipids* **2008**, *43*, 829–849. [CrossRef]

20. Guven, K.C.; Yazic, Z.; Akinci, S.; Okus, E. Fatty acids and sterols of Rapana venosa (Valenciennes, 1846). *J. Shellfish Res.* **1999**, *18*, 601–604.

21. Zolotarev, V. The Black Sea ecosystem changes related to the introduction of new mollusc species. *Mar. Ecol.* **1996**, *17*, 227–236. [CrossRef]

22. Merdzhanova, A.; Panayotova, V.; Dobreva, D.A.; Stancheva, R.; Peycheva, K. Lipid composition of raw and cooked Rapana venosa from the Black Sea. *Ovidius Univ. Ann. Chem.* **2018**, *29*, 49–55. [CrossRef]

23. Badiu, D.L.; Luque, R.; Dumitrescu, E.; Craciun, A.; Dinca, D. Amino acids from Mytilus galloprovincialis (L.) and Rapana venosa molluscs accelerate skin wounds healing via enhancement of dermal and epidermal neoformation. *Protein J.* **2010**, *29*, 81–92. [CrossRef] [PubMed]

24. Luo, F.; Xing, R.; Wang, X.; Yang, H.; Li, P. Antioxidant activities of Rapana venosa meat and visceral mass during simulated gastrointestinal digestion and their membrane ultrafiltration fractions. *Int. J. Food Sci. Technol.* **2018**, *53*, 395–403. [CrossRef]

25. Schmidt, M.M.; Dornelles, R.C.P.; Mello, R.O.; Kubota, E.H.; Mazutti, M.A.; Kempka, A.P.; Demiate, I.M. Collagen extraction process. *Int. Food Res. J.* **2016**, *23*, 913–922.

26. Nalinanon, S.; Benjakul, S.; Visessanguan, W.; Kishimura, H. Improvement of gelatin extraction from bigeye snapper skin using pepsin-aided process in combination with protease inhibitor. *Food Hydrocoll.* **2008**, *22*, 615–622. [CrossRef]

27. Shyni, K.; Hema, G.S.; Ninan, G.; Mathew, S.; Joshy, C.G.; Lakshmanan, P.T. Isolation and characterization of gelatin from the skins of skipjack tuna (Katsuwonus pelamis), dog shark (Scoliodon sorrakowah), and rohu (Labeo rohita). *Food Hydrocoll.* **2014**, *39*, 68–76. [CrossRef]

28. Lin, C.C.; Chiou, T.K.; Sung, W.C. Characteristics of gelatin from giant grouper (Epinephelus lanceolatus) skin. *Int. J. Food Prop.* **2015**, *18*, 2339–2348. [CrossRef]

29. Jamilah, B.; Harvinder, K.G. Properties of gelatins from skins of fish—black tilapia (Oreochromis mossambicus) and red tilapia (Oreochromis nilotica). *Food Chem.* **2002**, *77*, 81–84. [CrossRef]

30. Milovanovic, I.; Hayes, M. Marine Gelatine from rest raw materials. *Appl. Sci.* **2018**, *8*, 2407. [CrossRef]

31. Kittiphattanabawon, P.; Benjakul, S.; Visessanguan, W.; Shahidi, F. Effect of extraction temperature on functional properties and antioxidative activities of gelatin from shark skin. *Food Bioproc. Tech.* **2012**, *5*, 2646–2654. [CrossRef]

32. Gudipati, V. Fish gelatin: A versatile ingredient for the food and pharmaceutical industries. In *Marine Proteins and Peptides: Biological Activities and Applications*; Kim, S., Wijesekara, I., Eds.; Wiley-Blackwell: Hoboken, NJ, USA, 2013; pp. 271–295. ISBN 9781118375082.

33. Vallejos, N.; González, G.; Troncoso, E.; Zúñiga, R.N. Acid and enzyme-aided collagen extraction from the byssus of Chilean mussels (Mytilus Chilensis): Effect of process parameters on extraction performance. *Food Biophys.* **2014**, *9*, 322–331. [CrossRef]

34. Grossman, S.; Bergman, M. Process for the Production of Gelatin from Fish Skins. U.S. Patent No 5093474A, 3 March 1992.

35. Dos Reis, R.L. Method to Obtain Collagen/Gelatin from Marine Sponges. WO Patent WO2015151030A1, 8 October 2015.

36. Alfaro, A.T.; Biluca, F.C.; Marquetti, C.; Tonial, I.B.; de Souza, N.E. African catfish (Clarias gariepinus) skin gelatin: Extraction optimization and physical–chemical properties. *Food Res. Int.* **2014**, *65*, 416–422. [CrossRef]

37. Kittiphattanabawon, P.; Benjakul, S.; Visessanguan, W.; Shahidi, F. Comparative study on characteristics of gelatin from the skins of brownbanded bamboo shark and blacktip shark as affected by extraction conditions. *Food Hydrocoll.* **2010**, *24*, 164–171. [CrossRef]

38. Sotelo, C.G.; Comesaña, M.B.; Ariza, P.R.; Pérez-Martín, R.I. Characterization of collagen from different discarded fish species of the West coast of the Iberian Peninsula. *J. Aquat. Food Prod. T.* **2016**, *25*, 388–399. [CrossRef]

39. Blanco, M.; Vázquez, J.; Pérez-Martín, R.; Sotelo, C. Hydrolysates of fish skin collagen: An opportunity for valorizing fish industry byproducts. *Mar. Drugs* **2017**, *15*, 131. [CrossRef]

40. Cumming, M.H.; Hall, B.; Hofman, K. Isolation and characterization of major and minor collagens from hyaline cartilage of hoki (*Macruronus novaezelandiae*). *Mar. Drugs* **2019**, *17*, 223. [CrossRef]

41. Carvalho, A.; Marques, A.; Silva, T.; Reis, R. Evaluation of the potential of collagen from codfish skin as a biomaterial for biomedical applications. *Mar. Drugs* **2018**, *16*, 495. [CrossRef]

42. Lee, K.J.; Park, H.Y.; Kim, Y.K.; Park, J.I.; Yoon, H.D. Biochemical characterization of collagen from the starfish Asterias amurensis. *J. Korean Soc. Appl. Biol. Chem.* **2009**, *52*, 221–226. [CrossRef]

43. Mendis, E.; Rajapakse, N.; Byun, H.G.; Kim, S.K. Investigation of jumbo squid (Dosidicus gigas) skin gelatin peptides for their in vitro antioxidant effects. *Life Sci.* **2005**, *77*, 2166–2178. [CrossRef]

44. Di Paolo, N.C.; Shayakhmetov, D.M. Interleukin 1α and the inflammatory process. *Nat. Immunol.* **2016**, *17*, 906. [CrossRef]

45. Tanaka, T.; Narazaki, M.; Kishimoto, T. IL-6 in inflammation, immunity, and disease. *Cold Spring Harb. Perspect. Biol.* **2014**, *6*, a016295. [CrossRef]

46. Zhang, Q.; Dai, T.; Zhang, L.; Zhang, M.; Xiao, X.; Hu, H.; Huang, Y. Identification of potential biomarkers for predicting acute dermal irritation by proteomic analysis. *J. Appl. Toxicol.* **2011**, *31*, 762–772. [CrossRef] [PubMed]

47. Alves, A.; Marques, A.; Martins, E.; Silva, T.; Reis, R. Cosmetic potential of marine fish skin collagen. *Cosmetics* **2017**, *4*, 39. [CrossRef]

48. Edwards, C.A.; O'Brien, W.D., Jr. Modified assay for determination of hydroxyproline in a tissue hydrolyzate. *Clin. Chim. Acta* **1980**, *104*, 161–167. [CrossRef]

49. Laemmli, U.K. Cleavage of structural proteins during the assembly of the head of bacteriophage T4. *Nature* **1970**, *227*, 680. [CrossRef]

 *marine drugs*

Article

Serine Protease from *Nereis virens* Inhibits H1299 Lung Cancer Cell Proliferation via the PI3K/AKT/mTOR Pathway

Yanan Chen [1,2], Yunping Tang [1,*], Yanhua Tang [1], Zuisu Yang [1] and Guofang Ding [1,2,*]

[1] Zhejiang Provincial Engineering Technology Research Center of Marine Biomedical Products, School of Food and Pharmacy, Zhejiang Ocean University, Zhoushan 316022, China; cyanan2013@163.com (Y.C.); tyh19980126@163.com (Y.T.); abc1967@126.com (Z.Y.)

[2] Zhejiang Marine Fisheries Research Institution, Zhoushan 316021, China

* Correspondence: tangyunping1985@zjou.edu.cn (Y.T.); dinggf2007@163.com (G.D.); Tel.: +86-0580-229-9809 (G.D.); Fax: +86-0580-229-9866 (G.D.)

Received: 22 May 2019; Accepted: 18 June 2019; Published: 20 June 2019

Abstract: This study explores the in vitro anti-proliferative mechanism between Nereis Active Protease (NAP) and human lung cancer H1299 cells. Colony formation and migration of cells were significantly lowered, following NAP treatment. Flow cytometry results suggested that NAP-induced growth inhibition of H1299 cells is linked to apoptosis, and that NAP can arrest the cells at the G0/G1 phase. The ERK/MAPK and PI3K/AKT/mTOR pathways were selected for their RNA transcripts, and their roles in the anti-proliferative mechanism of NAP were studied using Western blots. Our results suggested that NAP led to the downregulation of p-ERK (Thr 202/Tyr 204), p-AKT (Ser 473), p-PI3K (p85), and p-mTOR (Ser 2448), suggesting that NAP-induced H1299 cell apoptosis occurs via the PI3K/AKT/mTOR pathway. Furthermore, specific inhibitors LY294002 and PD98059 were used to inhibit these two pathways. The effect of NAP on the downregulation of p-ERK and p-AKT was enhanced by the LY294002 (a PI3K inhibitor), while the inhibitor PD98059 had no obvious effect. Overall, the results suggested that NAP exhibits antiproliferative activity by inducing apoptosis, through the inhibition of the PI3K/AKT/mTOR pathway.

Keywords: lung cancer; nereis active protease; H1299 cells; PI3K/AKT/mTOR pathway

---

1. Introduction

Lung cancer is a highly malignant form of cancer that has some of the highest rates of morbidity and mortality worldwide [1,2]. Lung cancers caused the highest incidence of malignant tumors and led to 380,800 human deaths in Europe in 2018 [3]. Most lung cancer patients are in advanced stages at the time of diagnosis, which cannot be cured using surgery [4]. Currently, targeted drugs, such as tyrosine kinase inhibitors and immune checkpoint inhibitors, are used to treat cancers [5,6]. The advantage of targeted drugs is that they work primarily on cancer cells and have fewer side effects than chemotherapeutics. On the other hand, different cancer types exhibit different pathogenic behavior, and heterogeneity exists among cancer cells of the same type, so one targeted drug might be effective for one patient, but not for others [5,6]. Chemical drugs, such as cisplatin, carboplatin, and etoposide, are often used to treat advanced-stage lung cancers [7,8]. However, these drugs generally lead to cytotoxicity and side effects that limit their use as clinical treatments [9,10]. Therefore, improvements in the efficiency and toxicity profiles of anticancer drugs, are necessary for improving cancer treatments, which might be achievable through the use of natural products.

The earth's surface is covered by 70% ocean; this is a unique environment that is a resource for new drug discovery [11,12]. In recent years, the research on marine organisms has expanded, and numerous substances with anti-cancer activities have been found [13]. *Nereis succinea*, for example, has been used to extract an abundance of materials for pharmacological research [14–16]. In addition, a decapeptide from *Perinereies aibuhitensis* was shown to exhibit anti-proliferative activity on human lung cancer H1299 cells [14]. Ge et al. [15] also found that a serine protease from *Neanthes japonica* exhibits anti-cancer activity toward leukemia cells. In our previous study, a serine protease from *Nereis virens* (Nereis Active Protease (NAP)) exhibited anti-proliferative activity toward human lung cancer cells, including A549, 95C, SPC-A-1, and H1299 cells [16], however, the mechanism underlying this remains unclear.

The PI3K/AKT/mTOR and ERK/MAPK pathways are often used to elucidate anti-tumor mechanisms [17–21]. The PI3K/AKT/mTOR pathway plays an important role in pathological processes, including cell differentiation, survival, and proliferation. Therefore, this pathway is considered as a major regulator of cancer progression [17]. Continuous activation of this pathway causes continuous cell growth that can lead to the evolution of cancer cells [9,18,22]. Since this is a gradual process, pan PI3K blockers, subtype-specific PI3K blockers, PI3K/mTOR double blockers, AKT blockers, and mTOR blockers have been developed to counteract the pathway's influence on cancer formation [19]. In addition, the PI3K/AKT/mTOR pathway is connected to the ERK/MAPK pathway [20]. The activation of ERK is related to the continual growth of cells and affects the signal pathways related to cell proliferation. Previous studies suggest that apoptosis might be associated with the inhibition of the ERK/MAPK pathway [21,23,24]. Consequently, proteins in the PI3K/AKT/mTOR and ERK/MAPK signaling pathways could be good targets for cancer therapy.

As the NAP exhibited the strongest anti-proliferative activity toward H1299 cells, in this study, transcriptome sequencing was first used to identify the significant signal pathways related to the treatment of H1299 cells with *N. virens* NAP. Furthermore, the PI3K/AKT/mTOR and ERK/MAPK pathways were chosen to explore the anti-proliferative mechanism of NAP on H1299 cells. This research indicated that NAP inhibits H1299 cell proliferation via the PI3K/AKT/mTOR pathway. Therefore, NAP from *N. virens* demonstrates a strong potential as an anti-lung cancer drug candidate.

2. Results and Discussion

2.1. NAP Inhibits the Growth and Migration of H1299 Cells

Malignant cell proliferation is an uncontrolled process that increases the risk of carcinogenic factors that facilitate the dispersion and migration of cancer cells [25]. The inhibition of cancer cell growth and migration are effective ways to control tumor development [25]. In this work, the influence of NAP on the proliferation of individual H1299 cells was studied using a colony formation assay. The results indicated that the colony formation rate of H1299 cells significantly decreased after the NAP treatment (Figure 1A,B). The results were consistent with our previous studies [16], indicating that NAP could significantly inhibit the growth and proliferation of H1299 cells. Furthermore, a scratch wound assay was used to investigate the influence of NAP on the migrative ability of H1299 cells. Results revealed that NAP could inhibit wound healing through the inhibition of H1299 cell migration, after 24 h of treatment (Figure 1C,D). A similar phenomenon was reported by Song et al. [26], who found that a serine protease (*Trichosanthes kirilowii*) inhibited the proliferation of colorectal cancer cells.

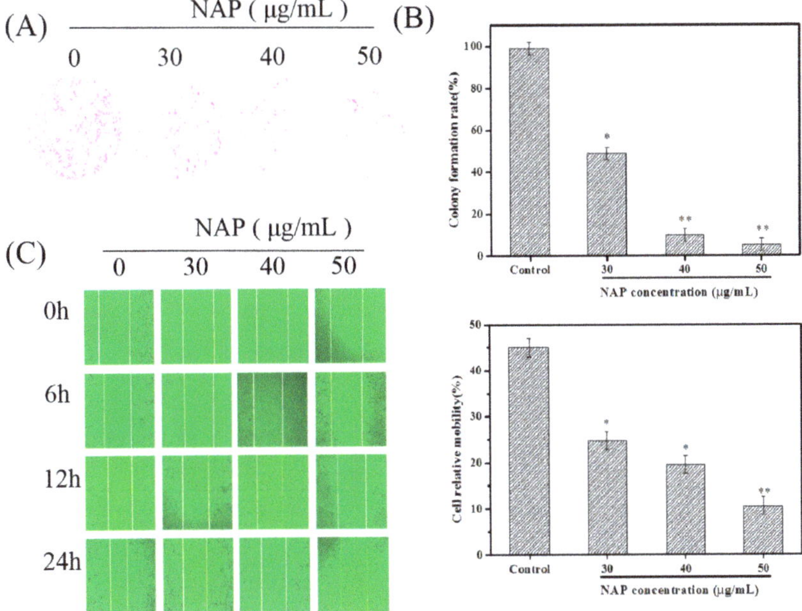

Figure 1. Nereis Active Protease (NAP)-based inhibition of the growth and migration of H1299 cells. (**A**) H1299 cells were treated with NAP (0, 30, 40, and 50 μg/mL) for 24 h and then cultured in RPMI-1640 complete culture medium for 12 days, to investigate the ability of single cells to form colonies through a colony formation assay. Magnification: 100 ×. (**B**) Colony formation rate of H1299 cells. (**C**) H1299 cells were treated with NAP (0, 30, 40, and 50 μg/mL) for 24 h and photographed under a microscope at 0, 6, 12, and 24 h. Magnification: 100 ×. (**D**) Wound healing rates after treatment of H1299 cells with NAP for one day. Significant results: * $p \leq 0.05$; ** $p \leq 0.01$ vs the blank group (0 μg/mL NAP).

2.2. NAP-Induced G0/G1 Phase Block in H1299 Cells

In the process of normal cell growth and proliferation, the cell cycle is divided into G0/G1, S and G2/M stages. G1 to S is a particularly important stage in the cell cycle [27]. During the period of complex and active molecular level changes, DNA replication is regulated by cyclin-dependent kinases (CDK), and cyclin D, and cyclin E proteins, which are easily affected by environmental conditions [28]. The regulation of G1 to S is thought to be of great significance for controlling the growth of tumors [29].

Flow cytometry was applied for subsequent investigation of the influence of NAP on the cell cycle. The percentages of cells blocked by NAP (0, 30, 40 and 50 μg/mL) at the G0/G1 phase were 60.7% ± 1.8, 68.9 ± 2.1%, 72.0 ± 1.9%, and 74.3 ± 1.5%, respectively (Figure 2A,B). In addition, the expression levels of CDK4, cyclin E1 and cyclin D1 proteins in the G0/G1 phase were examined [30,31]. Western blot results showed that CDK4, cyclin E1, and cyclin D1 proteins were down-regulated, indicating that NAP could have induced apoptosis by blocking the G0/G1 phase in H1299 cells (Figure 2C,D). These results were supported by results from Han et al. [30] who found that 8-Cetylcoptisine blocked A549 cells in the G0/G1 phase, and Zhang et al. [32] who demonstrated that the up-regulation of MiR-101-3p could significantly reduce the viability of H1299 cells and that MiR-101-3p up-regulation could block the cell cycle in the G0/G1 phase of H1299 cells.

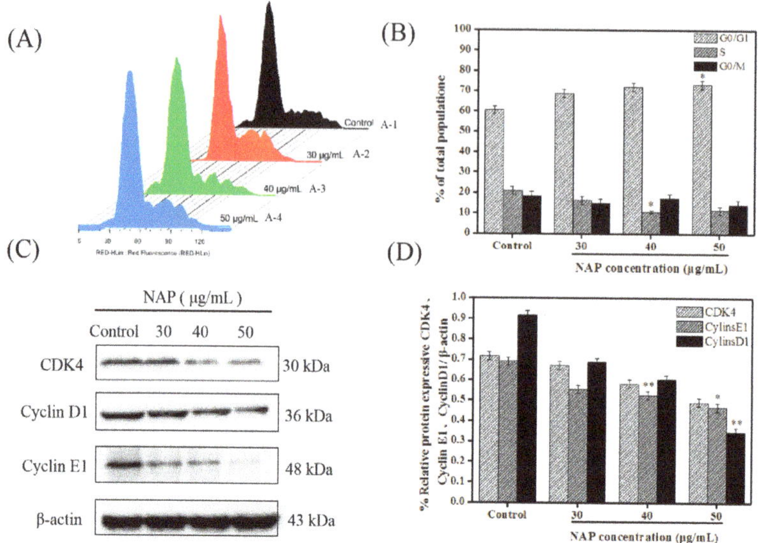

Figure 2. Flow cytometry of the cell cycle phase distribution reveals a NAP-induced G0/G1 stage block in H1299 cells. (**A**) Percentage of H1299 cells at each phase A1—control group; A2—30 μM NAP-treated group; A3—40 μg/mL NAP-treated group, and A4—50 μg/mL NAP-treated group. (**B**) Percentage of NAP-treated H1299 cells at the three stages of the cell cycle. (**C**) Western blot measurements of the cyclin-related proteins. (**D**) Ratio of CDK 4/β-actin, cyclin D1/β-actin, and cyclin E1/β-actin values. $* p \leq 0.05$, $** p \leq 0.01$ versus cells without NAP treatment.

2.3. Influence of NAP on the Transcriptome of H1299 Cells

Transcriptome sequencing is a major method used for studying gene expression. This method can highlight significant differentially expressed genes, which can be used to determine the major signaling pathways involved in biological processes [33]. In the process of RNA sequencing, a large concentration of cells was required. When the concentration of NAP is more than 30 μg/mL, cells will apoptosis and suspend in the medium. This is not conducive to collecting a large numbers of cells. Therefore, in the present study, 30 μg/mL of NAP was chosen for the transcriptome analysis. However, mRNA from H1299 cells with and without NAP treatment, was extracted. An mRNA library for the H1299 cells was constructed and a paired *t*-test was performed on three balanced experiments, with a *p*-value < 0.01 set as the significance level. Two hundred and nineteen differentially expressed genes were detected between the NAP-treated H1299 cells and the control group (Figure 3). Logarithmic ratios and *p*-values of the 219 genes were used to detect up- and down-regulated signaling pathways after the NAP treatment. Enrichment analysis of the KEGG pathways was performed for the 219 differentially expressed genes. Pathways with significant differences in pathway enrichment were selected for further study.

PI3K/AKT/mTOR and ERK/MAPK pathways are often related to the growth and proliferation of cancer cells [8,34,35]. For example, Li et al. [36] studied Tyr-Val-Pro-Gly-Pro (AAP-H), an active peptide extracted and purified from Anthopleura anjunae, which induces apoptosis in prostate cancer DU145 cells, via the PI3K/AKT/mTOR pathway. Liu et al. [37] demonstrated that 1, 4-naphthoquinone induces apoptosis in lung cancer cells by activating oxygen-dependent down-regulation of the proteins related to the MAPK cascade. Therefore, these two pathways were selected for investigation, for their role in NAP-induced apoptosis of H1299 cells.

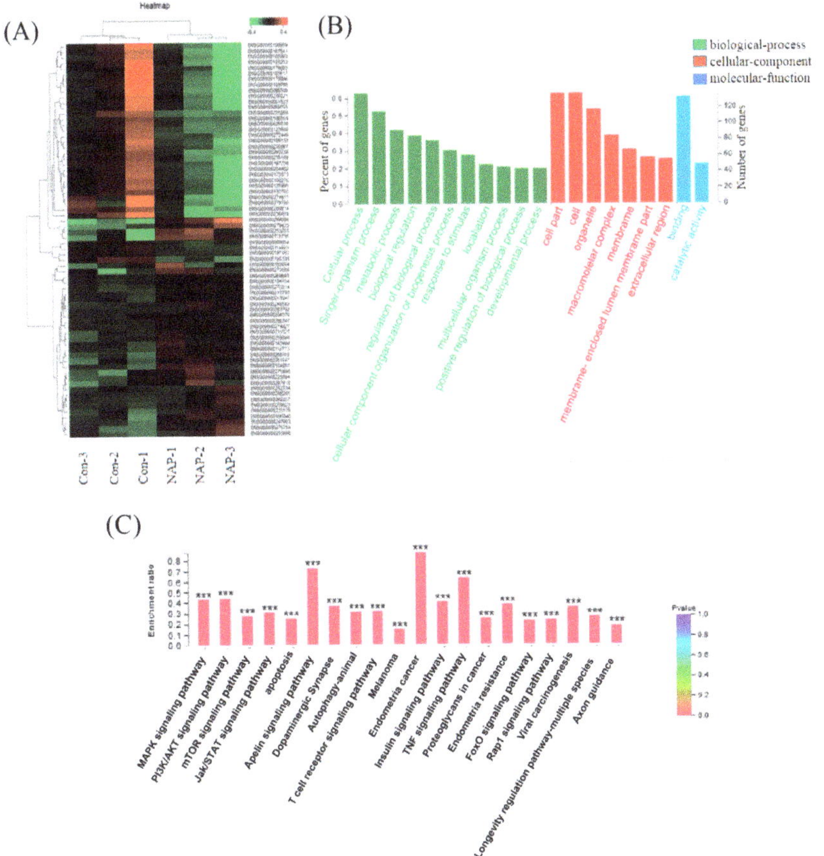

Figure 3. Influence of NAP on the transcriptome of H1299 cells. (**A**) Heat-map for 219 significantly different genes between the NAP-treated (NAP 1–3) and the control cells (Con 1–3). (**B**) Differential genes annotated according to the gene ontology (GO) classification criteria. (**C**) Annotations according to the enrichment of pathways and significant enrichment of 20 pathways.

2.4. Influence of NAP on the ERK/MAPK Pathway in H1299 Cells

Previous experiments have shown that the inhibitory activity of NAP on the H1299 cells is related to regulation of the cell cycle and the induction of apoptosis [16]. Transcriptome sequencing results have indicated that the PI3K/AKT/mTOR and ERK/MAPK pathways are associated with NAP's inhibitory effect on the H1299 cells. Johnson et al. [38] found that MAPK is closely related to the initiation of apoptosis and cell cycle quiescence, in various tumor cell lines. Therefore, the influence of NAP on the phosphorylation of the MAPK pathway-related kinases was investigated through Western blotting.

H1299 cells were treated with NAP and Western blots were used to visualize and measure ERK/MAPK pathway-related proteins (Figure 4). Results indicated that NAP treatment might have led to down-regulated p-ERK levels in H1299 cells, and the expression of p-P38 and p-MEK protein did not change significantly, when compared to cells without NAP treatment. Importantly, by inhibiting the activation of the PI3K/AKT/mTOR pathway, ERK phosphorylation might have been inhibited [39]. Therefore, we speculated that NAP might inhibit ERK phosphorylation by inhibiting the PI3K/AKT/mTOR pathway.

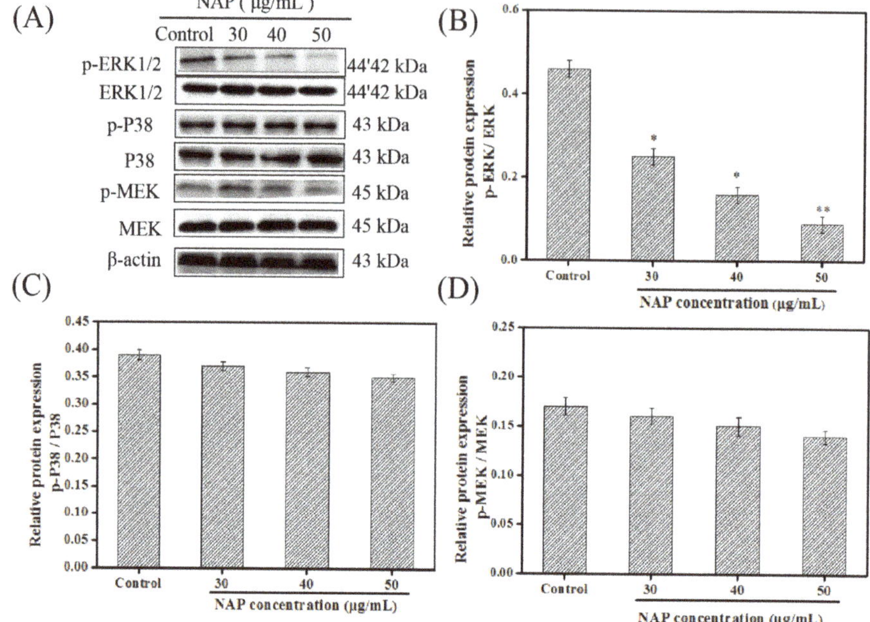

Figure 4. Influence of NAP (0, 30, 40 and 50 μg/mL) on the ERK/MAPK cascade in H1299 cells. (**A**) Western blot visualizations of the ERK/MAPK pathway-related proteins from the NAP-treated H1299 cells. (**B**) The ratio of phosphorylated ERK (p-ERK) to ERK in the NAP-treated H1299 cells. (**C**) The ratio of phosphorylated P38 (p-P38) to P38 in NAP-treated H1299 cells. (**D**) The ratio of phosphorylation MEK (p-MEK) to MEK in the NAP-treated H1299 cells. * $p \leq 0.05$, ** $p \leq 0.01$ versus cells without NAP treatment.

2.5. Influence of NAP on the PI3K/AKT/mTOR Pathway in H1299 Cells

The PI3K/AKT/mTOR pathway could activate or inhibit ERK, which has the same effect on cell proliferation and apoptosis [40]. The growth, proliferation, differentiation, and death of cells are regulated by PI3K and AKT protein kinases [41]. Research has indicated that this pathway plays a critical role in the abnormal activation of tumor cells [42]. When PI3K is activated, AKT is activated downstream through phosphorylation, due to the exposure of its phosphorylation site; this leads to the downstream activation of mTOR and other proteins, which plays an anti-apoptotic role within the cell [43,44].

To determine the effect of NAP concentrations (0, 30, 40, and 50 μg/mL) on the PI3K/AKT/mTOR pathway, PI3K, AKT, and mTOR were chosen for study, visualized, and analyzed, using Western blots (Figure 5). Results indicated that the NAP treatment led to the downregulation of p-mTOR, p-AKT, and p-PI3K levels in the H1299 cells, while PI3K, AKT, and mTOR levels did not change. Therefore, this suggests that the role of NAP in promoting apoptosis might be mediated via the inhibition of the PI3K/AKT/mTOR pathway in H1299 cells.

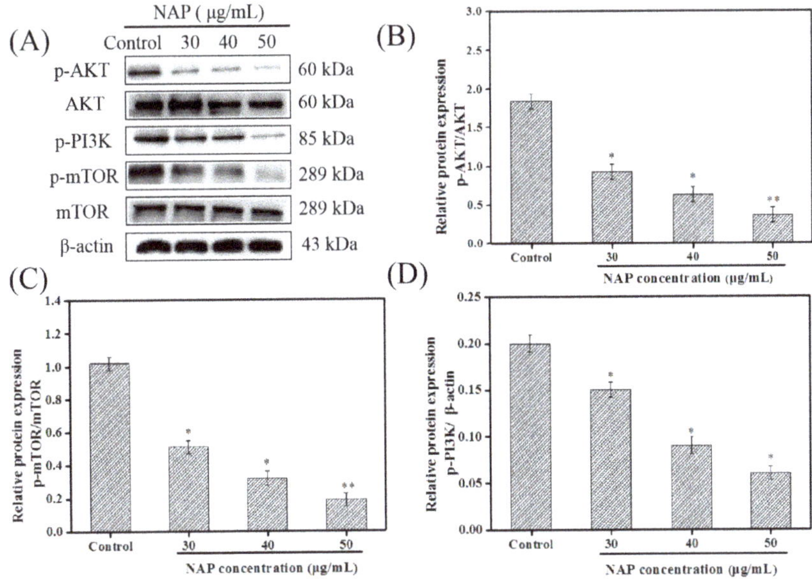

Figure 5. Influence of the NAP on PI3K/AKT/mTOR cascade pathway in H1299 cells. (**A**) Western blot visualizations of PI3K/AKT/mTOR pathway-related proteins from the NAP-treated H1299 cells. (**B**) The ratio of phosphorylated AKT (p-AKT) to AKT in NAP-treated H1299 cells. (**C**) The ratio of phosphorylated mTOR (p-mTOR) to mTOR, in the NAP-treated H1299 cells (**D**) The phosphorylation ratio of PI3K (p-PI3K) in the NAP-treated H1299 cells. * $p \leq 0.05$, ** $p \leq 0.01$ versus cells without NAP treatment.

2.6. NAP-Induced H1299 Cell Apoptosis Involves the PI3K/AKT/mTOR and ERK/MAPK Pathways

A further investigation was carried out to verify if the NAP-induced apoptosis is closely associated with the PI3K/AKT/mTOR and ERK/MAPK pathways in the H1299 cells. Cells were treated with 40 µg/mL NAP, then tested for changes in p-ERK, p-AKT, and cleaved PARP proteins, at different times [26,36]. The results indicated that NAP treatment led to the downregulation of p-AKT and p-ERK levels, and upregulation of the levels of cleaved-PARP (Figure 6). Importantly, the cleaved-PARP levels began to increase with concurrent decreases in levels of phosphorylated ERK and AKT. Results indicated that NAP could induce the apoptosis of H1299 cells via the PI3K/AKT/mTOR and ERK/MAPK pathways.

The inhibitors LY294002 and PD98059 were also applied to the cells, to inhibit these two pathways. Their effects on the NAP-induced apoptosis were assessed by Western blot analysis. Results indicated that treatment with LY294002 enhanced the NAP's effects on p-ERK and p-AKT proteins, while treatment with PD98059 did not enhance the NAP's effects on p-ERK and p-AKT proteins (Figure 7A–C). Therefore, our studies suggested that NAP could induce H1299 cell apoptosis by down-regulating the related proteins in the PI3K/AKT/mTOR pathway.

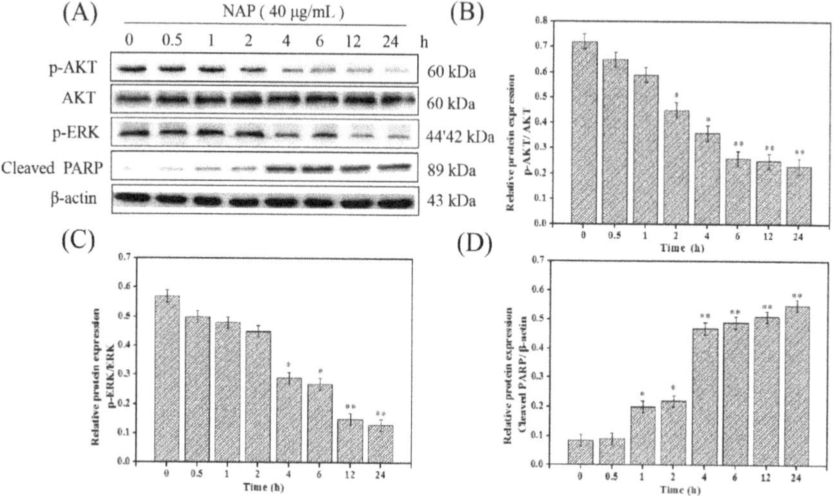

Figure 6. Influence of NAP on PI3K/AKT/mTOR and ERK/MAPK cascade at different time points. (**A**) Expression of p-AKT, AKT, p-ERK, and cleaved PARP proteins in H1299 cells treated with NAP at different time points. (**B**) The ratio of p-AKT (Ser473)/AKT in NAP-treated H1299 cells. (**C**) The ratio of p-ERK (Thr202/Tyr204)/β-actin in NAP-treated H1299 cells. (**D**) The ratio of cleaved-PARP/β-actin in NAP-treated H1299 cells. * $p \leq 0.05$, ** $p \leq 0.01$ versus cells without NAP treatment.

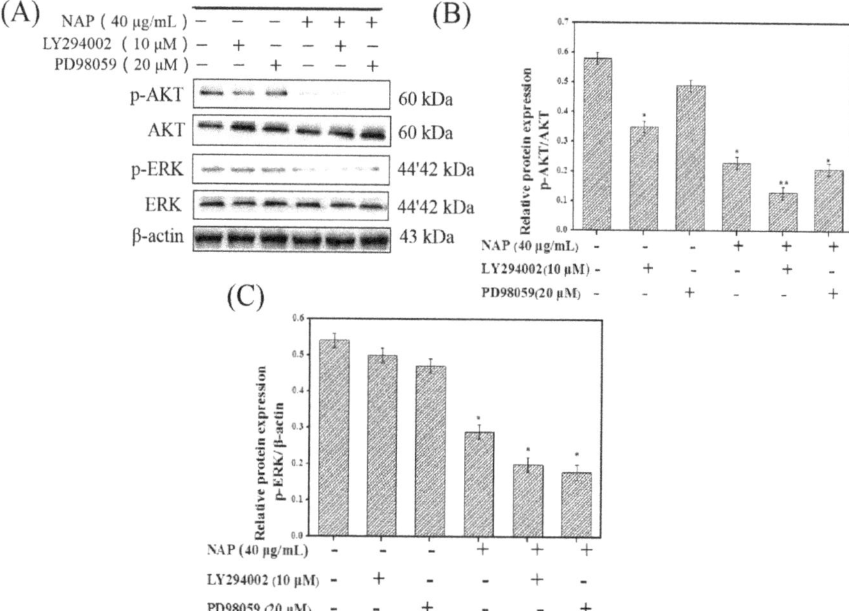

Figure 7. Influence of LY294002 and PD98059 on the PI3K/AKT/mTOR and ERK/MAPK pathway-related proteins. (**A**) Influence of LY294002 and PD98059 on ERK, AKT, p-ERK, and p-AKT proteins in H1299 cells. (**B**) p-AKT/AKT ratio expressed in H1299 cells. (**C**) Expression of p-ERK ratio expressed in H1299 cells.

3. Materials and Methods

3.1. Cell Culture

The H1299 cell lines were purchased from the Cell Bank of Chinese Academy of Sciences (Shanghai, China). The cells were incubated in RPMI-1640 medium containing 10% FBS, 1% 100 U/mL penicillin and streptomycin, and incubated in an incubator (Forma 3111, Thermo, Waltham, MA, USA) containing 5% CO_2 at 37 °C.

3.2. Material Sources

NAP was purified from *N. virens* and stored in our laboratory [16,45]. Fetal bovine serum (FBS) without mycoplasma, bacteriophage, and low endotoxin was purchased from Zhejiang Tianhang Biotechnology Co., Ltd. (Hangzhou, China). Antibodies against AKT, ERK, P38, MEK, mTOR, PARP, p-AKT, p-PI3K, p-mTOR, p-ERK, p-P38, p-MEK, cyclin B1, PARP, CDK4, and cyclin D1, were purchased from the Cell Signaling Technology (Boston, MA, USA). A cell cycle analysis kit was obtained from BestBio Science Co., (Shanghai, China).

3.3. Clonogenic Survival Assay

A clonogenic survival assay was applied for detecting single-cell survival, after treatment with NAP [46]. For adherent cells, 500 single cell suspensions of H1299 cells were inoculated into 6-well plates. After adherence, NAP (0, 30, 40, and 50 µg/mL) was added and removed after 24 h. Cells were then cultured in RPMI-1640 complete medium for 12 days, and the medium was washed over with PBS. Poly-formaldehyde liquid (4%) was applied for approximately 15 min to fix the cells, and the cells were then stained with 1% crystal violet for 15 min at 25 °C. The 6-well plates were rinsed gently with water and the number of clones (more than 50 cells) were determined using a microscope (OLYMPUS, Tokyo, Japan). Clone formation rate = (colonies/seed cells) × 100%.

3.4. Cell Migration (Scratch Wound) Assay

The migration of cells after the NAP-treatment was tested by wound scratch assay [47]. H1299 cells were inoculated in a 6-well culture plate and the appropriate medium was added until the cells covered 80% of the plate's surface, then, the cells were scraped to form a "wound" with a sterilized pipette tip. The 6-well plates were then washed twice with PBS, and photographed at 0, 6, 12, and 24 h, respectively. Finally, Image J software (NIH, Bethesda, MD, USA) was used to calculate the wound area.

$$Relative\ mobility\ (\%) = \frac{\left(Wound\ area_{(0\ h)} - Wound\ area_{(24\ h)}\right)}{Wound\ area_{(0\ h)}} \times 100$$

3.5. Cell Cycle Experiment

A cell cycle arrest test was used to determine the proportion of cells in different cell cycle phases after treatment with NAP [48]. Cells were inoculated in 6-well culture plates, and incubated until the cells covered 70% of the plate's surface. Then, the cells were treated with NAP (0, 30, 40, and 50 µg/mL) for 24 h, respectively. The steps for cell cycle arrest tests were in accordance with the method described by Li et al. [36]. Finally, flow cytometry was used to detect the cell cycle phase.

3.6. RNA Extraction and Integrity Test

H1299 cells in the log-phase growth were digested and dispersed to form single cells using trypsin, then incubated in an incubator for 24 h, treated with 0 µg/mL (blank control group), and 30 µg/mL NAP (medication group), for 24 h, and washed with PBS. The cells were lysed in 1 mL trizol solution to extract the total RNA. Transcriptome sequencing was performed by Sangon Biotech (Shanghai, China).

3.7. Western Blotting

The steps of Western blotting were in accordance with a method described by Li et al. [36]. A total of 40 µg of the samples were used for detection, in this study. Western blot membranes were probed with an enhanced chemiluminescence antibody and the images were obtained using an Alpha FluorChem FC3 imaging system (ProteinSimple, San Jose, CA, USA). The level of protein expression was quantified using Image J software. β-actin was used as the control.

3.8. Statistical Analysis

SPSS 19.0 software was used to conduct one-way ANOVA on the data; the results represent the mean with the standard deviation ($n = 3$). A p-value of ≤ 0.05 was considered to show a statistically significant difference between the two groups.

4. Conclusions

This research showed that NAP from *N. virens* exhibits strong anti-proliferative activity through the inhibition of proliferation and migration, and by blocking H1299 cells in the G0/G1 phase. Further studies showed that NAP exhibits anti-proliferative activity against human lung cancer H1299 cells, by inhibiting the PI3K/AKT/mTOR pathway. Therefore, NAP shows promising results that highlight its potential for treating lung cancer in the future.

Author Contributions: Y.T. and G.D. conceived and designed the experiments. Y.C., Y.T. (Yanhua Tang), and Z.Y. performed the statistical analysis of the data. Y.C. and Y.T. wrote the manuscript.

Funding: This work was financially supported by the National Natural Science Foundation of China (grant No. 81773629 and No. 41806153).

Conflicts of Interest: The authors declare no conflict of interest.

References

1. Mohandas, K.M. Colorectal cancer in India: Controversies, enigmas and primary prevention. *Indian J. Gastroenterol.* **2011**, *30*, 3–6. [CrossRef] [PubMed]
2. Liu, Z.; Zheng, Q.; Chen, W.; Man, S.; Teng, Y.; Meng, X.; Zhang, Y.; Yu, P.; Gao, W. Paris saponin I inhibits proliferation and promotes apoptosis through down-regulating AKT activity in human non-small-cell lung cancer cells and inhibiting ERK expression in human small-cell lung cancer cells. *RSC Adv.* **2016**, *6*, 70816–70824. [CrossRef]
3. Ferlay, J.; Colombet, M.; Soerjomataram, I.; Dyba, T.; Randi, G.; Bettio, M.; Gavin, A.; Visser, O.; Bray, F. Cancer incidence and mortality patterns in Europe: Estimates for 40 countries and 25 major cancers in 2018. *Eur. J. Cancer* **2018**, *103*, 356–387. [CrossRef] [PubMed]
4. Sarris, E.; Saif, M.; Syrigos, K. The biological role of PI3K pathway in lung cancer. *Pharmaceuticals* **2012**, *5*, 1236–1264. [CrossRef] [PubMed]
5. Wu, P.; Nielsen, T.E.; Clausen, M.H. FDA-approved small-molecule kinase inhibitors, trends pharmacol. *Sciences* **2015**, *36*, 422–439.
6. Brahmer, J.R.; Tykodi, S.S.; Chow, L.Q.M.; Hwu, W.-J.; Topalian, S.L.; Hwu, P.; Drake, C.G.; Camacho, L.H.; Kauh, J.; Odunsi, K.; et al. Safety and activity of anti-PD-L1 antibody in patients with advanced cancer. *N. Engl. J. Med.* **2012**, *366*, 2455–2465. [CrossRef]
7. Chapman, C.J.; Thorpe, A.J.; Murray, A.; Parsy-Kowalska, C.B.; Allen, J.; Stafford, K.M.; Chauhan, A.S.; Kite, T.A.; Maddison, P.; Robertson, J.F. Immunobiomarkers in small cell lung cancer: Potential early cancer signals. *Clin. Cancer Res.* **2011**, *17*, 1474–1480. [CrossRef]
8. Demedts, I.K.; Vermaelen, K.Y.; van Meerbeeck, J.P. Treatment of extensive-stage small cell lung carcinoma: Current status and future prospects. *Eur. Respir. J.* **2010**, *35*, 202–215. [CrossRef]
9. Taddia, L.; D'Arca, D.; Ferrari, S.; Marraccini, C.; Severi, L.; Ponterini, G.; Assaraf, Y.G.; Marverti, G.; Costi, M.P. Inside the biochemical pathways of thymidylate synthase perturbed by anticancer drugs: Novel strategies to overcome cancer chemoresistance. *Drug Resist. Updates* **2015**, *23*, 20–54. [CrossRef]

10. Memmott, R.M.; Dennis, P.A. The role of the Akt/mTOR pathway in tobacco carcinogen-induced lung tumorigenesis. *Clin. Cancer Res.* **2010**, *16*, 4–10. [CrossRef]
11. Kiuru, P.; D'Auria, M.V.; Muller, C.D.; Tammela, P.; Vuorela, H.; Yli-Kauhaluoma, J. Exploring marine resources for bioactive compounds. *Planta Med.* **2014**, *80*, 1234–1246. [CrossRef] [PubMed]
12. Simmons, T.L.; Andrianasolo, E.; McPhail, K.; Flatt, P.; Gerwick, W.H. Marine natural products as anticancer drugs. *Mol. Cancer Ther.* **2005**, *4*, 333–342. [PubMed]
13. Wu, Z.Z.; Ding, G.F.; Huang, F.F.; Yang, Z.S.; Yu, F.M.; Tang, Y.P.; Jia, Y.L.; Zheng, Y.Y.; Chen, R. Anticancer activity of anthopleura anjunae oligopeptides in prostate cancer DU-145 cells. *Mar. Drugs* **2018**, *16*, 125. [CrossRef] [PubMed]
14. Jiang, S.; Jia, Y.; Tang, Y.; Zheng, D.; Han, X.; Yu, F.; Chen, Y.; Huang, F.; Yang, Z.; Ding, G. Anti-proliferation activity of a decapeptide from perinereies aibuhitensis toward human lung cancer H1299 cells. *Mar. Drugs* **2019**, *17*, 122. [CrossRef] [PubMed]
15. Ge, X.; Bo, Q.; Hong, X.; Cui, J.; Jiang, X.; Hong, M.; Liu, J. A novel acidic serine protease, ASPNJ inhibits proliferation, induces apoptosis and enhances chemo-susceptibility of acute promyelocytic leukemia cell. *Leuk. Res.* **2013**, *37*, 1697–1703. [CrossRef] [PubMed]
16. Tang, Y.; Yu, F.; Zhang, G.; Yang, Z.; Huang, F.; Ding, G. A Purified serine protease from nereis virens and its impaction of apoptosis on human lung cancer cells. *Molecules* **2017**, *22*, 1123. [CrossRef] [PubMed]
17. Zhao, R.; Chen, M.; Jiang, Z.; Zhao, F.; Xi, B.; Zhang, X.; Fu, H.; Zhou, K. Platycodin-D induced autophagy in non-small cell lung cancer cells via PI3K/Akt/mTOR and MAPK signaling pathways. *J. Cancer* **2015**, *6*, 623–631. [CrossRef]
18. Li, X.; Wu, C.; Chen, N.; Gu, H.; Yen, A.; Cao, L.; Wang, E.; Wang, L. PI3K/Akt/mTOR signaling pathway and targeted therapy for glioblastoma. *Oncotarget* **2016**, *7*, 33440–33450. [CrossRef]
19. O'Donnell, J.S.; Massi, D.; Teng, M.; Mandala, M. PI3K-AKT-mTOR inhibition in cancer immunotherapy, redux. *Semin. Cancer Biol.* **2018**, *48*, 91–103. [CrossRef]
20. Fruman, D.A.; Rommel, C. PI3K and cancer: Lessons, challenges and opportunities. *Nat. Rev. Drug Discov.* **2014**, *13*, 140–156. [CrossRef]
21. Nagata, Y.; Todokoro, K. Requirement of activation of JNK and p38 for environmental stress-induced erythroid differentiation and apoptosis and of inhibition of ERK for apoptosis. *Blood* **1999**, *94*, 853–863. [PubMed]
22. Juric, D.; Krop, I.; Ramanathan, R.K.; Wilson, T.R.; Ware, J.A.; Sanabria Bohorquez, S.M.; Savage, H.M.; Sampath, D.; Salphati, L.; Lin, R.S. Phase I dose-escalation study of taselisib, an oral PI3K inhibitor, in patients with advanced solid tumors. *Cancer Discov.* **2017**, *7*, 704–715. [CrossRef] [PubMed]
23. Koo, H.M.; VanBrocklin, M.; McWilliams, M.J.; Leppla, S.H.; Duesbery, N.S.; Vande Woude, G.F. Apoptosis and melanogenesis in human melanoma cells induced by anthrax lethal factor inactivation of mitogen-activated protein kinase kinase. *Proc. Natl. Acad. Sci. USA* **2002**, *99*, 3052–3057. [CrossRef] [PubMed]
24. Jan, M.S.; Liu, H.S.; Lin, Y.S. Bad overexpression sensitizes NIH/3T3 cells to undergo apoptosis which involves caspase activation and ERK inactivation. *Biochem. Biophys. Res. Commun.* **1999**, *264*, 724–729. [CrossRef] [PubMed]
25. Vander Heiden, M.G.; Cantley, L.C.; Thompson, C.B. Understanding the warburg effect: The metabolic requirements of cell proliferation. *Science* **2009**, *324*, 1029–1033. [CrossRef] [PubMed]
26. Song, L.; Chang, J.; Li, Z. A serine protease extracted from trichosanthes kirilowii induces apoptosis via the PI3K/AKT-mediated mitochondrial pathway in human colorectal adenocarcinoma cells. *Food Funct.* **2016**, *7*, 843–854. [CrossRef]
27. Chang, F.; Lee, J.T.; Navolanic, P.M.; Steelman, L.S.; Shelton, J.G.; Blalock, W.L.; Franklin, R.A.; McCubrey, J.A. Involvement of PI3K/Akt pathway in cell cycle progression, apoptosis, and neoplastic transformation: A target for cancer chemotherapy. *Leukemia* **2003**, *17*, 590–603. [CrossRef]
28. Vermeulen, K.; Bockstaele DR, V.; Berneman, Z.N. The cell cycle: A review of regulation, deregulation and therapeutic targets in cancer. *Cell Prolif.* **2003**, *36*, 131–149. [CrossRef]
29. Bertoli, C.; Skotheim, J.M.; de Bruin, R.A. Control of cell cycle transcription during G1 and S phases. *Nat. Rev. Mol. Cell Biol.* **2013**, *14*, 518–528. [CrossRef]
30. Han, B.; Jiang, P.; Xu, H.; Liu, W.; Zhang, J.; Wu, S.; Liu, L.; Ma, W.; Li, X.; Ye, X. 8-Cetylcoptisine, a new coptisine derivative, induces mitochondria-dependent apoptosis and G0/G1 cell cycle arrest in human A549cells. *Chem. Biol. Interact.* **2019**, *299*, 27–36. [CrossRef]

31. Sun, P.; Wu, H.; Huang, J.; Xu, Y.; Yang, F.; Zhang, Q.; Xu, X. Porcine epidemic diarrhea virus through p53-dependent pathway causes cell cycle arrest in the G0/G1 phase. *Virus Res.* **2018**, *253*, 1–11. [CrossRef] [PubMed]

32. Zhang, X.; He, X.; Liu, Y.; Zhang, H.; Chen, H.; Guo, S.; Liang, Y. MiR-101-3p inhibits the growth and metastasis of non-small cell lung cancer through blocking PI3K/AKT signal pathway by targeting MALAT-1. *J. Hematol. Oncol.* **2017**, *93*, 1065–1073. [CrossRef] [PubMed]

33. Kanehisa, M.; Araki, M.; Goto, S.; Hattori, M.; Hirakawa, M.; Itoh, M.; Katayama, T.; Kawashima, S.; Okuda, S.; Tokimatsu, T.; et al. KEGG for linking genomes to life and the environment. *Nucleic Acids Res.* **2008**, *36*, 480–484. [CrossRef] [PubMed]

34. Morgensztern, D.; Mcleod, H. PI3K/Akt/mTOR pathway as a target for cancer therapy. J. *Anticancer Drugs*. **2005**, *16*, 797–803. [CrossRef]

35. Chang, L.; Graham, P.H.; Ni, J.; Hao, J.; Bucci, J.; Cozzi, P.J.; Li, Y. Targeting PI3K/Akt/mTOR signaling pathway in the treatment of prostate cancer radioresistance. *J. Crit. Rev. Oncol. Hematol.* **2015**, *96*, 507–517. [CrossRef] [PubMed]

36. Li, X.; Tang, Y.; Yu, F.; Sun, Y.; Huang, F.; Chen, Y.; Yang, Z.; Ding, G. Inhibition of prostate cancer DU-145 cells proliferation by anthopleura anjunae oligopeptide (YVPGP) via PI3K/AKT/mTOR signaling pathway. *Mar. Drugs* **2018**, *16*, 325. [CrossRef] [PubMed]

37. Liu, C.; Shen, G.N.; Luo, Y.H.; Piao, X.J.; Jiang, X.Y.; Meng, L.Q.; Wang, Y.; Zhang, Y.; Wang, J.R.; Wang, H.; et al. Novel 1,4-naphthoquinone derivatives induce apoptosis via ROS-mediated p38/MAPK, akt and STAT3 signaling in human hepatoma Hep3B cells. *Int. J. Biochem. Cell Biol.* **2018**, *96*, 9–19. [CrossRef]

38. Johnson, G.L.; Vaillancourt, R.R. Sequential protein kinase reactions controlling cell growth and differentiation. *Curr. Opin. Cell Biol.* **1994**, *6*, 230–238. [CrossRef]

39. Kiyatkin, A.; Aksamitiene, E.; Markevich, N.I.; Borisov, N.M.; Hoek, J.B.; Kholodenko, B.N. Scaffolding protein Grb2-associated binder 1 sustains epidermal growth factor-induced mitogenic and survival signaling by multiple positive feedback loops. *J. Biol. Chem.* **2006**, *281*, 19925–19938. [CrossRef]

40. Will, M.; Qin, A.C.; Toy, W.; Yao, Z.; Rodrik-Outmezguine, V.; Schneider, C.; Huang, X.; Monian, P.; Jiang, X.; de Stanchina, E.; et al. Rapid induction of apoptosis by PI3K inhibitors is dependent upon their transient inhibition of RAS-ERK signaling. *Cancer Discov.* **2014**, *4*, 334–347. [CrossRef]

41. Hollander, M.C.; Blumenthal, G.M.; Dennis, P.A. PTEN loss in the continuum of common cancers, rare syndromes and mouse models. *Nat. Rev. Cancer* **2011**, *11*, 289–301. [CrossRef] [PubMed]

42. Bader, A.G.; Kang, S.; Zhao, L.; Vogt, P.K. Oncogenic PI3K deregulates transcription and translation. *Nat. Rev. Cancer* **2005**, *5*, 921–929. [CrossRef] [PubMed]

43. Riquelme, I.; Tapia, O.; Espinoza, J.A.; Leal, P.; Buchegger, K.; Sandoval, A.; Bizama, C.; Araya, J.C.; Peek, R.M.; Roa, J.C. The gene expression status of the PI3K/AKT/mTOR pathway in gastric cancer tissues and cell lines. *Pathol. Oncol. Res.* **2016**, *22*, 797–805. [CrossRef] [PubMed]

44. Liu, P.; Cheng, H.; Roberts, T.M.; Zhao, J.J. Targeting the phosphoinositide 3-kinase pathway in cancer. *Nat. Rev. Drug Discov.* **2009**, *8*, 627–644. [CrossRef] [PubMed]

45. Zhang, Y.; Cui, J.; Zhang, R.; Wang, Y.; Hong, M. A novel fibrinolytic serine protease from the polychaete Nereis (Neanthes) virens (Sars): Purification and characterization. *Biochimie* **2007**, *89*, 93–103. [CrossRef] [PubMed]

46. Komina, O.; Wesierska-Gadek, J. Action of resveratrol alone or in combination with roscovitine, a CDK inhibitor, on cell cycle progression in human HL-60 leukemia cells. *Biochem. Pharmacol.* **2008**, *76*, 1554–1562. [CrossRef] [PubMed]

47. Lin, X.; Chen, Y.; Jin, H.; Zhao, Q.; Liu, C.; Li, R.; Yu, F.; Chen, Y.; Huang, F.; Yang, Z.; et al. Collagen extracted from bigeye tuna (*Thunnus obesus*) skin by isoelectric precipitation: Physicochemical properties, proliferation, and migration activities. *Mar. Drugs* **2019**, *17*, 261. [CrossRef] [PubMed]

48. Chen, J. The cell-cycle arrest and apoptotic functions of p53 in tumor initiation and progression. *Cold Spring Harb. Perspect. Med.* **2016**, *6*, a026104. [CrossRef]

Article

Enantioselective Hydrolysis of Styrene Oxide and Benzyl Glycidyl Ether by a Variant of Epoxide Hydrolase from *Agromyces mediolanus*

Huoxi Jin [1,*]**, Yan Li** [1]**, Qianwei Zhang** [1]**, Saijun Lin** [2]**, Zuisu Yang** [1] **and Guofang Ding** [1]

[1]   Zhejiang Provincial Engineering Technology Research Center of Marine Biomedical Products, School of Food and Pharmacy, Zhejiang Ocean University, Zhoushan 316022, China; m18625623399@163.com (Y.L.); zjouzqw@163.com (Q.Z.); abc1967@126.com (Z.Y.); dinggf2007@163.com (G.D.)

[2]   Hangzhou Institute for Food and Drug Control, Hangzhou 310019, China; saijunlin@126.com

*   Correspondence: jinhuoxi@zjou.edu.cn; Tel.: +86-580-2552-395

Received: 14 May 2019; Accepted: 19 June 2019; Published: 20 June 2019

Abstract: Enantiopure epoxides are versatile synthetic intermediates for producing optically active pharmaceuticals. In an effort to provide more options for the preparation of enantiopure epoxides, a variant of the epoxide hydrolase (vEH-Am) gene from a marine microorganism *Agromyces mediolanus* was synthesized and expressed in *Escherichia coli*. Recombiant vEH-Am displayed a molecular weight of 43 kDa and showed high stability with a half-life of 51.1 h at 30 °C. The purified vEH-Am exhibited high enantioselectivity towards styrene oxide (SO) and benzyl glycidyl ether (BGE). The vEH-Am preferentially converted (*S*)-SO, leaving (*R*)-SO with the enantiomeric excess (ee) >99%. However, (*R*)-BGE was preferentially hydrolyzed by vEH-Am, resulting in (*S*)-BGE with >99% ee. To investigate the origin of regioselectivity, the interactions between vEH-Am and enantiomers of SO and BGE were analyzed by molecular docking simulation. In addition, it was observed that the yields of (*R*)-SO and (*S*)-BGE decreased with the increase of substrate concentrations. The yield of (*R*)-SO was significantly increased by adding 2% (v/v) Tween-20 or intermittent supplementation of the substrate. To our knowledge, vEH-Am displayed the highest enantioselectivity for the kinetic resolution of racemic BGE among the known EHs, suggesting promising applications of vEH-Am in the preparation of optically active BGE.

Keywords: benzyl glycidyl ether; epoxide hydrolase; enantioselective; marine microorganism

---

1. Introduction

Because of significant differences in the pharmacological activities, metabolic processes, and toxicity of enantiomers in the human body, one isomer of chiral drugs may be effective while another may be ineffective or even harmful. Chiral drugs with high enantiopurity can not only improve the specificity of the drug by eliminating side effects caused by ineffective enantiomers but also reduce the amount of the drug needed to be taken by the patient. Due to these motivations, chiral drugs have become one of the main focuses of international drug research.

Chiral synthons play a key role in preparing chiral pharmaceuticals. As one of the important chiral synthons, chiral epoxides have been widely used to produce many chiral drugs due to their versatile reactivity. For example, chiral styrene oxide (SO) was an important chiral intermediate for the production of nematocide, levamisole, and hyperolactone C [1–3]. Chiral benzyl glycidyl ether (BGE) can be used for the synthesis of (+)-cryptocarya diacetate and synargentoide A, which have been found to show anti-tumor activity and have been used in the treatment of cancer pulmonary diseases [4,5].

One alternative method for preparing chiral epoxides is the enantioselective hydrolysis of racemic epoxides using epoxide hydrolase (EH, EC 3.3.2.3), which have been discovered in and cloned from many organisms such as plants, mammals, insects, bacteria, fungi, and yeasts [6–8]. To date, several EHs have been used to prepare chiral SO, but the enantiomeric excess (ee) or yields are always low. For example, the (*R*)-SO was obtained with an ee value of 91.6% and yield of 33.4% by using EH from *Vigna radiata* [9]. EH from *Gordonia* sp. H37 preferentially hydrolyzed (*R*)-SO, resulting in the preparation of an (*S*)-SO with ee > 99% and yield of 19.6% at 4 mM substrate concentration [10]. In addition, only a few EHs have been shown to be stereoselective for BGE. For example, EH from *bacillus alcalophilus* was used to provide (*S*)-benzyl glycidyl ether, but the ee value was only 30% [11]. EH from *Talaromyces flavus* resolved racemic BGE with moderate ee (< 98%) and selectivities (*E* < 15) [4]. The EH from *Aspergillus niger* M200 and its variants showed very low enantioselectivity (*E* < 5) for substrate BGE [12]. The limited number of EHs with high enantioselectivity for certain substrates demands studies that explore new EHs.

The ocean is rich in marine organisms that often have special metabolites and biosynthetic pathways because of their dramatically different environments. This poses potentially promising methods to screen highly for stereoselective EH from marine organisms. However, only a few EHs from marine organisms have been reported. To date, the EHs from marine microorganisms *Erythrobacter litoralis* HTCC2594, *Rhodobacterales bacterium* HTCC2654, *Erythrobacter* sp. JCS358, *Aspergillus sydowii*, and *Trichoderma* sp have been reported to produce chiral epoxides such glycidyl phenyl ether (GPE) and SO [13–16]. Xue et al. isolated a new marine microorganism *Agromyces mediolanus* from the coastal wetlands of Yancheng city, and found EH activity for epichlorohydrin (ECH), SO, and BGE [17]. However, the wild-type EH exhibited moderate enantioselectivity with only a 21.5% yield of (*S*)-ECH. Therefore, the author focused on improving enantioselectivity by site-saturation and site-directed mutagenesis of positions Ser207, Asn240 and Trp182 based on homologous modelling of the wild-type EH. Consequently, the variant (W182F/S207V/N240D) of this EH (vEH-Am) was found to enantioselectively hydrolyze racemic ECH with >99% ee and a 45.8% yield of (*S*)-ECH [18]. However, its application in the kinetic resolution of SO and BGE has still not been described. In general, the enzyme could exhibit different catalytic properties for different substrates because of the substrate specificity.

Due to the need for more information about the catalytic properties of vEH-Am for substrates SO and BGE, we synthesized the gene of vEH-Am and expressed it in *Escherichia coli* in the present study. The recombinant vEH-Am was used in the preparation of chiral SO and BGE (Scheme 1). We also focused on improving the yield of (*R*)-SO and (*S*)-BGE by adding surfactant or intermittent feeding of the substrate.

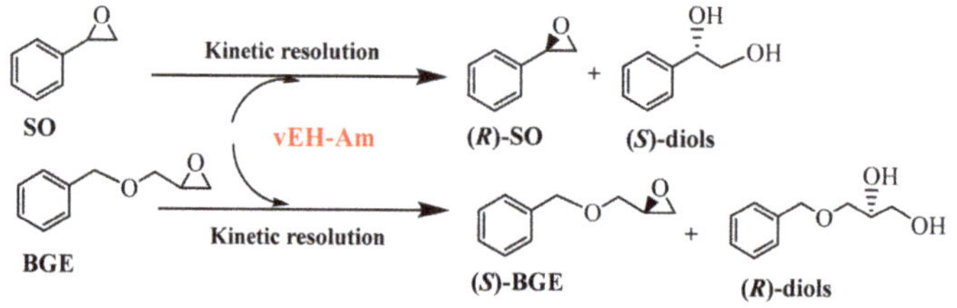

Scheme 1. Kinetic resolution of racemic styrene oxide (SO) and benzyl glycidyl ether (BGE) by a variant of the epoxide hydrolase gene from *Agromyces mediolanus* (vEH-Am).

2. Results and Discussion

2.1. Purification of the Recombinant vEH-Am

The recombinant *E. coli* BL21(DE3)/pET28a-vEH-Am was successfully constructed. After induction by 0.2 mM isopropyl β-D-thiogalactoside (IPTG), the expression of the recombinant vEH-Am was purified by His-tag affinity chromatography. The result of SDS-PAGE (Figure 1) showed a clear band with a molecular mass of approximately 43 kD in lane 2, which was consistent with the value predicted by the amino acid sequence of vEH-Am. The highest activity of vEH-Am was obtained at a pH of

8.0 and 35 °C (data not shown), which was the same as the original enzyme EH-Am [17]. This result indicated that there was no significant effect on the optimum reaction temperature and pH of the enzyme by the mutation treatment of W182F/S207V/N240D.

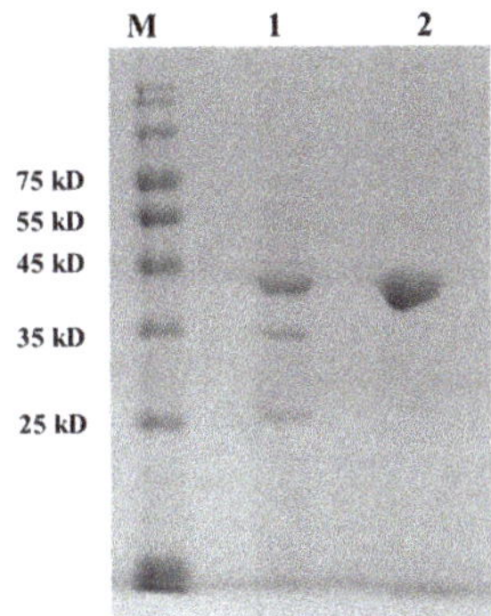

Figure 1. SDS-PAGE analysis of vEH-Am. Lane 1: cell extract; Lane 2: purified vEH-Am; Lane M: marker proteins.

2.2. Thermal Stability of vEH-Am

Enzyme stability is a very important indicator of enzyme properties and a very important parameter in catalytic processes. Although there are many factors affecting the stability of enzymes, including inhibitors, pH, organic solvents, etc., temperature is one of the most important factors affecting the stability of enzymes. To determine the thermal stability of vEH-Am, the enzyme solutions were pre-incubated at various temperatures (30, 37, and 50 °C) for different times (0–8 h) and then cooled immediately to measure the residual activity of vEH-Am at 30 °C. As shown in Figure 2, the residual activity of vEH-Am was more than 98% after 1 h at 30 °C, while only 81% was observed at 50 °C. After incubation for 8 h, 90% of activity was still obtained at 30 °C, but only 50% at 37 °C and 30% at 50 °C. The inactivation constant and half-life of vEH-Am at each temperature were calculated based on Equations (1) and (2). The results in Table 1 suggest that the half-life of vEH-Am is 51.1 h at 30 °C. As the temperature increased, the half-life decreases sharply. The half-life decreased to 8.0 h at 37 °C, and only 4.7 h at 50 °C. Therefore, although the highest activity of vEH-Am was obtained at 35 °C, 30 °C was selected as the reaction temperature for further studies in order to avoid a significant drop in enzyme activity during the reaction process.

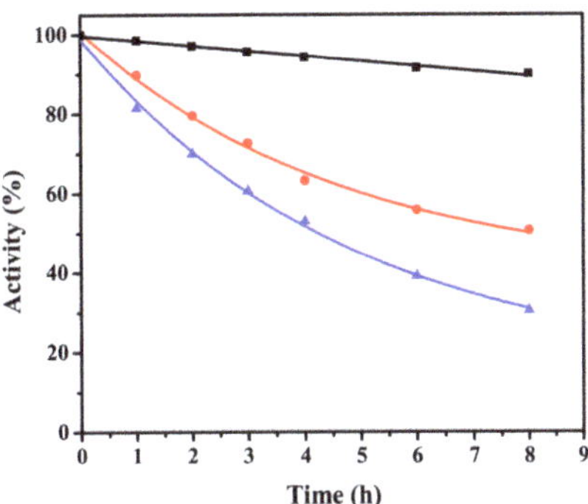

Figure 2. Thermal stability of vEH-Am at different temperatures. Symbols: 30 °C (■), 37 °C (●), 50 °C (▲).

Table 1. The inactivation constant and half-life of vEH-Am at different temperatures.

Temperature (°C)	Inactivation Constant (h^{-1})	Half-Life (h)
30	0.014	51.1
37	0.087	8.0
50	0.15	4.7

2.3. Hydrolysis of Racemic SO and BGE by vEH-Am

Enantiomerically pure epoxides can be directly prepared by the kinetic resolution of racemic epoxides with EH. However, the same enzyme shows different enantioselectivity for different substrates due to the substrate specificity. Under this context, SO and BGE were used to assess the vEH-Am's performance and selectivity.

The time courses of the kinetic resolution of racemic SO and BGE by vEH-Am were performed under optimal reaction conditions (pH 8.0, 30 °C). Figure 3A shows that the hydrolysis rate of (S)-SO is much faster than for (R)-SO. After 10 h of reaction, (S)-SO was hydrolyzed completely, and the ee of the remaining (R)-SO exceeded 99% with a yield of 25%. For substrate BGE, as shown in Figure 3B, vEH-Am preferentially converted (R)-BGE, and the remaining epoxide (S)-BGE ee was >99% when (R)-BGE was entirely consumed, leaving 34% of (S)-BGE. These results indicate that vEH-Am can enantioselectively hydrolyze both racemic SO and BGE, but results in different enantiomers and yields. It also seemed that BGE is more of a suitable substrate than SO for vEH-Am due to its higher activity and enantioselectivity.

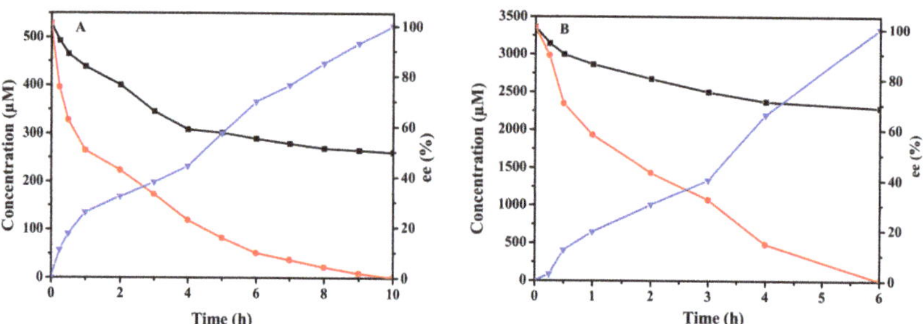

Figure 3. Time course of resolution of racemic SO (A) and BGE (B) by vEH-Am. Symbols: (*R*)-SO and (*S*)-BGE concentration (■); (*S*)-SO and (*R*)-BGE concentration (●); ee of (*R*)-SO and (*S*)-BGE (▼).

The kinetic resolution of racemic epoxides by EH was typically carried out in the aqueous medium, as described above. However, epoxides would spontaneously hydrolyze in this reaction system, resulting in a decrease in the yield of the optically pure epoxides. To overcome these disadvantages, the enzymatic reactions can also be performed in the presence of surfactants. Surfactants are amphoteric substances whose hydrophobic groups encapsulate hydrophobic substrates such as SO, reducing the occurrence of spontaneous hydrolysis [19]. Therefore, the hydrolysis of racemic SO and BGE by vEH-Am was investigated by adding 2% (v/v) Tween-20 to the reaction system. As shown in Figure 4, the kinetic resolution of racemic SO was also observed in the presence of Tween-20, but the hydrolysis rate of (R)-SO was significantly slower than that in Figure 3A (without Tween-20). After 10 h, the optically pure (R)-SO was obtained with a yield of 34%, which increased by 9 percentage points compared to the original 25% obtained without Tween-20. However, in BGE reactions compared to without Tween-20 (Figure 3B), the yield of (S)-BGE was not significantly improved by adding 2% (v/v) Tween-20 (data not shown). This result was probably attributed to the fact that the BGE was still mostly dissolved in the aqueous phase during the addition of the small amount of Tween-20.

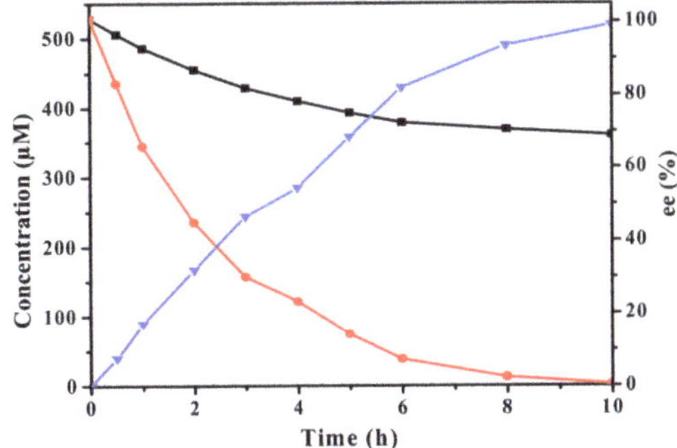

Figure 4. Effect of the Tween-20 on resolution of racemic SO by vEH-Am. Symbols: (*R*)-SO concentration (■); (*S*)-SO concentration (●); ee of (*R*)-SO (▼).

Although kinetic resolution of racemic SO by EH has been reported in many papers, there are only a few EHs that have been obtained from marine resources. EH from the marine fish (*Mugil cephalus*) exhibited (*R*)-preferred hydrolysis activity towards racemic SO, and (*S*)-SO was obtained with >99% ee and yield of 15.4% [8]. The kinetic resolution of racemic SO also was performed by EH from a marine microorganism (*Erythrobacter* spp), (*S*)-SO with an ee > 99%, but only an 8.5% yield was obtained [14]. In our study, (*R*)-SO was obtained with >99% ee and 34% yield from racemic SO by using vEH-Am from the marine microorganism *Agromyces mediolanus*, which was better than any previous EHs from marine resources.

Optically pure BGE is an important precursor for the synthesis of many drugs and natural products. However, as shown in Table 2, there are only a few studies in the literature for the preparation of chiral BGE by EH. (*R*)-BGE with an ee of 96% and *E* value of 13 was obtained by kinetic resolution of racemic BGE with EH from *Talaromyces flavus*. The EH from *Yarrowia lipolytica* enantioselectively hydrolyzed the racemic BGE with 95% ee and an *E* value of 10.4. In addition, the EH from *Bacillus alcalophilus*, *Aspergillus niger*, *Aspergillus sydowii*, *Streptomyces griseus*, and *Trichoderma* sp. were also used to produce chiral BGE, but ee values were all very low (<60%). Of course, enzyme catalysis is affected by many conditions, so it is possible that enzymes in Table 2 could display better performance if the conditions were altered. The vEH-Am from *Agromyces mediolanus* showed the highest ee and *E* value for substrate BGE in all the reported literature, suggesting the promising future of applications for vEH-Am.

Table 2. Kinetic resolution of racemic BGE with known epoxide hydrolases.

EH Source	Absolute Configuration	ee (%)	E-Values	Reference
Talaromyces flavus	R	96	13	[4]
Rhodotorula glutinis	R	98	6.5	[20]
Yarrowia lipolytica	R	95	10.4	[21]
Aspergillus niger	R	<60	3	[22]
Aspergillus sydowii	R	<46	/	[15]
Streptomyces griseus	R	7	1.5	[6]
Trichoderma sp.	S	<60	/	[15]
Bacillus alcalophilus	S	30	/	[11]
Agromyces mediolanus (W182F/S207V/N240D)	S	>99	14.5	This study

2.4. Kinetic Study of vEH-Am

To explore the detail of the catalytic properties of vEH-Am, kinetic studies for SO and BGE were performed using enantiopure as substrate. Kinetic parameters (V_m and K_m) of vEH-Am toward (*R*)-SO, (*S*)-SO, (*R*)-BGE, and (*S*)-BGE were measured, and the results are shown in Table 3. The V_m and K_m for the (*R*)-SO were 15.9 µmol·min^{-1}·mg^{-1} and 5.2 mM, whereas 3.4 µmol·min^{-1}·mg^{-1} and

0.9 mM were obtained for (*S*)-SO. The lower K_m for (*S*)-SO than (*R*)-SO indicated that (*S*)-SO had a higher affinity for vEH-Am and was preferentially hydrolyzed. However, the V_m of (*S*)-SO was also lower than that of (*R*)-SO, suggesting that (*R*)-SO was hydrolyzed with a much faster rate compared to (*S*)-SO when (*S*)-SO was completely converted. The V_m for (*R*)-BGE and (*S*)-BGE were 32.1 and 63.3 μmol·min^{-1}·mg^{-1}, respectively, which were significantly higher than those for (*R*)-SO and (*S*)-SO. The result indicated that vEH-Am showed a higher activity for substrate BGE than SO. In addition, the (*R*)-isomer of BGE was preferentially hydrolyzed because of the lower K_m, which was different from the preferential hydrolysis of (*S*)-isomer for SO. This result indicated that vEH-Am exhibits different stereoselectivity for different substrates.

Table 3. Kinetic parameters of hydrolysis of (*R*)-SO, (*S*)-SO, (*R*)-BGE, and (*S*)-BGE using the vEH-Am.

Kinetic Parameters	(*R*)-SO	(*S*)-SO	(*R*)-BGE	(*S*)-BGE
V_m (μmol·min^{-1}·mg^{-1})	15.9	3.4	32.1	63.3
K_m (mM)	5.2	0.9	9.3	36.5

2.5. Homology Structural Modeling and Substrate Docking

Homology modeling and molecular docking are the important methods for investigating the binding mode of the substrates and the origin of regioselectivity [9,23,24]. The 3D structure of vEH-Am was homologically modeled by Xue et al. [17,18]. In order to deeply understand the interactions between substrates and vEH-Am, the (*R*)-SO, (*S*)-SO, (*R*)-BGE, and (*S*)-BGE were antomatically docked into the active sites of vEH-Am using the Autodock 4.2 program (The Scripps Research Institute, USA). The binding modes of each enantiomer in the active site from molecular docking are shown in Figure 5. As in Figure 5A,B, both of (*R*)- and (*S*)-SO formed hydrogen bonds between Tyr 308 and the epoxide oxygen. The distance of the hydrogen bond (d_1) between Tyr 308 and the epoxide oxygen in (*S*)-SO was shorter than that in (*R*)-SO. However, it was observed that the distance of the hydrogen bond (d_2) between Asp181 oxygen and the attacked epoxide carbon in (*S*)-SO was slightly longer than that in (*R*)-SO. The total distance of hydrogen bonds, d ($d = d_1 + d_2$), was thought to be of particular importance for the catalytic efficacy. The d value should be shorter for the preferred enantiomer [25]. The differences in the modeled distance, Δd, for the (*R*)-enantiomer and (*S*)-enantiomer were well consistent with regioselectivity. The d value for (*S*)-SO was 5.1 Å, which was shorter than 6.7 Å for (*R*)-SO, indicating that (*S*)-SO was preferably hydrolyzed by vEH-Am. For the substrate BGE (Figure 5C,D), the d value for (*R*)-BGE was 5.5 Å, while it was 7.3 Å for (*S*)-BGE, suggesting that (*R*)-BGE was preferentially hydrolyzed. Furthermore, the Δd value between (*R*)-BGE and (*S*)-BGE was 1.8, which was longer than 1.6 Å between (*R*)-SO and (*S*)-SO, showing that vEH-Am revealed a higher regioselectivity for the kinetic resolution of racemic BGE than SO. The results were consistent with those obtained in Figure 3.

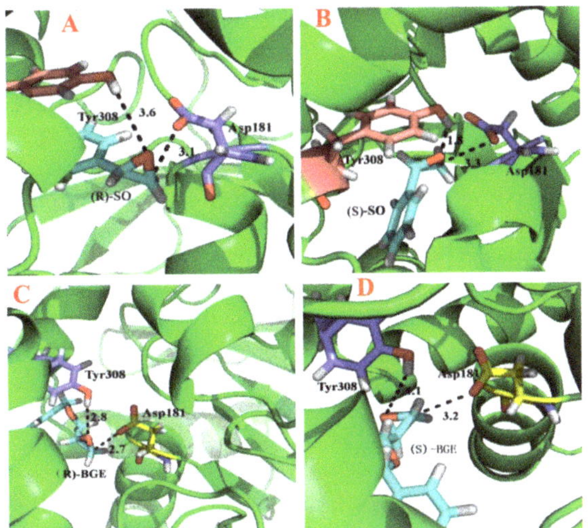

Figure 5. Molecular docking simulations between vEH-Am and (*R*)-SO (**A**), (*S*)-SO (**B**), (*R*)-BGE (**C**), and (*S*)-BGE (**D**).

2.6. Biocatalytic Synthesis of (R)-SO and (S)-BGE

To test the feasibility of vEH-Am and its potential application in the synthesis of (*R*)-SO or (*S*)-BGE, the effect of the substrate concentration on the reaction was investigated. As shown in Figure 6, >99% ee of (*R*)-SO or (*S*)-BGE was obtained in all tested substrate concentrations, but the yields of them were both decreased with increasing substrate concentrations. For the substrate SO (Figure 6A), the yield of (*R*)-SO was 30.2% at the 180 μM of racemic SO, but decreased to 27.6% at concentration of 700 μM. When the concentration of racemic SO reached 2110 μM, the yield of (*R*)-SO was only 24.8%. Furthermore, a similar trend was observed for the substrate BGE (Figure 6B). With 670 μM racemic BGE as substrate, the optical purity of (*S*)-BGE was obtained with yield of 40.1%. When kinetic resolution of racemic BGE was performed at concentration of 5380 μM, the yield of (*S*)-BGE sharply decreased to 25.1%. If the concentration of racemic BGE continued to increase to 10750 μM, the yield of (*S*)-BGE was only 21.2%. The phenomenon that the yield decreased with increasing substrate concentration was also observed in kinetic resolution of epoxides by other EHs [26,27]. The result may be ascribed to enzyme inactivation caused by high concentrations of the substrate.

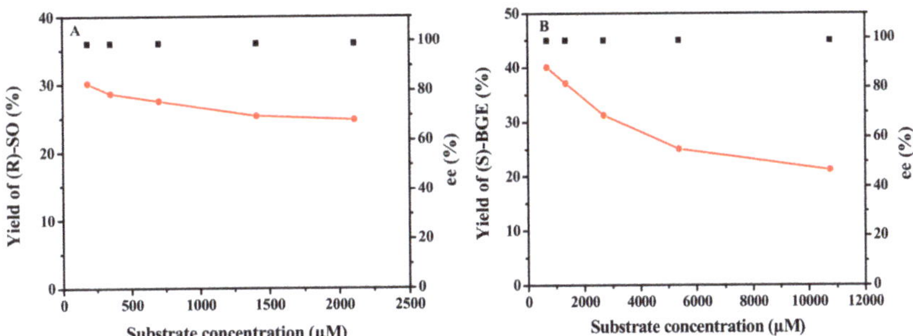

Figure 6. The effect of substrate concentration on the production of (*R*)-SO (**A**) and (*S*)-BGE (**B**). Symbols: yield of (*R*)-SO or (*S*)-BGE (■); ee of (*R*)-SO or (*S*)-BGE (●).

Substrate inhibition has always been observed in many reactions by enzyme catalysis. It is well known that the substrate inhibition can be alleviated or even eliminated by intermittent feeding of the substrate. For example, in the kinetic resolution of racemic epichlorohydrin by recombinant EH from *Agrobacterium radiobacter*, the substrate concentration increased from 320 mM to 448 mM by intermittent feeding of the substrate [28]. We therefore performed kinetic resolutions of racemic SO and BGE by intermittent feeding of the substrate to increase the yield of (*R*)-SO and (*S*)-BGE. The kinetic resolution of racemic SO was initiated at 50 mL phosphate buffer (0.2 M, pH 8.0) with 1056 μM of the initial concentration substrate and 0.5 mg of vEH-Am. A volume of 1.5 μL of racemic SO was added to the reaction system every 30 min until the final concentration was 2110 μM. As shown in Figure 7A, during the addition of the substrate, the concentration of (*R*)-SO was always higher than the concentration of (*S*)-SO, and this gap gradually increased with the extension of the reaction time. When the addition of racemic SO was halted after 2 h, the (*S*)-SO hydrolyzed rapidly, while the rate of hydrolysis of (*R*)-SO was slower. Ultimately, 701 μM of (*R*)-SO remained from 2110 μM of racemic SO after (*S*)-SO was completely consumed at 12 h of reaction. The yield of (*R*)-SO with >99% ee was 33.2%, which was significantly higher than the 24.8% obtained by adding substrate only once, as shown in Figure 6A.

The reaction of racemic BGE was performed at an initial concentration of 5380 μM, and 20 μL of racemic BGE was added every 30 min until the final concentration was 10,750 μM. Finally, (*S*)-BGE with ee > 99% was obtained at a yield of 26.9% from 10750 μM racemic BGE (Figure 7B). Compared to 21.2% obtained without addition of substrate (Figure 6B), the yield of (*S*)-BGE was significantly improved by intermittent feedings. However, it was observed that the extent of improvement in yield of (*S*)-BGE (increasing 5.7%) was not as significant as (*R*)-SO (increasing by 8.4%). This similar result was also observed in biosynthesis of (*R*)-SO and (*S*)-BGE by addition of Tween-20, as shown in Figure 4. These results indicate that the substrate inhibition of SO is stronger than that of BGE, which may be mainly attributed to the more hydrophobic tendency of SO than BGE. Another promising way to overcome substrate inhibition is to perform the reaction in an aqueous/organic two-phase system, which had been used in previous research to improve substrate concentration [28,29]. In the future, we would try to use a two-phase reaction system to increase the substrate concentration and yield of the chiral epoxide.

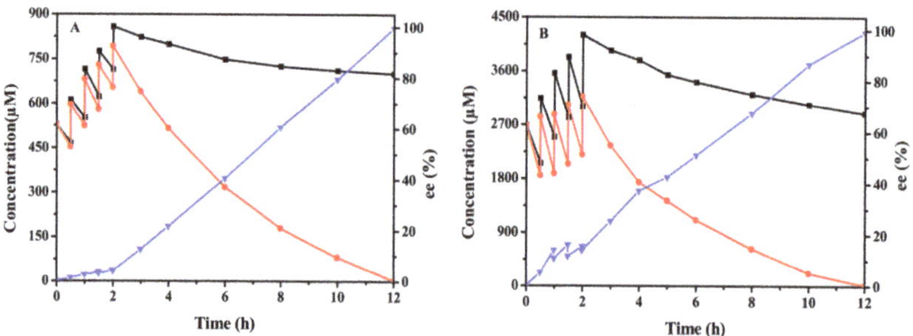

Figure 7. Production of (*R*)-SO (**A**) and (*S*)-BGE (**B**) by intermittent feeding of the substrate. Symbols: (*R*)-SO or (*S*)-BGE concentration (■); (*S*)-SO or (*R*)-BGE concentration (●); ee of (*R*)-SO or (*S*)-BGE (▼).

3. Materials and Methods

3.1. Materials

(*R,S*)-SO, (*R*)-SO, (*S*)-SO, (*R,S*)-BGE, (*R*)-BGE, and (*S*)-BGE were obtained from Aladdin Bio-Chem Technology Co., Ltd. (Shanghai, China). *E. coli* BL21(DE3) and pET-28a were used for the expression of vEH-Am. All other chemicals were of analytical grade from commercial sources.

3.2. Cell Culture and Protein Expression

The vEH-Am sequence was synthesized according to the gene of EH from *Agromyces mediolanus* ZJB120203 (GenBank accession no. JX467176) and mutant sites (W182F/S207V/N240D) using the PCR method [30]. Six His-tags were added at the 3′ end of the vEH-Am gene. The synthesized gene was inserted into pET28a between NcoI and XhoI sites. The recombinant *E. coli* was obtained by transforming the pET28a-vEH-Am into *E. coli* BL21(DE3). The strain was inoculated into 50 mL of LB liquid medium containing 50 µg/mL kanamycin and cultured overnight at 37 °C as a seed solution. A volume of 0.5 mL of seed solution was then transferred into 50 mL of the same LB medium and incubated at 37 °C at 200 rpm until the OD_{600} of the fermentation liquid was about 0.7. Isopropyl β-D-thiogalactoside (IPTG) was added with the final concentration of 0.2 mM. The fermentation liquid was further cultured for 8 h at 28 °C and then centrifuged at 10,000 rpm for 15 min to obtain cells.

3.3. Purification of vEH-Am

Four grams of wet cells were suspended in 40 mL of phosphate buffer (20 mM, pH 8.0) and broken by a 30 min ultrasound treatment. The disrupted solution was centrifuged at 10,000 rpm for 20 min to remove cell debris. The supernatant solution (crude enzyme solution) was collected for subsequent separation and purification. The purification column was Nickel–nitrilotriacetic (Ni-NTA), and the packed volume was 10 mL. The Ni-NTA column was equilibrated with buffer A (20 mM phosphate buffer, 500 mM NaCl and 20 mM imidazole, pH 8.0) for 10 min. The crude enzyme solution was then inflowed into the column at a rate of 1 mL/min. After the unbound proteins were eluted by buffer A, the target protein vEH-Am was collected by elution at a rate of 3 mL/min with buffer B (20 mM phosphate buffer, 500 mM NaCl and 500 mM imidazole, pH 8.0). The enzyme solution was dialyzed overnight in a phosphate buffer of 20 mM (pH 8.0), and the purified enzyme was analyzed by 12% sodium dodecyl sulfate polyacrylamide gel electrophoresis (SDS-PAGE).

3.4. Thermostability of vEH-Am

The temperature stability of vEH-Am was determined in 10 ml of phosphate buffer (0.2 mM, pH 8.0) containing 0.1 mg of vEH-Am. The mixture was pre-incubated for different times at 30, 37, and 50 °C, and the remaining activity was assayed by adding 10 µL of racemic BGE. The inactivation constant k_d and half-life $t_{1/2}$ were calculated based on the following formula:

$$\ln\left(\frac{[E]}{[E_0]}\right) = -k_d \times t \tag{1}$$

$$t_{1/2} = 0.693/k_d \tag{2}$$

3.5. Activity Assay and Analytical Methods

The racemic substrate (1 µL SO or 10 µL BGE) and 0.1 mg vEH-Am were mixed in 10 mL of phosphate buffer (0.2 M, pH 8.0). The reaction was performed for 15 min at 30 °C at 150 rpm. The 1.0 mL reaction solution was then taken and added to 2 mL of hexane. After 3 min, 1 mL of the organic layer was separated and dried with anhydrous sodium sulfate. The reaction solution was treated with a 0.2 µm filter membrane after centrifugation at 4000 rpm for 3 min and then analyzed by HPLC to determine the enzyme activity and ee value. The concentrations of the substrates were analyzed using an Agilent LC 1260 with a CHIRALPAK AS-H column (5 µm, 4.6 × 250 mm). Detection conditions were as follows: 20 µL injection volume, mobile phase n-hexane/isopropanol (95:5), 210 nm detection wavelength, 30 °C column temperature, 1 mL/min flow rate. Retention times: 5.3 min for (*R*)-SO, 5.6 min for (*S*)-SO, 8.3 min for (*S*)-BGE, and 9.8 min for (*R*)-BGE. The ee was calculated from the concentrations of the two enantiomers based on the formula: ee (%) = $(S - R)/(S + R) \times 100$. The enantiomeric ratio (*E* value) was calculated based on the ee of the remaining epoxide and the conversion (*C*) of racemate according to equation: $E = \ln[(1 - C)(1 - ee)]/\ln[(1 - C)(1 + ee)]$. One unit of enzyme activity was defined as the amount of enzyme required to convert 1 µmol of substrates at 30 °C.

3.6. Determination of Kinetic Properties

The kinetic study of the vEH-Am was performed by measuring the initial rate at different concentrations of (R)-SO, (S)-SO, (R)-BGE, and (S)-BGE. The kinetic parameters V_m (maximum reaction rate) and K_m (Michaelis constant) were calculated based on the Michaelis–Menten equation: $1/v = K_m/V_m[S] + 1/V_m$, where v is the initial velocity and [S] is the substrate concentration.

3.7. Homology Modeling and Docking

The three-dimensional structure of vEH-Am was generated by Modeller 9.12 in Discovery studio (DS) 2.1 (Accelrys Software, San Diego, CA, USA) based on the crystal structure of EH (PDB accession no. 4i19) [18]. The docking studies were performed by Autodock 4.2 (The Scripps Research Institute, USA).

3.8. vEH-Am Hydrolysis of Racemic SO and BGE

Hydrolysis of racemic SO and BGE was performed in 50 mL 0.2 M sodium phosphate buffer (pH 8.0) containing 0.5 mg vEH-Am. The reactions were initiated by adding various concentrations of racemic substrates at 30 °C and 200 rpm. The 1.0 mL mixtures were taken out regularly at different times intervals and added to 2 mL of hexane to stop the reaction. The hydrolysis progressions of racemic SO and BGE were analyzed by measuring the concentrations of each enantiomer.

4. Conclusions

In this study, we report on the kinetic resolution of racemic SO and BGE by the particular variant of epoxide hydrolase from *Agromyces mediolanus* (vEH-Am). The vEH-Am enantioselectively hydrolyzed (S)-SO and (R)-BGE, leaving (R)-SO and (S)-BGE with >99% ee. Moreover, the yield of optically pure (S)-BGE reached 34%, which was the highest among all reported epoxide hydrolases. Molecular docking simulations showed that the hydrogen bonds between vEH-Am and (S)-SO or (R)-BGE were shorter than the other enantiomers. The concentration of the substrate had a negative impact on the yield of (R)-SO or (S)-BGE, which could be significantly improved by adding Tween-20 or intermittent feeding of the substrate. This study lays the theoretical foundation for the application of vEH-Am in the preparation of enantiopure SO and BGE.

Author Contributions: H.J. designed the experiments and wrote the paper. Y.L. and Q.Z. performed the experiments. S.L., Z.Y. and G.D. provided experimental materials and equipment.

Funding: The work was supported by the National Natural Science Foundation of China (NSFC) (no. 81773629).

Acknowledgments: We are grateful to Feng Xue (Yancheng Institute of Technology) for providing technical assistance of the molecular docking simulations.

Conflicts of Interest: The authors declare no conflict of interest.

References

1. Chang, D.; Wang, Z.; Heringa, M.F.; Wirthner, R.; Witholt, B.; Li, Z. Highly enantioselective hydrolysis of alicyclic meso-epoxides with a bacterial epoxide hydrolase from *Sphingomonas* sp. HXN-200: Simple syntheses of alicyclic vicinal trans-diols. *Chem. Commun.* **2003**, *8*, 960–961. [CrossRef]
2. Wu, S.; Li, A.; Chin, Y.S.; Li, Z. Enantioselective Hydrolysis of Racemic and Meso-Epoxides with Recombinant *Escherichia coli* Expressing Epoxide Hydrolase from *Sphingomonas* sp. HXN-200: Preparation of Epoxides and Vicinal Diols in High ee and High Concentration. *Acs Catal.* **2013**, *3*, 752–759. [CrossRef]
3. Hu, D.; Wang, R.; Shi, X.L.; Ye, H.H.; Wu, Q.; Wu, M.C.; Chu, J.J. Kinetic resolution of racemic styrene oxide at a high concentration by recombinant *Aspergillus usamii* epoxide hydrolase in an n-hexanol/buffer biphasic system. *J. Biotechnol.* **2016**, *236*, 152–158. [CrossRef] [PubMed]

4. Wei, C.; Chen, Y.; Shen, H.; Wang, S.; Chen, L.; Zhu, Q. Biocatalytic resolution of benzyl glycidyl ether and its derivates by *Talaromyces flavus*: effect of phenyl ring substituents on enantioselectivity. *Biotechnol. Lett.* **2012**, *34*, 1499–1503. [CrossRef] [PubMed]

5. Sabitha, G.; Gopal, P.; Reddy, C.N.; Yadav, J.S. First stereoselective synthesis of synargentolide A and revision of absolute stereochemistry. *Tetrahedron Lett.* **2009**, *50*, 6298–6302. [CrossRef]

6. Saini, P.; Kumar, N.; Wani, S.I.; Sharma, S.; Chimni, S.S.; Sareen, D. Bioresolution of racemic phenyl glycidyl ether by a putative recombinant epoxide hydrolase from *Streptomyces griseus* NBRC 13350. *World J. Microbiol. Biotechnol.* **2017**, *33*, 82. [CrossRef]

7. Woo, J.H.; Lee, E.Y. Enantioselective hydrolysis of racemic styrene oxide and its substituted derivatives using newly-isolated *Sphingopyxis* sp. exhibiting a novel epoxide hydrolase activity. *Biotechnol. Lett.* **2014**, *36*, 357–362. [CrossRef]

8. Lee, S.J.; Kim, H.S.; Kim, S.J.; Park, S.; Kim, B.J.; Shuler, M.L.; Lee, E.Y. Cloning, expression and enantioselective hydrolytic catalysis of a microsomal epoxide hydrolase from a marine fish, *Mugil cephalus*. *Biotechnol. Lett.* **2007**, *29*, 237–246. [CrossRef]

9. Hu, D.; Tang, C.; Li, C.; Kan, T.; Shi, X.; Feng, L.; Wu, M. Stereoselective Hydrolysis of Epoxides by reVrEH3, a Novel *Vigna radiata* Epoxide Hydrolase with High Enantioselectivity or High and Complementary Regioselectivity. *J. Agric. Food Chem.* **2017**, *65*, 9861–9870. [CrossRef]

10. Woo, J.H.; Kwon, T.H.; Kim, J.T.; Kim, C.G.; Lee, E.Y. Identification and characterization of epoxide hydrolase activity of polycyclic aromatic hydrocarbon-degrading bacteria for biocatalytic resolution of racemic styrene oxide and styrene oxide derivatives. *Biotechnol. Lett.* **2013**, *35*, 599–606. [CrossRef]

11. Bala, N.; Kaur, K.; Chimni, S.S.; Saini, H.S.; Kanwar, S.S. Bioresolution of benzyl glycidyl ether using whole cells of *Bacillus alcalophilus*. *J. Basic Microbial.* **2012**, *52*, 383–389. [CrossRef] [PubMed]

12. Kotik, M.; Stepanek, V.; Kyslik, P.; Maresova, H. Cloning of an epoxide hydrolase-encoding gene from *Aspergillus niger* M200, overexpression in *E. coli*, and modification of activity and enantioselectivity of the enzyme by protein engineering. *J. Biotechnol.* **2007**, *132*, 8–15. [CrossRef] [PubMed]

13. Woo, J.H.; Hwang, Y.O.; Kang, S.G.; Lee, H.S.; Cho, J.C.; Kim, S.J. Cloning and characterization of three epoxide hydrolases from a marine bacterium, *Erythrobacter litoralis* HTCC2594. *Appl. Microbial. Biotechnol.* **2007**, *76*, 365–375. [CrossRef] [PubMed]

14. Hwang, Y.O.; Kang, S.G.; Woo, J.H.; Kwon, K.K.; Sato, T.; Lee, E.Y.; Han, M.S.; Kim, S.J. Screening enantioselective epoxide hydrolase activities from marine microorganisms: detection of activities in *Erythrobacter* spp. *Mar. Biotechnol.* **2008**, *10*, 366–373. [CrossRef] [PubMed]

15. Martins, M.P.; Mouad, A.M.; Boschini, L.; Regali Seleghim, M.H.; Sette, L.D.; Meleiro Porto, A.L. Marine fungi *Aspergillus sydowii* and Trichoderma sp. catalyze the hydrolysis of benzyl glycidyl ether. *Mar. Biotechnol.* **2011**, *13*, 314–320. [CrossRef] [PubMed]

16. Woo, J.H.; Kang, J.H.; Hwang, Y.O.; Cho, J.C.; Kim, S.J.; Kang, S.G. Biocatalytic resolution of glycidyl phenyl ether using a novel epoxide hydrolase from a marine bacterium, *Maritimibacter alkaliphilus* KCCM 42376. *J. Biosci. Bioeng.* **2010**, *109*, 539–544. [CrossRef]

17. Xue, F.; Liu, Z.Q.; Zou, S.P.; Wan, N.W.; Zhu, W.Y.; Zhu, Q.; Zheng, Y.G. A novel enantioselective epoxide hydrolase from *Agromyces mediolanus* ZJB120203: Cloning, characterization and application. *Process Biochem.* **2014**, *49*, 409–417. [CrossRef]

18. Xue, F.; Liu, Z.Q.; Wan, N.W.; Zhu, H.Q.; Zheng, Y.G. Engineering the epoxide hydrolase from *Agromyces mediolanus* for enhanced enantioselectivity and activity in the kinetic resolution of racemic epichlorohydrin. *RSC Adv.* **2015**, *5*, 31525–31532. [CrossRef]

19. Yoo, S.S.; Park, S.; Lee, E.Y. Enantioselective resolution of racemic styrene oxide at high concentration using recombinant Pichia pastoris expressing epoxide hydrolase of *Rhodotorula glutinis* in the presence of surfactant and glycerol. *Biotechnol. Lett.* **2008**, *30*, 1807–1810. [CrossRef]

20. Weijers, C.A.G.M. Enantioselective hydrolysis of aryl, alicyclic and aliphatic epoxides by *Rhodotorula glutinis*. *Tetrahedron Asymm.* **1997**, *8*, 639–647. [CrossRef]

21. Bendigiri, C.; Harini, K.; Yenkar, S.; Zinjarde, S.; Sowdhamini, R.; RaviKumar, A. Evaluating Ylehd, a recombinant epoxide hydrolase from *Yarrowia lipolytica* as a potential biocatalyst for the resolution of benzyl glycidyl ether. *RSC Adv.* **2018**, *8*, 12918–12926. [CrossRef]

22. Kotik, M.; Kyslik, P. Purification and characterisation of a novel enantioselective epoxide hydrolase from *Aspergillus niger* M200. *Biochim. Biophys. Acta* **2006**, *1760*, 245–252. [CrossRef] [PubMed]

23. Zheng, H.; Reetz, M.T. Manipulating the stereoselectivity of limonene epoxide hydrolase by directed evolution based on iterative saturation mutagenesis. *J. Am. Chem. Soc.* **2010**, *132*, 15744–15751. [CrossRef] [PubMed]

24. Saenz-Mendez, P.; Katz, A.; Perez-Kempner, M.L.; Ventura, O.N.; Vazquez, M. Structural insights into human microsomal epoxide hydrolase by combined homology modeling, molecular dynamics simulations, and molecular docking calculations. *Proteins* **2017**, *85*, 720–730. [CrossRef] [PubMed]

25. Reetz, M.T.; Bocola, M.; Wang, L.W.; Sanchis, J.; Cronin, A.; Arand, M.; Zou, J.; Archelas, A.; Bottalla, A.L.; Naworyta, A.; et al. Directed evolution of an enantioselective epoxide hydrolase: Uncovering the source of enantioselectivity at each evolutionary stage. *J. Am. Chem. Soc.* **2009**, *131*, 7334–7343. [CrossRef] [PubMed]

26. Wang, R.; Hu, D.; Zong, X.; Li, J.; Ding, L.; Wu, M.; Li, J. Enantioconvergent hydrolysis of racemic styrene oxide at high concentration by a pair of novel epoxide hydrolases into (R)-phenyl-1,2-ethanediol. *Biotechnol. Lett.* **2017**, *39*, 1917–1923. [CrossRef]

27. Zou, S.P.; Zheng, Y.G.; Wu, Q.; Wang, Z.C.; Xue, Y.P.; Liu, Z.Q. Enhanced catalytic efficiency and enantioselectivity of epoxide hydrolase from *Agrobacterium radiobacter* AD1 by iterative saturation mutagenesis for (R)-epichlorohydrin synthesis. *Appl. Microbial. Biotechnol.* **2018**, *102*, 733–742. [CrossRef]

28. Jin, H.X.; Liu, Z.Q.; Hu, Z.C.; Zheng, Y.G. Biosynthesis of (R)-epichlorohydrin at high substrate concentration by kinetic resolution of racemic epichlorohydrin with a recombinant epoxide hydrolase. *Eng. Life Sci.* **2013**, *13*, 385–392. [CrossRef]

29. Chen, W.J.; Lou, W.Y.; Yu, C.Y.; Wu, H.; Zong, M.H.; Smith, T.J. Use of hydrophilic ionic liquids in a two-phase system to improve Mung bean epoxide hydrolases-mediated asymmetric hydrolysis of styrene oxide. *J. Biotechnol.* **2012**, *162*, 183–190. [CrossRef]

30. Liu, Z.Q.; Zhang, L.; Sun, L.H.; Li, X.J.; Wan, N.W.; Zheng, Y.G. Enzymatic production of 5′-inosinic acid by a newly synthesised acid phosphatase/phosphotransferase. *Food Chem.* **2012**, *134*, 948–956. [CrossRef]

Article

Collagen Extracted from Bigeye Tuna (*Thunnus obesus*) Skin by Isoelectric Precipitation: Physicochemical Properties, Proliferation, and Migration Activities

Xinhui Lin [1,†], Yinyue Chen [1,†], Huoxi Jin [1,†], Qiaoling Zhao [2], Chenjuan Liu [1], Renwei Li [3], Fangmiao Yu [1], Yan Chen [1,*], Fangfang Huang [1], Zuisu Yang [1], Guofang Ding [1] and Yunping Tang [1,4,*]

[1] Zhejiang Provincial Engineering Technology Research Center of Marine Biomedical Products, School of Food and Pharmacy, Zhejiang Ocean University, Zhoushan 316022, China; linxinhui1995@163.com (X.L.); zhdchenyingyue@163.com (Y.C.); jinhuoxi@163.com (H.J.); 17805800624@163.com (C.L.); fmyu@zjou.edu.cn (F.Y.); gracegang@126.com (F.H.); abc1967@126.com (Z.Y.); dinggf2007@163.com (G.D.)

[2] Zhoushan Institute for Food and Drug Control, Zhoushan 316000, China; zql850410@126.com

[3] Zhejiang Ocean Family CO., LTD, Zhoushan 316022, China; lirw6886@163.com

[4] Zhejiang Changxing Pharmaceutical CO., LTD, Huzhou 313108, China

* Correspondence: cyancy@zjou.edu.cn (Y.C.); tangyunping1985@zjou.edu.cn (Y.T.); Tel.: +86-0580-229-9809 (Y.T.); Fax: +86-0580-229-9866 (Y.T.)

† These authors contributed equally to this work.

Received: 22 April 2019; Accepted: 30 April 2019; Published: 1 May 2019

Abstract: Collagen was extracted from bigeye tuna (*Thunnus obesus*) skins by salting-out (PSC-SO) and isoelectric precipitation (PSC-IP) methods. The yield of the PSC-IP product was approximately 17.17% (dry weight), which was greater than the yield obtained from PSC-SO (14.14% dry weight). Sodium dodecyl sulfate-polyacrylamide gel electrophoresis analysis indicated that collagen from bigeye tuna skin belongs to collagen type I. Inductively coupled plasma mass spectrometry results indicate that the heavy metal abundance in PSC-IP was lower than the maximum acceptable amounts according to Chinese regulatory standards. In addition, results from a methylthiazolyldiphenyl-tetrazolium bromide assay and an in vitro scratch assay demonstrated that PSC-IP could promote the proliferation and migration of NIH-3T3 fibroblasts. Overall, results suggest PSC-IP could be used to rapidly extract collagen from marine by-products instead of traditional salting-out methods. Collagen from bigeye tuna skin may also have strong potential for cosmetic and biomedical applications.

Keywords: *Thunnus obesus*; collagen; isoelectric precipitation; physicochemical properties; proliferation and migration

---

1. Introduction

Rapid developments of seafood products in the marine product processing industry have led to a large number of marine by-products being discarded without treatment, such as fish skin, bones, scales, and swimming bladders [1,2]. By-products are a cause of environmental issue and are wastes of biological resources, therefore the utilization of by-products to produce high-added-value compounds is an urgent problem, such as the blue granary scientific and technological innovation plan in China. Collagen, which accounts for approximately thirty percent of the total protein content, is the most abundant fibrous protein in animals [3,4]. Recently, marine collagen has been extracted from sponges, octopus, jellyfishes, squids, and fish offal such as skins, bones, fins, and scales [5–8]. Marine collagen has attracted an increasing interest for applications in the biomedical, and pharmaceutical, cosmetic and food industries, as it has no religious limitations, and shows low immunogenicity and non-cytotoxicity [9–11]. Therefore, the extraction of collagen from marine by-products would be a suitable way to utilize them, which could produce a high-added-value compound.

Neutral salt solubilization, acid solubilization, and pepsin solubilization are three major methods for extracting collagen in the collagen extraction phase [6]. However, the recovery phase is also an extremely important step during industrial collagen production. The salting-out method was often used for precipitating collagen during the recovery process [12–14]. The main principle of the salting-out method is that the charge carried by salt ions in solution neutralizes the surface charge of collagen molecules, which decreases the electrostatic interactions between collagen molecules and leads to their gradual precipitation. The salting-out method offers high recovery rates of collagen and does not affect its triple helix structure; therefore, it has often been used in previous studies [14,15]. However, the addition of high concentrations of salt can greatly prolong the downstream dialysis process. Generally, it takes three to four days of dialysis after salting-out to ensure that the final product is not affected by the salt and acid [3,15]. Moreover, dialysis of collagen extract containing a high concentration of salt produces a large amount of wastewater. Therefore, the purification process could be improved by the development of a highly efficient, highly productive, and sustainable method to take full advantage of collagen-rich fish by-products.

Isoelectric precipitation has become a popular method for protein purification due to its high efficiency and strong specificity; it also does not often require a lengthy downstream dialysis for the product [16–18]. Isoelectric precipitation is based on the principle that proteins have their lowest solubility at their isoelectric point and that different proteins have different isoelectric points. Isoelectric precipitation has been used for the separation and purification of proteins such as PSE-like chicken protein [19], walnut protein [20], Chinese quince seed protein [21], and ovalbumin [22]. However, there are few reports demonstrating the use of isoelectric precipitation for the recovery and purification of collagen.

Tuna are an economically important fish worldwide; in 2018 the estimated global harvest of tuna was 7.5 million tons [23]. Because the white meat of tuna is only used for sashimi or canned, the tuna industry produces a lot of waste or by-products, which includes fish heads, bones, skins, scales, internal organs, and dark meat, which account for approximately 50% to 70% of the total mass [24]. However, there are no reports of the use of isoelectric precipitation to extract collagen from bigeye tuna (*Thunnus obesus*) skins. Therefore, the aim of this study is to use isoelectric precipitation to extract collagen from bigeye tuna skin. The physiochemical properties of collagen obtained from isoelectric precipitation, and its activity on NIH-3T3 proliferation and migration are also evaluated to assess its suitability for biomedical and cosmetic applications

2. Results and Discussion

2.1. Determination of the Isoelectric Point of Collagen

The determination of the collagen isoelectric point is necessary for using isoelectric precipitation to recover and purify collagen. As shown in Figure 1, pepsin-solubilized collagen (PSC) from bigeye tuna skin had the best solubility in the range of acidic pH values (≤4.0) and had lower solubility in the range of neutral or alkaline pH values. Similarity, the PSC from *Nibea japonica* skin and the PSC from silver carp showed the maximum solubility in the pH range of 1.0–4.0, and lower solubility in the neutral or slightly alkaline pH range [9,25]. Our result was consistent with these studies. When the pH value is above or below the isoelectric point of a protein, the net charge, repulsive forces, and interaction capacity with water, increase. However, if the protein has no net charge at the isoelectric point, the protein will aggregate and precipitate due to hydrophobic–hydrophobic interactions. In the present study, PSC showed the lowest solubility at pH 7.0, therefore, this value was selected for collagen extraction from bigeye tuna skin.

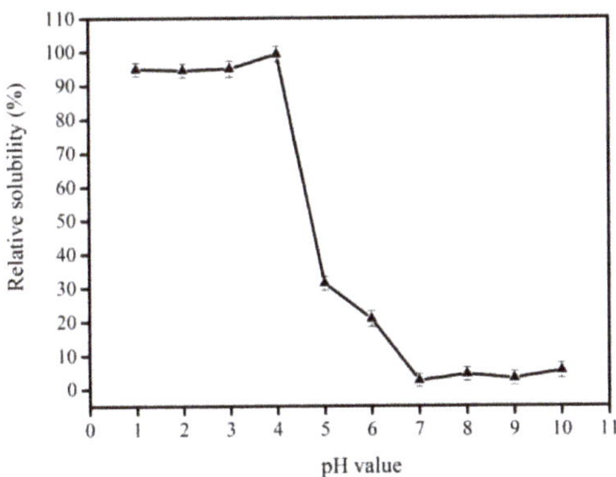

Figure 1. Determination of the isoelectric point of pepsin-solubilized collagen (PSC) from bigeye tuna skin.

2.2. Collagen Yield from Bigeye Tuna Skin

The PSC from the bigeye tuna skin was precipitated using the salting-out method (PSC-SO) and isoelectric precipitation method (PSC-IP) separately. The yield of PSC-SO and PSC-IP was 14.14% (dry weight) and 17.17% (dry weight), respectively (Figure 2). Due to the high concentration of NaCl (1.5 M) in PSC-CO, it took four days of dialysis after salting-out to ensure that PSC-SO was not affected by NaCl and acetic acid. In addition, large amounts of wastewater were produced due to the use of high salt concentrations. Our results were consistent with previous studies that demonstrated the use of NaCl (0.5–1.5 M) to precipitate collagen [3,15]. Due to the low salt concentration required for PSC-IP, dialysis only took two days after the process and less waste water was produced than for PSC-SO. Since collagen yields were greater from PSC-IP, and a shorter time of dialysis was required than for PSC-SO, the PSC-IP method was selected for extracting PSC from bigeye tuna skin.

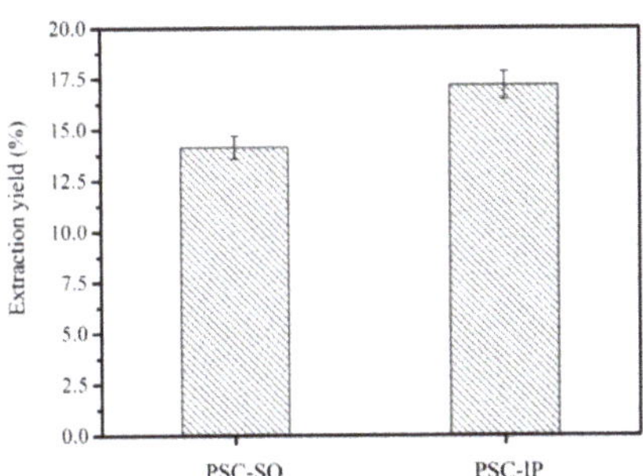

Figure 2. Comparison of extraction yield of collagen by the salting out method and isoelectric precipitation method (dry weight).

2.3. Sodium Dodecyl Sulfate-Polyacrylamide Gel Electrophoresis (SDS-PAGE) Analysis

Figure 3 indicates the SDS-PAGE patterns of PSC-SO and PSC-PI products extracted from bigeye tuna skin, along with bovine collagen type I for comparison. The band patterns of PSC-SO and PSC-PI were highly identical, which contained two clear bands attributed to two different types of α-chains (α_1 and α_2) in accordance with bovine collagen type I (Figure 3). The density of α_1-chain band was approximately 2-fold greater than α_2-chain band, indicating that collagen type I was the main collagen in bigeye tuna skin. In addition, high molecular weight compounds, including β compounds as well as a small number of γ compounds were also observed in PSC-SO and PSC-PI products. Ahmed et al. [26] explored bacterial collagen protease to extracted collagen from bigeye tuna, and their SDS-PAGE results indicated that collagen from bigeye tuna consisted of two different types of α-chains (α_1 and α_2), which was consistent with our result. Our results were also consistent with previous results from PSC extracted from other marine fish skins, including *Scomberomorous niphonius* [27], *Evenchelys macrura* [28], and *Aluterus monocerous* [29], which also belongs to collagen type I.

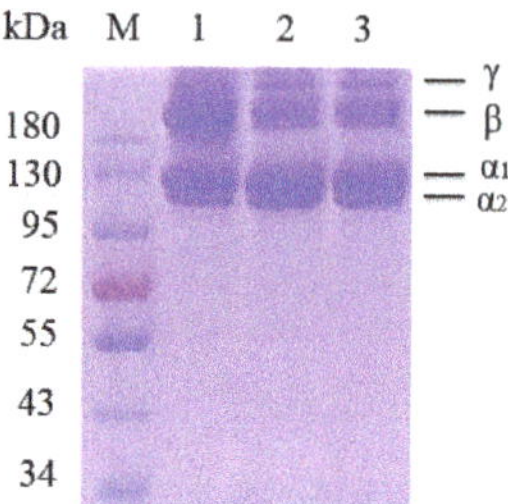

Figure 3. Sodium Dodecyl Sulfate-Polyacrylamide Gel Electrophoresis (SDS-PAGE) analysis of PSC-SO and PSC-PI product from bigeye tuna skin. M: Protein markers; Lane 1: Bovine collagen type I; lane 2: PSC-SO from bigeye tuna skin; lane 3: PSC-PI from bigeye tuna skin.

2.4. Amino Acid Contents

The amino acid contents of the PSC-IP product are shown in Figure 4. The major amino acid in the PSC-IP product is glycine, which constitutes approximately 24% of the total amino acid contents. Other amino acids that were in high proportions were proline (10.86%), hydroxyproline (9.56%), alanine (9.04%), glutamic acid (9.13%), and arginine (7.49%). Cysteine was not detected in the PSC-IP product from bigeye tuna skin. The amino acid contents in PSC-IP product were consistent with the collagen from bigeye tuna in the previous study (glycine (22.2–22.7%), proline (14.8–15.1%), alanine (9.7–9.9%), glutamic acid (9.8–9.9%), hydroxyproline (8.0–8.2%)) [26]. In addition, the high content of glycine, proline, and hydroxyproline in the PSC-IP product is consistent with the high frequency of occurrence of the glycine-proline-hydroxyproline sequence in collagen, which is essential for triple helical formation [4,5]. The content of amino acids (proline and hydroxyproline) in PSC-IP product is 20.42%, which was similar to jellyfish *Acromitus hardenbergi* collagen (19.50%) [2], crap scales or bones collagen (19.2%) [30].

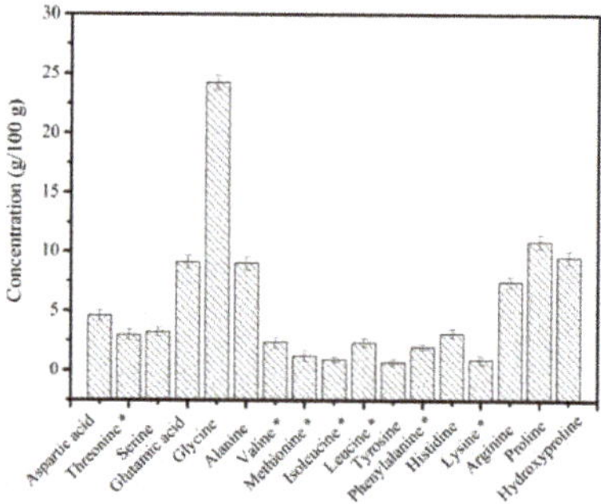

Figure 4. Amino acid contents of PSC-IP product extracted from skin of bigeye tuna. Note: * Essential amino acid.

2.5. Fourier Transform Infrared Spectroscopy (FTIR) Analysis

Figure 5 indicates the FTIR spectra of the PSC-IP product from bigeye tuna skin. The five main absorption bands of the PSC-IP product were located in the amide zone, containing a peak for amide A (3425. 57 cm^{-1}), B (2930.97 cm^{-1}), I (1646.26 cm^{-1}), II (1550.75 cm^{-1}), and III (1238.94 cm^{-1}). The amide A wavenumber of the PSC-IP product was located at 3425.57cm^{-1}, which fits the common wavenumber of free N–H vibrations as an indication of hydrogen bonds [31]. The wavenumber of the amide B band of the PSC-IP product was 2930.97 cm^{-1}, indicating the existence of the asymmetrical stretch of CH$_2$. The amide I band is associated with C=O stretching vibration on the main polypeptide chain or the hydrogen bond coupled with COO$^-$ [4]. The amide I band from the PSC-IP sample is supported by strong absorbance shown in the range of 1600–1700 cm^{-1}. The amide II band represents the N–H bending vibration couples with C–N stretching vibration, and the measured wavenumbers of PSC-IP are within the range of 1550–1600 cm^{-1}. In addition, the appearance of the PSC-IP sample amide III band corresponds to the helical arrangement in the PSC sample.

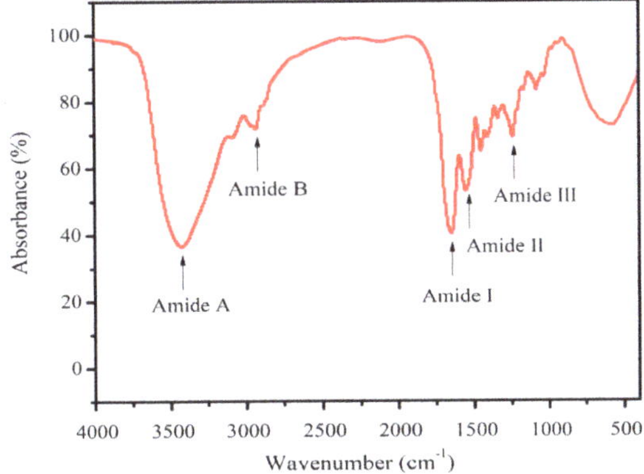

Figure 5. FTIR spectra of PSC-IP product from bigeye tuna skin.

2.6. Inductively Coupled Plasma Mass Spectrometry (ICP-MS)

Heavy metal ions such as As, Pb, and Hg were often used for evaluating the extracted collagen for cosmetic and biomedical applications [4,14]. For example, Tang et al. [4] used ICP-MS to detect As, Pb, and Hg content in collagen from *Nibea japonica* skin and found the content of these heavy metal ions was significantly lower than Chinese regulatory standards. Zhang et al. [14] used ICP-MS to detect these heavy metal ions in collagen from frog skin, which was also lower than Chinese regulatory standards. Thus, the contents of As, Pb, and Hg in PSC-IP product were analyzed using ICP-MS to verify the possibility of applying PSC-IP for cosmetic and biomedical applications (Table 1). ICP-MS results indicate that the heavy metal abundance in PSC-IP was lower than the maximum acceptable amounts according to Chinese regulatory standards (GB 6783-2013). Our results indicate that the heavy metal ions As, Pb, and Hg did not accumulate during the PSC-IP extraction process. Thus, the PSC-IP process is safe for use on bigeye tuna skin for cosmetic and biomedical applications.

Table 1. Elemental analysis of PSC-IP product from bigeye tuna skin by ICP-MS.

Collagen	Element	Content (mg/kg)	National Standard of Edible Gelatin (GB 6783-2013 in China) (mg/kg)
PSC-IP	As	0.51 ± 0.04	≤1.0
	Pb	0.17 ± 0.02	≤1.5
	Hg	1.18 ± 0.07	

2.7. Cytotoxic and Allergenic Tests

The cytotoxicity and sensitization of the PSC-IP product from bigeye tuna skin were detected using a methylthiazolyldiphenyl-tetrazolium bromide (MTT) assay and lactate dehydrogenase (LDH) toxicity assay. The MTT assay was used to assess the cell compatibility of the PSC-IP product, and results are shown in Figure 6A. Results indicate that after treatment with increasing concentrations of the PSC-IP product, the viability of NIH-3T3 fibroblasts did not decrease after 24 h incubation. In addition, the PSC-IP product promoted the growth of NIH-3T3 fibroblasts. Therefore, results demonstrate that collagen from bigeye tuna skin has no significant cytotoxic effect in vitro. Our finding agreed with that of Jeong et al. [32], where pre-osteoblast (MC3T3-E1) cells were used for biocompatibility evaluation of collagen from *Thunnus obesus* bone. Their results revealed that collagen scaffolds from *Thunnus obesus* bone were biocompatible and non-toxic in vitro. The cytoplasm in every human tissue contains LDH, and the disruption of cell membrane integrity leads to an increase in LDH concentration in the surrounding matrix. Thus, due to its close association with allergic reactions and inflammation, LDH release has been utilized as a criterion to evaluate allergenicity [33,34]. As shown in Figure 6B, the LDH release of cells in the presence of the PSC-IP product was relatively low when compared with untreated cells. The results indicate that PSC-IP extract from bigeye tuna skin could be considered as a non-cytotoxic and hypoallergenic biomaterial for cosmetic and biomedical applications.

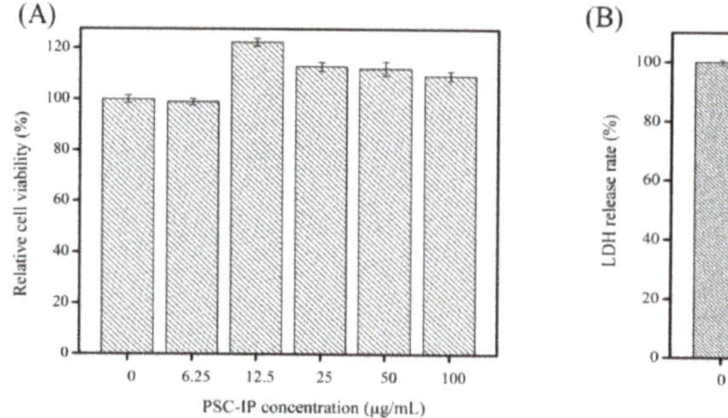
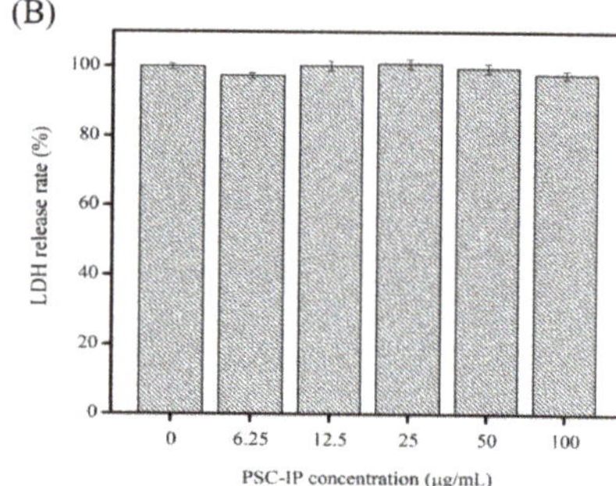

Figure 6. Relative cell viability (**A**) and lactate dehydrogenase (LDH) release (**B**) as affected by 24 h treatment of PSC-IP product (0, 6.25, 12.5, 25, 50, and 100 µg/mL) from bigeye tuna skin.

2.8. Morphological Examination

Morphological examination of cells with collagen solutions could also reveal its biocompatibility and non-toxicity [2,4]. For example, 3T3 F442A cells were treated collagen solutions from jellyfish and there were no observable changes among cells when compared with control group [2]. In the present study, NIH-3T3 cells were treated with PSC-IP (12, 25, and 50 µg/mL) and examined for morphological changes. Treated cells showed no significant change in comparison with the untreated cells, as cells in all groups grew uniformly and presented normal morphologies (Figure 7). The PSC-IP product facilitated the growth of NIH-3T3 cells, which is in agreement with our MTT results. This phenomenon also suggests that collagen from bigeye tuna skin is non-toxic and has the potential to be utilized in biomaterials, for example as a humectant agent and in wound dressing.

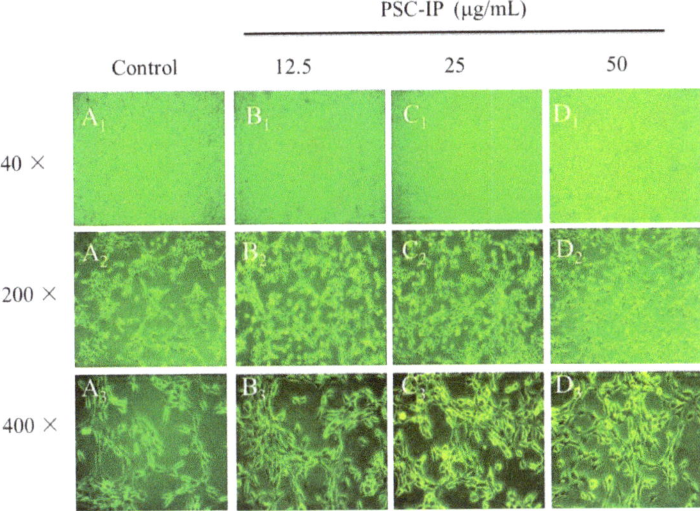

Figure 7. Morphological changes of NIH-3T3 cells treated with 0, 12.5, 25, and 50 µg/mL of PSC-IP product from bigeye tuna skin, respectively (40×, 200×, 400×). A_1–A_3: Untreated cells; B_1–B_3: Treated cells with 12.5 µg/mL of PSC-IP; C_1–C_3: Treated cells with 25 µg/mL of PSC-IP; D_1–D_3: Treated cells with 50 µg/mL of PSC-IP.

2.9. In Vitro Scratch Wound Closure

PSC extracted from various marine organisms have been shown to be beneficial for wound healing, as they can reinforce the adhesion and proliferation of the fibroblasts, and influence inflammatory cytokines such as IL-1β [35]. Fibroblast migration can accelerate the process of wound re-epithelialization and promote wound closure during wound healing [36]. In the previous studies, in vitro scratch test was often used to simulate wound healing [36,37]. Thus, in our study, the effects of the PSC sample on wound healing were estimated via an in vitro scratch test (Figure 8). The scratch closure rate at different time points was recorded and calculated. Figure 8A shows that the wound scratch area decreased remarkably in a dose-dependent manner when treated with the PSC-IP product in comparison with the negative control group area. In addition, the scratch closure rates of the PSC-IP product treated groups were apparently greater than the control group (Figure 8B). It is worth noting, that the scratch nearly closed in the experimental groups, and showed a stronger healing effect similar to bovine collagen after 24 h of treatment. Overall, in vitro wound healing results indicate that collagen from bigeye tuna skin can effectively promote cell migration and has the potential for wound healing. However, the molecular mechanism of collagen promoting cell migration and proliferation is also unclear. In our further study, some biochemical/cellular signaling pathways will be chosen for investigating, such as AKT/mTOR signaling or nuclear factor kappa enhancer binding protein (NF-kB) signaling pathway [36].

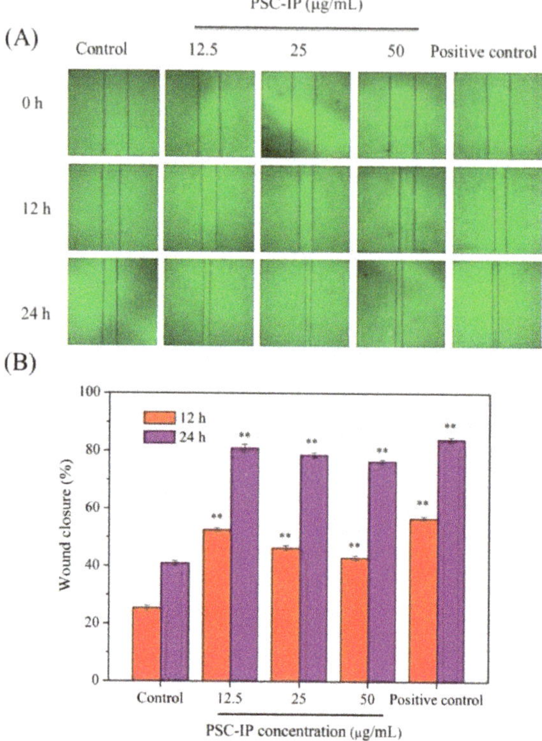

Figure 8. Effects of PSC-IP product from bigeye tuna skin on the scratch closure rate. (**A**) PSC-IP promoted cell migration was evaluated using a scratch wound healing assay. (**B**) Wound closure rate. * $p < 0.05$ and ** $p < 0.001$ vs. control.

3. Materials and Methods

3.1. Raw Materials

Bigeye tuna skin was provided by Zhejiang Ocean Family CO., LTD (Zhoushan, China). The NIH-3T3 fibroblasts were stored in the laboratory [4]. Bovine collagen type I (cat. no. C8060) was purchased from Solarbio (Beijing, China). The LDH cytotoxicity assay kit (cat. no. C0016) and the high molecular weight markers (cat. no. P0068) was obtained from Beyotime Biotechnology (Shanghai, China). The MTT cell proliferation and cytotoxicity assay kit (cat. no. AR1156) was obtained from Boster Biological Technology Co., Ltd (Wuhan, China). All other reagents were of analytical grade.

3.2. Extraction of Collagen from Bigeye Tuna Using the Salting-Out Method

The extraction procedure of PSC extraction from bigeye tuna skin was performed at 4 °C according to Tang et al. [4]. Fish skins that had been removed from non-collagenous protein and defatted were cut into small pieces and incubated in 0.5 M acetic acid (1:50, *w/v*) and 1200 U/g pepsin to extract PSC. The extract was then filtered and the supernatants were salted-out using 1.5 M NaCl. After 24 h, the precipitate was harvested by centrifugation (10,000× g, 15 min) and dissolved in acetic acid (0.5 M) subsequently. The PSC was dialyzed with deionized water until the pH was neutral and the silver nitrate method was used to detect the presence of chloride ions. The resultant suspensions (PSC-SO) were lyophilized and stored at −20 °C for further study.

3.3. Extraction of Collagen from Bigeye Tuna by Using Isoelectric Precipitation

The small pieces of fish skins mentioned above were incubated in 0.5 M acetic acid (1:50, *w/v*) and 1200 U/g pepsin to extract PSC. The extract was then filtered and the obtained supernatants were named as collagen stock solution. Then, the pH of collagen stock solution (3 mg/mL) was adjusted to 1.0, 2.0, 3.0, 4.0, 5.0, 6.0, 7.0, 8.0, 9.0, or 10.0 with HCl (6 N) or NaOH (6 N), allowing to stand for approximately 3 h. The supernatant was obtained after centrifugation (10,000 × g, 15 min) and used to determine the protein concentration according to the instructions of the BCA protein detection kit. Then, the collagen stock solution was adjusted to the isoelectric point of collagen. The precipitate was harvested by centrifugation (10,000× g, 15 min), and dissolved in acetic acid (0.5 M). The samples were dialyzed with deionized water until the pH value was neutral. The resultant suspensions (PSC-IP) were lyophilized and stored at –20 °C for further study.

3.4. SDS-PAGE Analysis

The SDS-PAGE method was performed according to Laemmli et al. [38]. Protein (30 μg) obtained from PSC-SO and PSC-IP was loaded in each well of pre-made 8.0% SDS-PAGE gels. The molecular weights of collagen samples were then estimated by inferring from the high molecular weight markers. Bovine collagen type I was used as a positive control.

3.5. Amino Acid Analysis

PSC-SO and PSC-IP products (0.02 g each) were hydrolyzed in 6 N HCl at 110 °C for approximately 24 h. The hydrolysates were diluted and then analyzed using an amino acid analyzer (Hitachi L-8800, Tokyo, Japan). The chloramine T method was used to analyze the content of hydroxyproline [15].

3.6. FTIR Analysis

FTIR spectra of PSC-IP product were determined using a Bruker Tensor 27 FTIR spectrometer (Bruker, Rheinstetten, Germany) under dry conditions. The infrared spectra were recorded in the 4000–500 cm^{-1} range at 1 cm^{-1} resolution for a single scan.

3.7. ICP-MS

The contents of heavy metals in the PSC-IP product were analyzed using ICP-MS (Agilent, CA, USA). The PSC-IP (0.5 mg/mL) was dissolved in deionized water and cooled to room temperature before detection [14].

3.8. Cytotoxic and Allergenic Properties of PSC-IP

NIH-3T3 fibroblasts were used to determine the cytotoxic and allergenic properties of PSC-IP product according to the MTT assay kit instructions. Cells were inoculated in 96-well plates (1×10^5 cells/well) and incubated in a 5% CO_2 incubator for 24 h at 37 °C. Then, cells were dealt with PSC-IP (0, 6.25, 12.5, 25, 50 and 100 μg/mL) and cultured for another 24 h. The cytotoxic possibility of the PSC-IP product was determined by MTT assay and the absorbance values were determined at 490 nm. Cell growth inhibition (%) was obtained according to the MTT assay kit instructions.

The allergenic properties of the PSC-IP product were determined using the LDH release assay. Cells were inoculated in 96-well plates (1×10^5 cells/well) and cultured for 24 h. The cells were dealt with PSC-IP (0, 6.25, 12.5, 25, 50, and 100 μg/mL) and cultured for another 24 h. The LDH release rate (%) was calculated according to the LDH cytotoxicity assay kit instructions.

3.9. Morphological Changes

NIH-3T3 fibroblasts (1×10^5 cells/mL) were suspended and cultured in a 6-well flat bottom plate with a cover glass (20 × 24 mm) for 24 h. Cells were dealt with PSC-IP (0, 12.5, 25, and 50 μg/mL).

After 24 h incubation, the changes in cell morphology were assessed using an inverted microscope (Olympus, Tokyo, Japan).

3.10. In Vitro Scratch Closure Assay

NIH-3T3 cells were inoculated into 6-well plates (2×10^5 cells/well) and cultured for 24 h at 37 °C in 5% CO_2 to reach 80–90% cell confluency. A scratch wound was created using a 200 μL pipette tip and the wound debris was washed away using PBS. PSC-IP product (0, 12.5, 25, and 50 μg/mL) was added and cultured for a further 12 or 24 h. The phase-contrast microscope (CKX41-A32PH, Olympus, Tokyo, Japan) was used to observed the scratch closure and the scratch area was obtained using Image J software.

The scratch closure rate (%) was calculated as follows:

$$\text{Scratch closure rate (\%)} = (A_0 - A_t)/A_0 \times 100\%$$

where A_0 represents the scratch area at 0 h and A_t represents the scratch area at the designated time point.

3.11. Statistical Analysis

All tests were expressed as mean ± standard deviation (SD, $n = 3$). Data were analyzed by analysis of variance (ANOVA) using IBM SPSS 19.0 software (Ehningen, Germany). A difference was considered statistically significant when * $p < 0.05$ and ** $p < 0.001$.

4. Conclusions

In the present study, PSC was extracted from bigeye tuna skin using salt outing (PSC-SO) and isoelectric precipitation (PSC-IP) methods. Considering the high yield and short time for extracting PSC, the isoelectric precipitation method was chosen for obtaining PSC-IP product from bigeye tuna skin. SDS-PAGE analysis indicated that PSC-IP product from bigeye tuna is a collagen type I. Furthermore, ICP-MS analysis showed that the PSC-IP product was free of heavy metals. The effects of the PSC-IP on proliferation and migration PSC-IP indicates that collagen from bigeye tuna skin has good potential for use in cosmetic and biomedical applications.

Author Contributions: Y.T. and Y.C. (Yan Chen) conceived and designed the experiments. X.L., Y.C. (Yinyue Chen), H.J., Q.Z., and C.L. performed the experiments. R.L., F.Y., F.H., Z.Y., and G.D. performed the statistical analysis of the data. X.L. and Y.T. wrote the manuscript.

Funding: This work was financially supported by the National Natural Science Foundation of China (grant No. 41806153, No. 81773629 and No. 21808208), the Natural Science Foundation of Zhejiang Province (No. LQ18B060004), the National Undergraduate Training Program for Innovation and Entrepreneurship (grant No. 201810340015), and the Zhejiang Xinmiao Talents Program (grant No. 2018R411052).

Conflicts of Interest: The authors declare no conflict of interest.

References

1. Blanco, M.; Vazquez, J.A.; Perez-Martin, R.I.; Sotelo, C.G. Hydrolysates of fish skin collagen: An opportunity for valorizing fish industry byproducts. *Mar. Drugs* **2017**, *15*, 131. [CrossRef] [PubMed]
2. Khong, N.M.H.; Yusoff, F.M.; Jamilah, B.; Basri, M.; Maznah, I.; Chan, K.W.; Armania, N.; Nishikawa, J. Improved collagen extraction from jellyfish (*Acromitus hardenbergi*) with increased physical-induced solubilization processes. *Food Chem.* **2018**, *251*, 41–50. [CrossRef] [PubMed]
3. Chen, J.; Li, M.; Yi, R.; Bai, K.; Wang, G.; Tan, R.; Sun, S.; Xu, N. Electrodialysis extraction of pufferfish skin (*Takifugu flavidus*): A promising source of collagen. *Mar. Drugs* **2019**, *17*, 25. [CrossRef] [PubMed]
4. Tang, Y.; Jin, S.; Li, X.; Li, X.; Hu, X.; Chen, Y.; Huang, F.; Yang, Z.; Yu, F.; Ding, G. Physicochemical properties and biocompatibility evaluation of collagen from the skin of giant croaker (*Nibea japonica*). *Mar. Drugs* **2018**, *16*, 222. [CrossRef] [PubMed]

5. Berillis, P. Marine collagen: Extraction and applications. In *Research Trends in Biochemistry, Molecular Biology and Microbiology*; smgebooks: Dover, DE, USA, 2015; pp. 1–13.

6. Silva, T.; Moreira-Silva, J.; Marques, A.; Domingues, A.; Bayon, Y.; Reis, R. Marine origin collagens and its potential applications. *Mar. Drugs* **2014**, *12*, 5881–5901. [CrossRef]

7. Rahman, M.A. Collagen of extracellular matrix from marine invertebrates and its medical applications. *Mar. Drugs* **2019**, *17*, 118. [CrossRef]

8. Ehrlich, H.; Wysokowski, M.; Żółtowska-Aksamitowska, S.; Petrenko, I.; Teofil Jesionowski, T. Collagens of poriferan origin. *Mar. Drugs* **2018**, *16*, 79. [CrossRef]

9. Yu, F.; Zong, C.; Jin, S.; Zheng, J.; Chen, N.; Huang, J.; Chen, Y.; Huang, F.; Yang, Z.; Tang, Y.; et al. Optimization of extraction conditions and characterization of pepsin-solubilised collagen from skin of giant croaker (*Nibea japonica*). *Mar. Drugs* **2018**, *16*, 29. [CrossRef] [PubMed]

10. Rastian, Z.; Putz, S.; Wang, Y.J.; Kumar, S.; Fleissner, F.; Weidner, T.; Parekh, S.H. Type I Collagen from jellyfish catostylus mosaicus for biomaterial applications. *ACS Biomater. Sci. Eng.* **2018**, *4*, 2115–2125. [CrossRef]

11. Avila Rodríguez, M.I.; Rodríguez Barroso, L.G.; Sanchez, M.L. Collagen: A review on its sources and potential cosmetic applications. *J. Cosmet. Dermatol.* **2018**, *17*, 20–26. [CrossRef]

12. Wang, J.; Pei, X.; Liu, H.; Zhou, D. Extraction and characterization of acid-soluble and pepsin-soluble collagen from skin of loach (*Misgurnus anguillicaudatus*). *Int. J. Biol. Macromol.* **2018**, *106*, 544–550. [CrossRef] [PubMed]

13. Li, J.; Wang, M.C.; Qiao, Y.Y.; Tian, Y.Y.; Liu, J.H.; Qin, S.; Wu, W.H. Extraction and characterization of type I collagen from skin of tilapia (*Oreochromis niloticus*) and its potential application in biomedical scaffold material for tissue engineering. *Process Biochem.* **2018**, *74*, 156–163. [CrossRef]

14. Zhang, J.; Duan, R. Characterisation of acid-soluble and pepsin-solubilised collagen from frog (*Rana nigromaculata*) skin. *Int. J. Biol. Macromol.* **2017**, *101*, 638–642. [CrossRef]

15. Sun, L.; Hou, H.; Li, B.; Zhang, Y. Characterization of acid- and pepsin-soluble collagen extracted from the skin of Nile tilapia (*Oreochromis niloticus*). *Int. J. Biol. Macromol.* **2017**, *99*, 8–14. [CrossRef]

16. Bramaud, C.; Aimar, P.; Daufin, G. Whey protein fractionation: Isoelectric precipitation of alpha-lactalbumin under gentle heat treatment. *Biotechnol. Bioeng.* **1997**, *56*, 391–397. [CrossRef]

17. Mireles DeWitt, C.A.; Nabors, R.L.; Kleinholz, C.W. Pilot plant scale production of protein from catfish treated by acid solubilization/isoelectric precipitation. *J. Food. Sci.* **2007**, *72*, 351–355. [CrossRef]

18. Salcedo-Chavez, B.; Osuna-Castro, J.A.; Guevara-Lara, F.; Dominguez-Dominguez, J.; Paredes-Lopez, O. Optimization of the isoelectric precipitation method to obtain protein isolates from amaranth (*Amaranthus cruentus*) seeds. *J. Agric. Food Chem.* **2002**, *50*, 6515–6520. [CrossRef]

19. Zhao, X.; Xing, T.; Wang, P.; Xu, X.; Zhou, G. Oxidative stability of isoelectric solubilization/precipitation-isolated PSE-like chicken protein. *Food Chem.* **2019**, *283*, 646–655. [CrossRef]

20. Zhao, X.; Liu, H.; Zhang, X.; Zhu, H. Comparison of structures of walnut protein fractions obtained through reverse micelles and alkaline extraction with isoelectric precipitation. *Int. J. Biol. Macromol.* **2019**, *125*, 1214–1220. [CrossRef] [PubMed]

21. Deng, Y.; Huang, L.; Zhang, C.; Xie, P.; Cheng, J.; Wang, X.; Li, S. Physicochemical and functional properties of Chinese quince seed protein isolate. *Food Chem.* **2019**, *283*, 539–548. [CrossRef] [PubMed]

22. Geng, F.; Xie, Y.; Wang, J.; Li, S.; Jin, Y.; Ma, M. Large-scale purification of ovalbumin using polyethylene glycol precipitation and isoelectric precipitation. *Poult. Sci.* **2019**, *98*, 1545–1550. [CrossRef]

23. FAO. *The State of World Fisheries and Aquaculture 2018—Meeting the Sustainable Development Goals*; FAO: Roma, Italy, 2018; p. 1121.

24. Herpandi, N.H.; Rosma, A.; Nadiah, W.A.W. The tuna fishing industry: A new outlook on fish protein hydrolysates. *Compr. Rev. Food Sci. Food Safty* **2011**, *10*, 195–207. [CrossRef]

25. Abdollahi, M.; Rezaei, M.; Jafarpour, A.; Undeland, I. Sequential extraction of gel-forming proteins, collagen and collagen hydrolysate from gutted silver carp (*Hypophthalmichthys molitrix*), a biorefinery approach. *Food Chem.* **2018**, *242*, 568–578. [CrossRef] [PubMed]

26. Ahmed, R.; Getachew, A.T.; Yeon-Jin Cho, Y.J.; Chun, B.S. Application of bacterial collagenolytic proteases for the extraction of type I collagen from the skin of bigeye tuna (*Thunnus obesus*). *LWT-Food SCI. Technol.* **2018**, *89*, 44–51. [CrossRef]

27. Li, Z.R.; Wang, B.; Chi, C.; Zhang, Q.H.; Gong, Y.; Tang, J.J.; Luo, H.; Ding, G. Isolation and characterization of acid soluble collagens and pepsin soluble collagens from the skin and bone of Spanish mackerel (*Scomberomorous niphonius*). *Food Hydrocolloid.* **2013**, *31*, 103–113. [CrossRef]

28. Veeruraj, A.; Arumugam, M.; Balasubramanian, T. Isolation and characterization of thermostable collagen from the marine eel-fish (*Evenchelys macrura*). *Process Biochem.* **2013**, *48*, 1592–1602. [CrossRef]

29. Mehraj, A.; Soottawat, B. Extraction and characterisation of pepsin-solubilised collagen from the skin of unicorn leatherjacket (*Aluterus monocerous*). *Food Chem.* **2010**, *120*, 817–824.

30. Duan, R.; Zhang, J.; Du, X.; Yao, X.; Konno, K. Properties of collagen from skin, scale and bone of carp (*Cyprinus carpio*). *Food Chem.* **2009**, *112*, 702–706. [CrossRef]

31. Doyle, B.B.; Bendit, E.G.; Blout, E.R. Infrared spectroscopy of collagen and collagen-like polypeptides. *Biopolymers* **1975**, *14*, 937–957. [CrossRef] [PubMed]

32. Jeong, H.S.; Venkatesan, J.; Kim, S.K. Isolation and Characterization of Collagen from Marine Fish (*Thunnus obesus*). *Biotechnol. Bioproc. Eng.* **2013**, *18*, 1185–1191. [CrossRef]

33. Cincin, Z.B.; Unlu, M.; Kiran, B.; Bireller, E.S.; Baran, Y.; Cakmakoglu, B. Molecular mechanisms of quercitrin-induced apoptosis in non-small cell lung cancer. *Arch. Med. Res.* **2014**, *45*, 445–454. [CrossRef] [PubMed]

34. Chen, H.; Xu, Y.; Wang, J.; Zhao, W.; Ruan, H. Baicalin ameliorates isoproterenol-induced acute myocardial infarction through iNOS, inflammation and oxidative stress in rat. *Int. J. Clin. Exp. Patho.* **2015**, *8*, 10139–10147.

35. Castillo-Briceño, P.; Bihan, D.; Nilges, M.; Hamaia, S.; Meseguer, J.; García-Ayala, A.; Farndale, R.W.; Mulero, V. A role for specific collagen motifs during wound healing and inflammatory response of fibroblasts in the teleost fish gilthead seabream. *Mol. Immunol.* **2011**, *48*, 826–834. [CrossRef] [PubMed]

36. Park, Y.R.; Sultan, M.T.; Park, H.J.; Lee, J.M.; Ju, H.W.; Lee, O.J.; Lee, D.J.; Kaplan, D.L.; Park, C.H. NF-kappa B signaling is key in the wound healing processes of silk fibroin. *Acta Biomater.* **2018**, *67*, 183–195. [CrossRef] [PubMed]

37. Liu, H.; Mu, L.X.; Tang, J.; Shen, C.B.; Gao, C.; Rong, M.Q.; Zhang, Z.Y.; Liu, J.; Wu, X.Y.; Yu, H.N.; et al. A potential wound healing-promoting peptide from frog skin. *Int. J. Biochem. Cell Biol.* **2014**, *49*, 32–41. [CrossRef] [PubMed]

38. LaemmLi, U.K. Cleavage of structural proteins during the assembly of bacteriophage T4. *Nature* **1970**, *227*, 680–685. [CrossRef]

MDPI

St. Alban-Anlage 66

4052 Basel

Switzerland

Tel. +41 61 683 77 34

Fax +41 61 302 89 18

www.mdpi.com

Marine Drugs Editorial Office

E-mail: marinedrugs@mdpi.com

www.mdpi.com/journal/marinedrugs

www.ingramcontent.com/pod-product-compliance
Lightning Source LLC
LaVergne TN
LVHW071242170726
843515LV00006BA/1311